D1088426

Exotic Atoms '79

Fundamental Interactions and Structure of Matter

ETTORE MAJORANA INTERNATIONAL SCIENCE SERIES

Series Editor:
Antonino Zichichi
European Physical Society
Geneva, Switzerland

(PHYSICAL SCIENCES)

Volume 1 INTERACTING BOSONS IN NUCLEAR PHYSICS
Edited by F. Iachello

Volume 2 HADRONIC MATTER AT EXTREME ENERGY DENSITY
Edited by Nicola Cabibbo and Luigi Sertorio,

Volume 3 COMPUTER TECHNIQUES IN RADIATION TRANSPORT AND DOSIMETRY
Edited by Walter R. Nelson and T. M. Jenkins

Volume 4 EXOTIC ATOMS '79: Fundamental Interactions and Structure of Matter
Edited by Kenneth Crowe, Jean Duclos, Giovanni Fiorentini, and Gabriele Torelli

Exotic Atoms '79
Fundamental Interactions and Structure of Matter

Edited by

Kenneth Crowe
University of California
Berkeley, California

Jean Duclos
CEN de Saclay
Gif sur Yvette, France

Giovanni Fiorentini and **Gabriele Torelli**
Italian National Institute for Nuclear Physics (INFN)
Pisa, Italy

Plenum Press · New York and London

Library of Congress Cataloging in Publication Data

International School of Physics of Exotic Atoms, 2d, Erice, Italy, 1979.
Exotic atoms '79.

(Ettore Majorana international science series: Physical sciences; v. 4)
"Proceedings of the second course of the International School of Physics of Exotic Atoms, held in Erice, Trapani, Sicily March 25–April 5, 1979."
Includes index.
1. Nuclear reactions—Congresses. 2. Matter—Constitution—Congresses. 3. Nuclear chemistry—Congresses. 4. Muons—Congresses. I. Crowe, Kenneth. II. Title. III. Series.
QC793.9.I595 1979 539.7'21 79-23072
ISBN 0-306-40322-6

Proceedings of the Second Course of the International School of Physics of Exotic Atoms, held in Erice, Trapani, Sicily, March 25–April 5, 1979.

A Division of Plenum Publishing Corporation
227 West 17th Street, New York, N.Y. 10011

Printed in the United States of America

PREFACE

The second course of the International School on the Physics of Exotic Atoms took place at the "Ettore Majorana" Center for Scientific Culture, Erice, Sicily, during the period from March 25 to April 5, 1979. It was attended by 40 participants from 23 institutes in 8 countries.

The purpose of the course was to review the various aspects of the physics of exotic atoms, with particular emphasis on the results obtained in the last two years, i.e., after the first course of the School (Erice, April 24-30, 1977). The course dealt with two main topics, A) Exotic atoms and fundamental interactions and B) Applications to the study of the structure of matter.

One of the aims of the course was to offer an opportunity for the exchange of experiences between scientists working in the two fields. In view of this, the lectures in the morning discussed the more general arguments in a common session, whereas the more specialized topics were treated in the afternoon, in two parallel sections.

Section A was organized around four main subjects, briefly positronium and muonium, quarkonium, baryonium and neutral currents in atomic physics. In addition various progresses were reported in muon and antiproton physics.

In section B a comprehensive account of the field of muon spin rotation was given. The topics covered were: muon motion in solids, muons in semiconductors and ferromagnetic materials, isotope effects of Hydrogen diffusion in metals, the chemistry of muonic radicals, problems of site determination, muonium induced quadrupole moments, muons in semimetals and quantum tunnelling.

The lecturers were kind enough to edit their notes so that we are now able to present the proceedings of the course; we would like to express our appreciation and gratitude to all of them for the effort they put into both the preparation and presentation of these lectures.

The Center for Scientific Culture was a very efficient and pleasant host for this course, and our sincere thanks go to its Director, Prof. A. Zichichi.

The Editors
Pisa, July, 1979

CONTENTS

I. FUNDAMENTAL INTERACTIONS

II. QUARK ATOMS

III. THE CHEMICAL PHYSICS OF MESIC ATOMS AND MOLECULES

IV. MUON SPIN ROTATION

Part I
Fundamental Interactions

MUONIUM

Vernon W. Hughes

Yale University
Physics Department, Gibbs Laboratory
New Haven, CT 06520 U.S.A.

1. INTRODUCTION

Muonium is the hydrogen-like atom consisting of a positive muon and an electron. From the viewpoint of elementary particle physics and fundamental atomic physics, the motivations for studying muonium are to determine the properties of the muon and to measure the muon-electron interaction in this bound atomic state. Indeed muonium provides an ideal simple system for testing modern muon electrodynamics and for searching for effects of weak, strong, or unknown interactions on the electron-muon bound state.

From the viewpoint of atomic and molecular physics and solid state physics, muonium should be considered as an isotope of hydrogen. Because of the unique method of muon spin resonance applicable to the study of muonium, new unique information, involving the microsecond time scale and the muon magnetism, can be obtained about the atomic collisions, molecular binding, and behaviour in condensed matter of a hydrogen-like atom. Recent reviews on muonium research as it applies to atomic and solid state physics have been given by Brewer et al., 1975; Brewer and Crowe, 1978.

Quantum electrodynamics in its modern theoretical formulation has been available for about 30 years, and experimental tests of quantum electrodynamics, including the behaviour of the muon as a heavy electron, have been active over this period. Theory and experiment are in excellent agreement to a level of high precision for a wide range of phenomena and energies which involve the electromagnetic interaction of leptons and hadrons. This agreement constitutes one of the major successes of modern physics and justifies the hope that quantum field theory may provide the correct foundation for understanding all physical phenomena.

A central aim of modern elementary particle physics is a unified theory of the electromagnetic, strong, weak, and gravitational interactions. Modern experiments which measure principally the electromagnetic interaction of particles, and hence test quantum electrodynamics, seek also to detect the effects of weak or strong interactions and hence provide information about phenomena where the interplay of several different interactions is important. One example is the precise measurement of the g value of the muon, to which the strong interactions contribute through hadron vacuum polarization (Bailey, et al., 1979). A second example is the discovery of parity nonconservation in the high energy scattering of longitudinally polarized electrons by nucleons (Prescott et al., 1978, 1979), which is believed to be caused by an interference between the electromagnetic interaction and the weak interaction mediated by a neutral vector boson (Weinberg, 1979).

Muonium is an ideal system to study the electromagnetic interaction of two different leptons and hence to test muon electrodynamics, which includes the viewpoint that the muon and electron are elementary, structureless particles; and also to determine the electromagnetic properties of the muon, including its spin, magnetic moment and mass. In addition, a precise value of the fine structure constant α is determined from the hyperfine structure interval of muonium. Also certain aspects of the weak interactions of the muon and electron can be tested with muonium. A precise study of a basic system like muonium, for which the measured quantities can be calculated accurately, provides criteria for testing new speculative theories.

A review of muon electrodynamics, including a discussion of muonium research, up to 1977 has been given by Hughes and Kinoshita (1977). The present short article will present recent developments in muonium research relevant to muon electrodynamics, and certain speculations about future directions for muonium research. The topics covered are the hyperfine structure of the ground state, the muon magnetic moment, muonium-antimuonium conversion, the excited n=2 state of muonium, and the muonic helium atom.

2. HYPERFINE STRUCTURE AND ZEEMAN EFFECT IN GROUND STATE

Important advances have been made in both our experimental and theoretical knowledge of the hyperfine structure interval $\Delta\nu$ in the ground n=1 state of muonium. Closely related improvement has been achieved in the precision of measurement of the magnetic moment of the positive muon.

The classic muonium microwave magnetic resonance experiment has recently been redone at the Los Alamos Meson Physics Facility

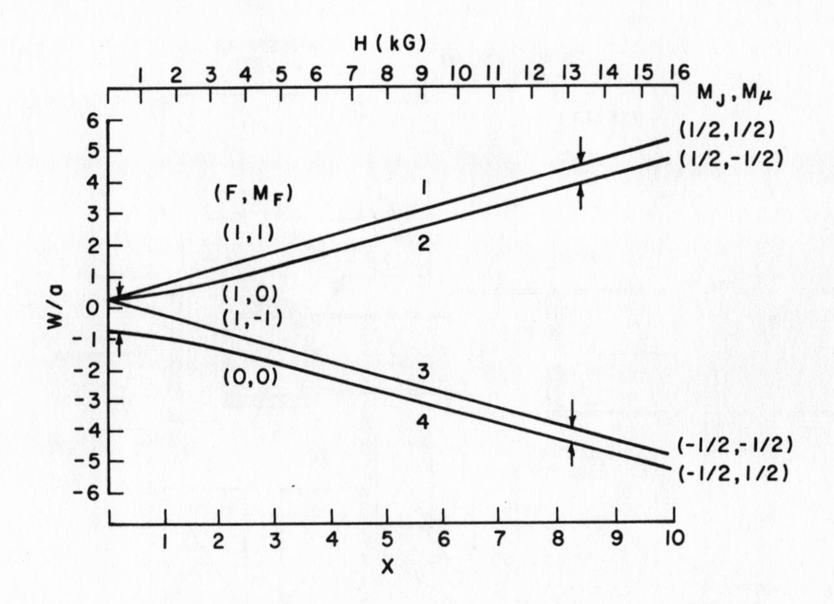

Fig. 1. Energy levels in ground n=1 state of muonium based on the Hamiltonian

$$\mathcal{H} = a\vec{I}_\mu \cdot \vec{J} + \mu_B^e g_J \vec{J}\cdot\vec{H} - \mu_B^\mu g'_\mu \vec{I}_\mu \cdot \vec{H}$$

$$X = (g_J \mu_B^e + g'_\mu \mu_B^\mu) H / (h\Delta\nu)$$

$$a \simeq 4\ 463 \text{ MHz}$$

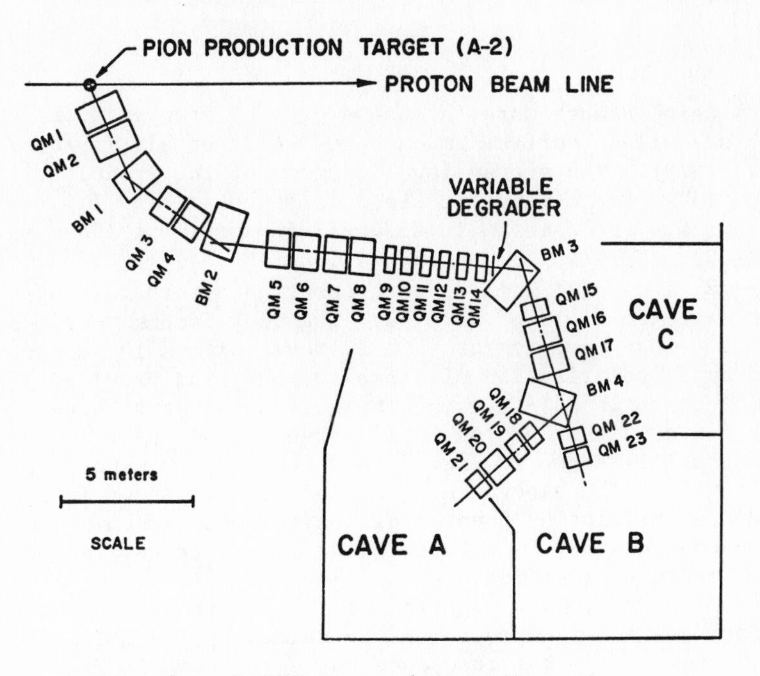

Fig. 2. LAMPF Stopped Muon Channel

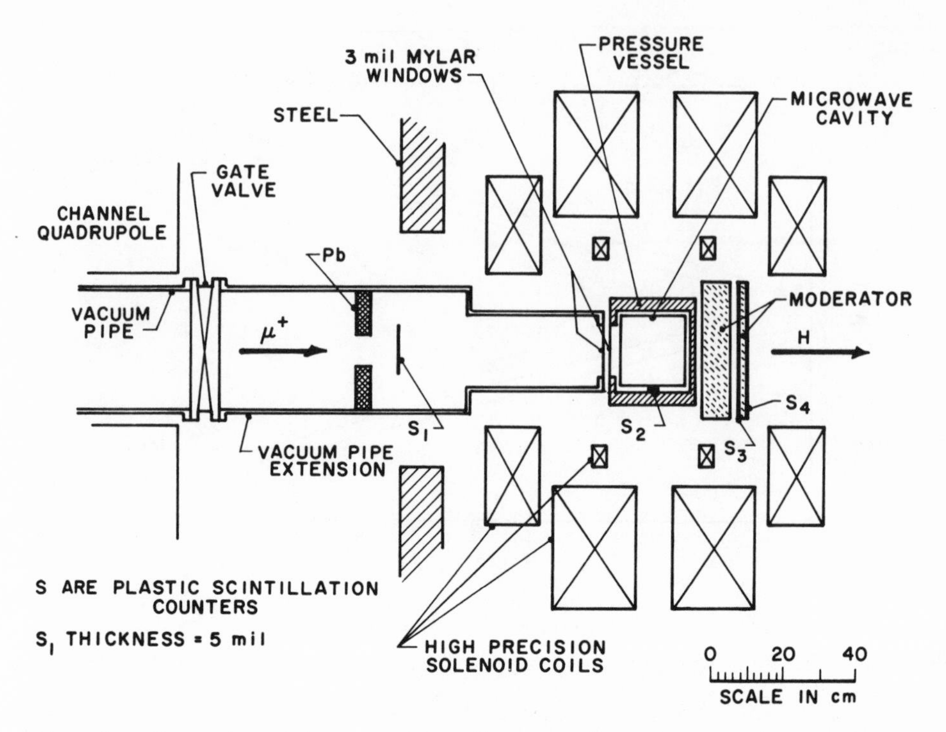

Fig. 3. Experimental apparatus for muonium resonance experiment using surface muons.

(LAMPF) using a much more intense source of stopped muons obtained from a so-called "surface" muon beam (Reist et al., 1978; Thompson et al., 1979). The energy level diagram of the ground state of muonium in a static magnetic field is shown in Fig. 1. The general technique was the same as in an earlier experiment (Casperson et al., 1977). The two Zeeman transitions $(M_J,M_\mu)=(1/2,1/2)\leftrightarrow(1/2,-1/2)$ and $(-1/2,-1/2)\leftrightarrow(-1/2,+1/2)$ were observed alternately at a strong static magnetic field of 13.6 kG. The very intense surface muon beam had a mean momentum of p_μ = 28 MeV/c with a 10% momentum spread and a polarization of close to 100%, and provided μ^+ stopping rates of up to 10^6 s^{-1} average in a 0.5 atm Kr gas target 20 cm in length for a primary proton beam current of 300 μA. The stopped muon channel is shown in Fig. 2, and the muonium experiment is set up in cave B. The experimental arrangement is shown in Fig. 3. The thin scintillation counter S1 monitors the incoming muon beam and the counters S3, S4 detect the decay electrons. A central element in this experiment is the high precision solenoid electromagnet which provides a magnetic field over the region of the microwave cavity which is homogeneous to several ppm and is stable to about 1 ppm. A typical resonance curve is shown in Fig. 4. It was obtained in a period of about 2 hrs by varying the magnetic field

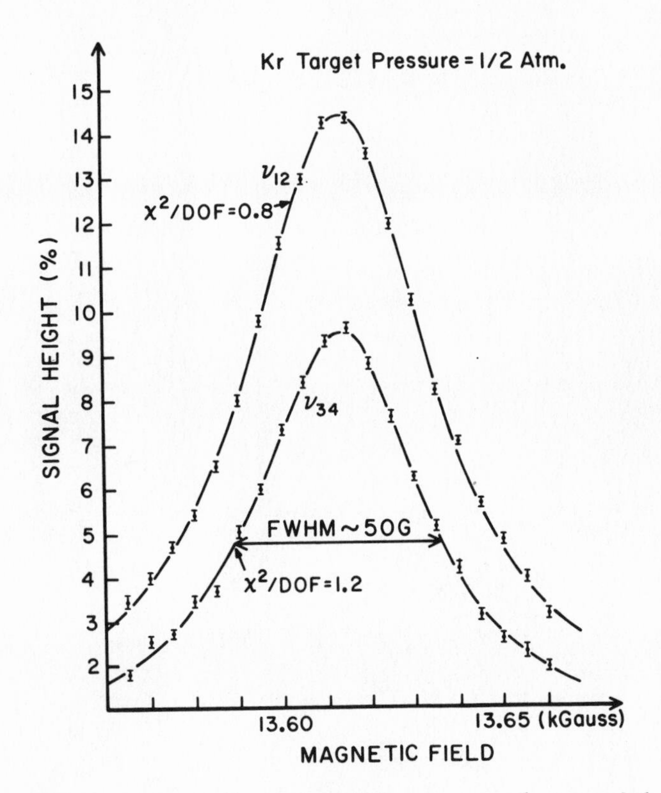

Fig. 4. Resonance curve for transition ν_{12} and ν_{34} with fitted theoretical line shapes.

under computer control together with the NMR signal. For each field point the microwave frequency is changed from the ν_{12} to the ν_{34} frequency and also the microwave power is turned on and off with a modulation period of several minutes. The fitted theoretical line shapes are shown. The preliminary result (Egan et al., 1979) of our data analysis for the hfs interval $\Delta\nu$ and for the ratio μ_μ/μ_p of muon to proton magnetic moments is

$$\Delta\nu = 4\ 463\ 302.90 \pm 0.27\ (0.10)\ \text{kHz}\ [0.06\ \text{ppm}] \tag{1}$$

$$\mu_\mu/\mu_p = 3.183\ 347\ 8 \pm 0.000\ 002\ 6\ (0.000\ 000\ 8)\ [0.8\ \text{ppm}] \tag{2}$$

in which one standard deviation total errors are given. The errors given in () parentheses are statistical errors. Further analysis of our systematic errors should allow us to quote total errors about 1/2 of those given here.

The present theoretical value for $\Delta\nu$ is given in Table 1. There are two principal contributions to the estimated error in

Table 1. Theoretical Value for Muonium $\Delta\nu$ and Comparison with Experiment

$$\Delta\nu_{th} = [\frac{16}{3}\alpha^2 cR_\infty(\mu_\mu/\mu_B^e)][1 + (m_e/m_\mu)]^{-3}$$

$$\times [1 + \frac{3}{2}\alpha^2 + a_e + \varepsilon_1 + \varepsilon_2 + \varepsilon_3 - \delta'_\mu]$$

$$a_e = \frac{\alpha}{2\pi} - 0.32848\frac{\alpha^2}{\pi^2} + 1.184\ (7)\frac{\alpha^3}{\pi^3}\ ;\ \varepsilon_1 = \alpha^2(\ln 2 - 5/2)$$

$$\varepsilon_2 = -\frac{8\alpha^3}{3\pi}\ln\alpha(\ln\alpha - \ln 4 + \frac{281}{480});\ \varepsilon_3 = \frac{\alpha^3}{\pi^3}(18.4 \pm 5)$$

$$\delta'_\mu = \frac{m_e}{m_\mu}\{\frac{3\alpha}{\pi}[1-(\frac{m_e}{m_\mu})^2]^{-1}\ln\frac{m_\mu}{m_e} + 2\alpha^2\ln\alpha\,[1 + (\frac{m_e}{m_\mu})]^{-2} + 2\frac{\alpha^2}{\pi^2}(\ln\frac{m_\mu}{m_e})^2\}$$

$\alpha^{-1} = 137.035\ 963\ (15)\ (0.11\ \text{ppm})$ [a]

$R_\infty = 1.097\ 373\ 147\ 6\ (32) \times 10^5\ \text{cm}^{-1}\ (0.003\ \text{ppm})$ [a]

$c = 2.997\ 924\ 58\ (1.2) \times 10^{10}\ \text{cm/sec}\ (0.004\ \text{ppm})$ [a]

$\mu_\mu/\mu_B^e = (\mu_\mu/\mu_p)\ (\mu_p/\mu_B^e);\ \mu_p/\mu_B^e = 1.521\ 032\ 209\ (16)\ (0.01\ \text{ppm})$ [b]

$m_\mu/m_e = 206.768\ 26\ (10)\ (0.5\ \text{ppm})$ [c]

$\mu_\mu/\mu_p = 3.183\ 345\ 5\ (17)\ (0.5\ \text{ppm})$

$\Delta\nu_{th} = (\mu_\mu/\mu_p)\alpha^2\ [2.632\ 945\ 55 \pm 0.6\ \text{ppm}] \times 10^{10}\ \text{kHz}$

$\Delta\nu_{th} = (\mu_\mu/\mu_p)\ \{1.402\ 079\ 91 \pm 0.7\ \text{ppm}\} \times 10^6\ \text{kHz}$

$\Delta\nu_{th} = 4\ 463\ 304.8\ (4.5)\ (1.0\ \text{ppm})\ \text{kHz}$

$\Delta\nu_{exp} = 4\ 463\ 302.90\ (27)\ (0.06\ \text{ppm})\ \text{kHz}$

$\Delta\nu_{th} - \Delta\nu_{exp} = 1.9\ (4.5)\ \text{kHz}$

Determination of α:

$$\Delta\nu_{th} = \alpha^2[8.381\ 575\ 37 \pm 1.0\ \text{ppm}] \times 10^{10}\ \text{kHz}$$

$$\alpha^{-1} = 137.035\ 99\ (7)\ (0.5\ \text{ppm})$$

---

a. (Williams and Olsen, 1979); b. (Hughes and Kinoshita, 1977); c. (Obtained from μ_μ/μ_p in this Table).

$\Delta\nu_{th}$. One is inaccuracy in the calculation of the radiative correction term ε_3 and uncalculated radiative - recoil terms of relative order $(m_e/m_\mu)(\frac{\alpha}{\pi})^2 \ln(m_\mu/m_e)$ and $(m_e/m_\mu)\alpha^2$. Together these contribute an estimated error of about 0.6 ppm. (Caswell and Lepage, 1978; Kinoshita, 1979). The second is the experimental uncertainty in μ_μ/μ_p. The value given in Table 1 is the average value from four experiments: the value given in Eq. 2, an earlier value obtained from muonium (Casperson et al., 1977), a recent value from muon spin resonance in liquid bromine (Camani et al., 1978) and an older μSR value in H_2O (Crowe et al., 1972).

The agreement of the theoretical and experimental values of $\Delta\nu$ is well within the error of the theoretical value. This agreement provides one of the most sensitive tests of modern muon electrodynamics.

The modern unified gauge theories of the weak and electromagnetic interactions, in particular the standard theory of Weinberg and Salam, predict a contribution of the weak interaction to $\Delta\nu$ arising from an axial vector-axial vector coupling of the electron and muon (Beg and Feinberg, 1974) which amounts to about -0.07 kHz. This weak interaction contribution is comparable to the present experimental precision in the determination of $\Delta\nu$, but is about a factor of 100 less than the error in the theoretical value of $\Delta\nu$.

The suggestion that there might be a weak interaction directly coupling muonium $M \equiv \mu^+e^-$ to antimuonium $(\bar{M} \equiv \mu^-e^+)$ was made some 20 years ago (Pontecorvo, 1957; Feinberg and Weinberg, 1959, 1961; Glashow, 1961) in connection with theories of the relation of the electron and the muon. A four-fermion weak interaction of the universal V-A form was suggested.

$$\mathcal{H}_{M-\bar{M}} = \frac{G}{\sqrt{2}} \bar{\psi}_\mu \gamma_\mu (1 + \gamma_5) \psi_e \bar{\psi}_\mu \gamma^\lambda (1 + \gamma_5) \psi_e + \text{h.c.} \qquad (3)$$

in which the field operators ψ and γ matrices have their usual meanings and G is the coupling constant. The matrix element between M and $\bar{M}$ ground states is

$$\langle \bar{M} | \mathcal{H}_{M-\bar{M}} | M \rangle = 1.0 \times 10^{-12} \frac{G}{G_F} \text{ eV}, \qquad (4)$$

in which G_F is the Fermi vector coupling constant of β decay. This interaction would contribute to the hfs interval the amount of 4 kHz for $G=G_F$. The present agreement of $\Delta\nu_{th}$ and $\Delta\nu_{exp}$ to about 1 ppm thus sets an upper limit on G of about G_F.

Active theoretical work is now in progress on $\Delta\nu$ to evaluate further radiative and recoil contributions to the level of a few parts in 10^7. On the experimental side, apart from the improvement of about a factor of 2 in $\Delta\nu$ and in μ_μ/μ_p anticipated when the

Table 2. Muon Properties

Mass

$$\frac{m_{\mu^+}}{m_e} = 206.768\ 26\ (10)\ (0.5\ \text{ppm})$$

$$\frac{m_{\mu^-}}{m_e} = 206.765\ (20)\ (100\ \text{ppm})$$

Spin

$$I_\mu = 1/2$$

Magnetic Moment

$$\frac{\mu_{\mu^+}}{\mu_p} = 3.183\ 345\ 5\ (17)\ (0.5\ \text{ppm})$$

$$\frac{\mu_{\mu^-}}{\mu_p} = 3.183\ 4\ (9)\ (300\ \text{ppm})$$

g-Value

$$(g_{\mu^+} - 2)/2 = 1\ 165\ 911\ (11) \times 10^{-9}\ (10\ \text{ppm})$$

$$(g_{\mu^-} - 2)/2 = 1\ 165\ 937\ (12) \times 10^{-9}\ (10\ \text{ppm})$$

Electric Dipole Moment

$$\mu_e \leq 8 \times 10^{-19}\ \text{e-cm}\ (95\%\ \text{confidence level})$$

Statistics

Fermi - Dirac

analysis of the recent LAMPF experiment is completed (Egan et al., 1979), the next important advance will probably involve the use of line-narrowing techniques in the measurement of strong field Zeeman transitions. At LAMPF this will require a pulsed muon beam with a time structure of the order of the muon lifetime of 2μs, which might be obtained either by pulsing the primary proton beam or by pulsing the muon beam in the muon channel.

Table 2 summarizes the properties of the muon.

3. MUONIUM IN VACUUM

Two further researches on muonium that would be of great interest are, firstly, the measurement of the fine structure, Lamb shift, and hyperfine structure intervals in the n=2 excited state, and, secondly, a direct search for the conversion of muonium to antimuonium.

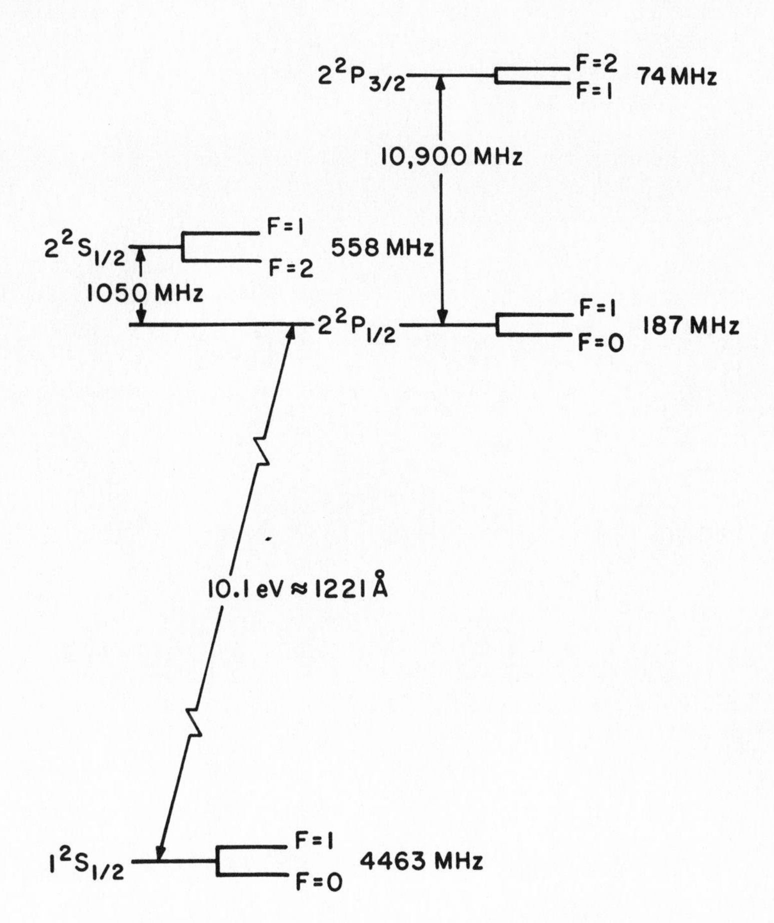

Fig. 5. Energy level diagram for the n=1 and n=2 states of muonium.

The energy level diagram for the n=1 and n=2 states of muonium is shown in Fig. 5. It would be particularly valuable to make a precise measurement of the Lamb shift, or $2^2S_{1/2}$ to $2^2P_{1/2}$ energy difference. Although the Lamb shift in H is known experimentally to high precision (Lundeen and Pipkin, 1975; Mohr, 1977), its theoretical interpretation is limited by uncertainties associated with proton structure. Muonium is an ideal system in which to test the quantum electrodynamic theory of the Lamb shift since both the muon and the electron are believed to be structureless particles, unlike the proton which has a structure associated with the strong interactions.

Microwave spectroscopy experiments to measure the fine structure, Lamb shift, and hyperfine structure of the n=2 state can be done in principle by detecting the change in decay positron angular distribution associated with a change in muon spin direction. However, such an experiment would seem to require that the metastable 2S state of muonium be obtained in a collision-free or vacuum environment. It is well known (Kass and Williams, 1971) that the cross section for quenching the 2S state of H in atomic collisions is large ($\sim 10^{-14}$ cm^2) at thermal energies. Muonium will have approximately the same collision quenching cross section as hydrogen, and hence, for example, in 1 atmosphere of Ar thermal muonium in the 2S state would have a lifetime of about 10^{-11}s. This lifetime is too short to allow a microwave spectroscopy experiment.

A search for a direct transition of muonium to antimuonium with a sensitivity adequate to detect a branching ratio corresponding to a value of the coupling constant G in the Hamiltonian of Eq. (3) equal to or less than G_F would be of great interest. Some forms of the unified gauge theories of weak and electromagnetic interactions predict that the $M \to \bar{M}$ transition may occur at a rate corresponding to $G \simeq 0.1\ G_F$ and be associated with a massive Higgs boson (Derman, 1978, 1979). Such a transition would of course violate the additive law of muon number conservation, but would be allowed by a multiplicative law of muon number conservation.

If muonium M is formed at time t=0, the $\bar{M}$ component of the system wavefunction develops with time due to the presence of $\mathcal{H}_{M-\bar{M}}$, and the probability that the muon will decay from the $M-\bar{M}$ system as a μ^- rather than as a μ^+ is

$$\begin{aligned} P(\bar{M}) &= \int_0^\infty \gamma e^{-\gamma t} |\langle \bar{M}|\ \psi(t)\rangle|^2\, dt \\ &= \delta^2/2\hbar^2\gamma^2 = 2.5 \times 10^{-5}(G/G_F)^2 \end{aligned} \tag{5}$$

where γ is the free muon decay rate, $4.5 \times 10^5\ \text{sec}^{-1}$. One experiment has been done (Amato et al., 1969) to search for the $M-\bar{M}$ transi-

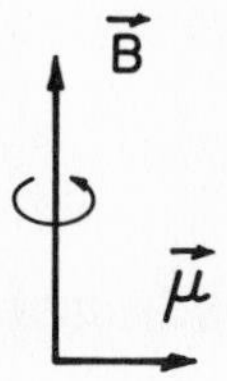

$$f_L = \frac{\mu B}{Ih}$$

$$f_{L\mu} = \frac{\mu_\mu B}{(1/2)h} = 13.5 \text{ B kHz} \quad \text{(free muon)}$$

$$f_{LM} = \frac{\mu_M B}{h} = 1.40 \text{ B MHz} \quad (\text{muonium } (F,M_F)=(1,1))$$

Fig. 6. Figure illustrating Larmor precession of free muon and muonium.

tion by forming M in 1 atm of Ar gas, and obtained the limit $G \leq 5800\, G_F$ at the 95% confidence level. Collisions of M with Ar remove the degeneracy of M and $\bar{M}$ and severely inhibit the development of the $\bar{M}$ component in the M-$\bar{M}$ system wavefunction, and largely determined the sensitivity of this early experiment. Greatly increased sensitivity in this sort of experiment requires that muonium be in vacuum. In a storage ring experiment an upper limit to the cross section for the reaction $e^- + e^- \rightarrow \mu^- + \mu^-$ was established (Barber et al., 1969), which implied $G \leq 610\ G_F$.

Because of current interest in the muonium to antimuonium conversion and in the Lamb shift in the n=2 state of muonium, two similar experiments have been done recently to try to obtain muonium in vacuum. Positive muons are stopped in a target consisting of thin foils in vacuum with adjacent foils spaced by distances of several mm. The Larmor precession frequency signals characteristic of free muons and of muonium (see Fig. 6) are observed. The first experiment (Barnett et al., 1977) was done at SREL and reported the abundant formation of thermal muonium by gold foils in vacuum with a formation probability of 0.28 ± 0.05. The second experiment, of higher sensitivity, was done at LAMPF (Beer et al., 1979) and was unable to confirm the results of the earlier experiment; indeed no thermal muonium signal was observed and a formation probability of less than 0.06 at the 95% confidence level was reported.

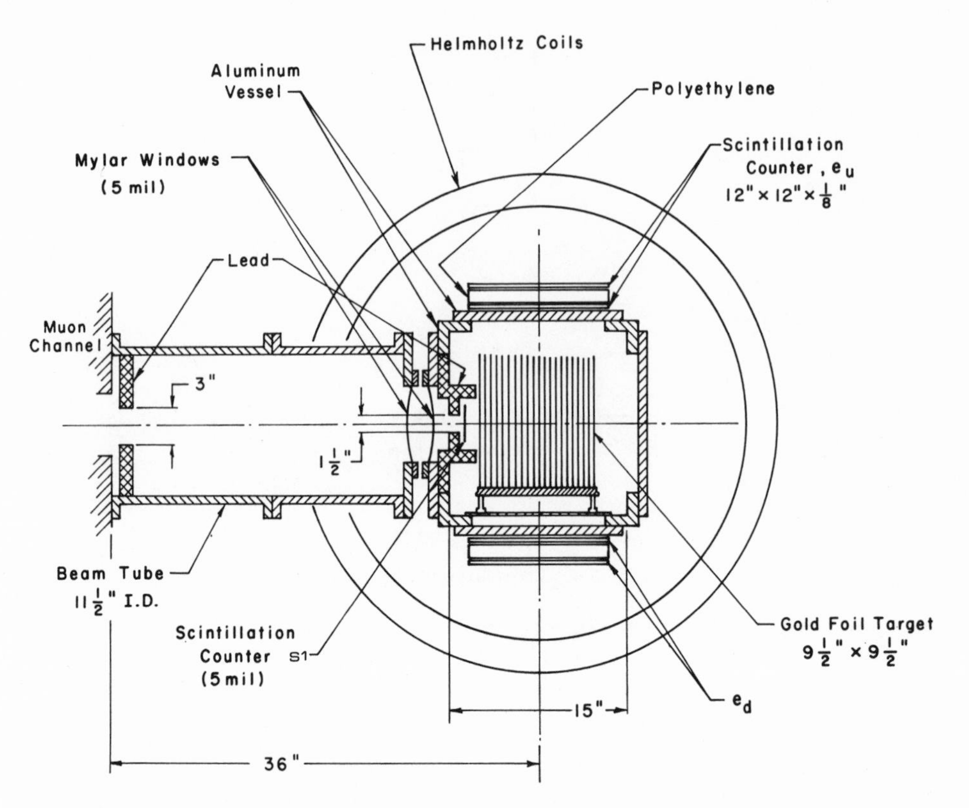

Fig. 7. Experimental apparatus to search for muonium formation in a thin foil target.

A brief discussion of the LAMPF experiment will be given. A diagram of the experimental apparatus with the target of thin (1000A°) gold foils is shown in Fig. 7. The range curve for the surface muon beam is shown in Fig. 8 where it is seen that most of the μ^+ stop in the thin foil target. Free muon precession signals were observed but no muonium signals were observed for the various types of foil targets used. The following upper limits (95% confidence level) for the fraction of stopped muons which form free muonium in the thermal energy range have been reported: 6% for gold foils, 9% for MgO coated gold foils, 12% for SiO (SiO_2) coated gold foils, and 12% for aluminum (surface oxides) foils. Although the discrepancy between the two experiments might conceivably be due to different surface conditions on the Au foils in the two experiments, it seems more likely that the more sensitive LAMPF experiment is correct and that no thermal muonium is formed in vacuum between the foils.

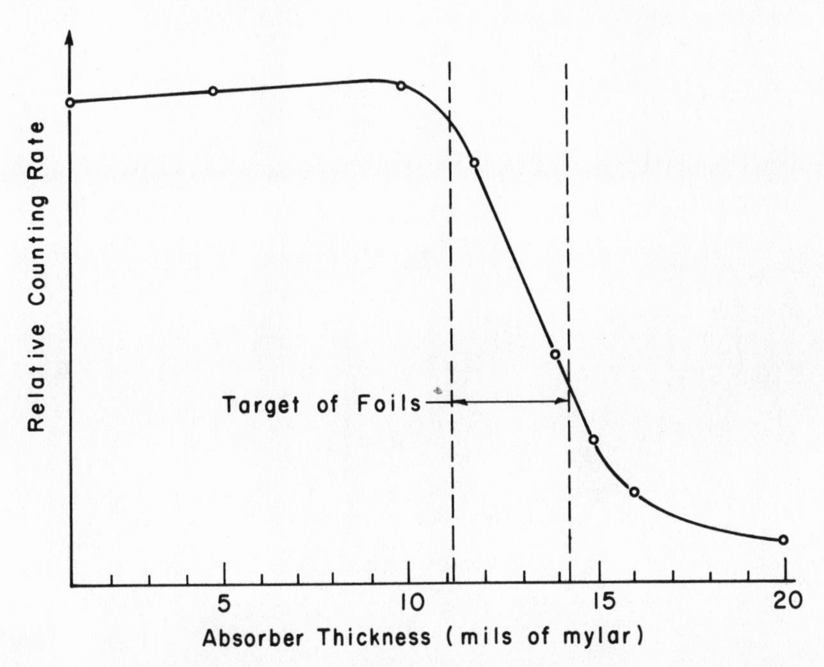

Fig. 8. Range curve for positive muon beam used in experiment of figure 7.

Further searches for thermal muonium formation from the surfaces of foils would seem to require a deeper scientific approach. The physical processes occurring include the stopping of μ^+ in the foil, its diffusion in the foil, and finally its behaviour at the surface of the foil. These processes are of interest to current μSR studies in solids and are being actively pursued.

Although we do not know how to form thermal muonium in a vacuum, we can be quite sure that more energetic muonium will emerge from the surface of a foil through which a relatively energetic μ^+ has passed. A large body of information is available (Bashkin, 1976) from the field of beam foil spectroscopy on the charge states and atomic states produced when protons pass through foils. In general we can expect that the neutral atom H will be formed abundantly when the proton velocity is comparable to the velocities of the outer shell atomic electrons. For ground state (or 2S state) muonium formation this implies muon and hence muonium kinetic energies of the order of 5 keV. It would seem promising to try to form energetic muonium in vacuum in this way and to consider the possibility of investigating the $M \to \bar{M}$ conversion and the Lamb shift in muonium.

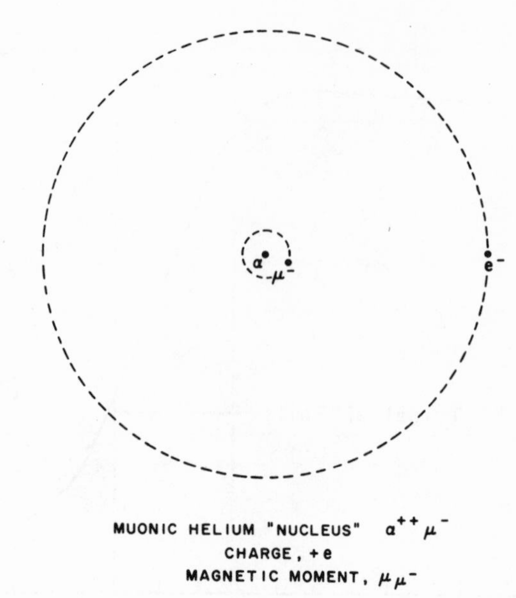

Fig. 9. Bohr orbits of muonic helium atom, $^4He\mu^-e^-$.

MUONIC HELIUM ATOM

Some years ago the muonic helium atom ($^4He\mu^-e^-$) was discovered (Souder et al., 1975) from the observation of its characteristic, muonium-like, Larmor precession frequency. The Bohr picture of this amusing atom is indicated in Fig. 9. In a first approximation it can be considered to consist of an inner core or pseudo-nucleus consisting of $^4He\mu^-$ and of an outer electron. The pseudo nucleus has a unit positive charge and a magnetic moment equal to that of μ^-. Hence the muonic helium atom is analogous to muonium and would be expected to have a hyperfine structure interval similar in magnitude to that of μ^+e^- but inverted because of the negative magnetic moment of μ^-.

The hyperfine structure transition of $^4He\mu^-e^-$ has recently been observed at SIN in a microwave magnetic resonance experiment at zero static magnetic field (Arnold et al., 1979). The resonance signal is relatively weak because of the low residual polarization of μ^- in $^4He\mu^-e^-$. The value reported for $\Delta\nu$ is 4464.9(5) MHz, which is close to the value of $\Delta\nu$ for μ^+e^- which is 4463.3 MHz. Much improved precision in the determination of $\Delta\nu$ should be possible from the zero field experiment. A strong magnetic field measurement is in progress at LAMPF which should determine both $\Delta\nu$ and μ_μ/μ_p.

The theoretical value of $\Delta\nu$ for $^4He\mu^-e^-$ can be written:

(Huang, 1974; Huang and Hughes, 1979)

$$\Delta\nu = \Delta\nu_F[1 + \delta^{rel} + \delta^{rad} + \delta^{rec}] \quad (6)$$

in which $\Delta\nu_F$ is the leading Fermi term, and δ^{rel}, δ^{rad} and δ^{rec} refer to relativistic, radiative, and recoil contributions. Evaluation of $\Delta\nu_F$ with high accuracy is rather difficult because the most important aspect of the wavefunction involves the correlation of the muon and electron whereas the binding energy of the atom, the quantity to which a variational calculation applies, is determined principally by the muon part of the wavefunction. Two recent calculations of $\Delta\nu$ have been done. One uses a second order perturbation approach to compute $\Delta\nu_F$ (Lakdawala and Mohr, 1979) and the other uses a variational approach with a 496 term wavefunction (Huang and Hughes, 1979). These two calculations give, respectively, $\Delta\nu$ = (4462.6 ± 3) MHz and $\Delta\nu$ = (4465.1 ± 1.0) MHz, in which the estimated errors arise from calculation of $\Delta\nu_F$. Both theoretical values are in satisfactory agreement with the experimental value. Improvement in the accuracy of the value of $\Delta\nu_F$ is particularly needed.

Research supported in part by the U.S. Department of Energy under contract No. EY-76-C-02-3075 and The John Simon Guggenheim Memorial Foundation.

REFERENCES

Amato, J.J., et al., (1968), Phys. Rev. Lett. 21, 1709.

Arnold, K.P., et al., (1979), "Measurement of the Muonic Helium ($\alpha\mu^-e^-$) Hyperfine Structure Interval." Abstract submitted to the 8th International Conference on High Energy Physics and Nuclear Structure, August 13-17, 1979, University of British Columbia Vancouver, British Columbia, Canada.

Bailey, J., et al., (1979), Nucl. Phys. B150, 1.

Barber, W.C., et al., (1969), Phys. Rev. Lett. 22, 902.

Barnett, B.A., et al., (1977), Phys. Rev. A15, 2246.

Bashkin, S., (1976), Beam-Foil Spectroscopy, (Springer-Verlag, Berlin, Heidelberg).

Beer, W., et al., (1979), "Search for Free Muonium in Vacuum Produced with Thin Foil Targets." Abstract submitted to the 8th International Conference on High Energy Physics and Nuclear Structure, August 13-17, 1979, University of British Columbia Vancouver, British Columbia, Canada; Bull. Am. Phys. Soc. 24, 675.

Beg, M.A., and Feinberg, G., (1974), Phys. Rev. Lett. 33, 606.

Brewer, J.H., et al., (1975), Muon Physics, Vol. III, ed. V.W. Hughes and C.S. Wu (Academic Press, New York) p. 3.

Brewer, J.H., and Crowe, K.M., (1978), Ann. Rev. Nucl. Part. Sci. 28, 239.

Camani, M., et al., (1978), Phys. Lett. 77B, 326.

Casperson, D.E., et al., (1977), Phys. Rev. Lett. 38, 956; 1504.

Caswell, W.E. and Lepage, G.P., (1978), Phys. Rev. Lett. 41, 1092.

Caswell, W.E., and Lepage, G.P., (1978), Phys. Rev. A18, 810.
Crowe, K.M., et al., (1972), Phys. Rev. D5, 2145.
Derman, E., (1978), Phys. Lett. 78B, 497
Derman, E., (1979), Phys. Rev. 19D, 317.
Egan, P.O., et al., (1979), "Higher Precision Measurement of the HGS Interval of Muonium and of the Muon Magnetic Moment." Abstract submitted to the 8th International Conference on High Energy Physics and Nuclear Structure, August 13-17, 1979, University of British Columbia, Vancouver, British Columbia, Canada; Bull. Am. Phys. Soc. 24, 26.
Feinberg, G., and Weinberg, S., (1959), Nuovo Cimento 14, 571.
Feinberg, G., and Weinberg, S., (1961), Phys. Rev. 123, 1439.
Glashow, S.L., (1961), Phys. Rev. Lett. 6, 196.
Huang, K.-N., (1974), Ph.D. Thesis, Yale University.
Huang, K.-N., and Hughes, V.W., (1979), "Theoretical Hyperfine Structure of Muonic Helium." accepted for publication in Phys. Rev. A; Abstract submitted to the 8th International Conf. on High Energy Physics and Nuclear Structure, August 13-17, 1979, University of British Columbia, Vancouver, British Columbia, Canada.
Hughes, V.W., and Kinoshita, T., (1977), Muon Physics, Vol. I, ed. V.W. Hughes and C.S. Wu (Academic Press, New York) p. 11.
Kass, R.S., and Williams, W.L., (1971), Phys. Rev. Lett. 27, 473.
Kinoshita, T., (1979), Proceedings of the 19th International Conf. on High Energy Physics Tokyo, 1978 (Physical Society of Japan), p. 571
Lakdawala, S.D., and Mohr, P.J., (1979), "Hyperfine Structure in Muonic Helium." Preprint.
Lepage, G.P., (1977), Phys. Rev. A16, 863.
Lundeen, S.R., and Pipkin, F.M., (1975), Phys. Rev. Lett. 34, 1368.
Mohr, P.J., (1977), Atomic Physics 5, eds. R. Marrus, M. Prior, and H. Shugart (Plenum Press, New York) p. 37.
Pontecorvo, B., (1957), Zh. Eksp. Teoret. Fiz. 33, 549.
Prescott, C.Y., et al., (1978), Phys. Lett. 77B, 347.
Prescott, C.Y., et al., (1979), SLAC-PUB-2319. (May 1979).
Reist, H.W., et al., (1978), Nucl. Inst. Meth. 153, 61.
Souder, P.A., et al., (1975), Phys. Rev. Lett. 34, 1417.
Thompson, P.A., et al., (1979), Nucl. Inst. Meth. 161, 391.
Weinberg, S., (1979), Proceedings of the 19th International Conf. on High Energy Physics Tokyo, 1978 (Physical Society of Japan) p. 907.
Williams, E.R., and Olsen, P.T., (1979), Phys. Rev. Lett. 42, 1575.

POSITRONIUM

Vernon W. Hughes

Yale University
Physics Department, Gibbs Laboratory
New Haven, CT 06520 U.S.A.

The general motivation for studying positronium (e^+e^-) is similar to that discussed in the accompanying short article on muonium. Positronium is the bound state of a particle (e^-) and its antiparticle (e^+) which interact through the electromagnetic interaction. Perhaps the most special feature about positronium is that its quantitative understanding requires the full use of a relativistic two-body bound state equation; because the positron and electron masses are equal, recoil contributions are of central important to positronium. Several review articles on fundamental studies of positronium have appeared in recent year (Hughes 1969, 1973; Mills et al., 1977).

A summary of our present experimental information on energy intervals of positronium is given in Fig. 1. Two recent measurements of $\Delta\nu$ in the n=1 state have been done at Yale (Egan et al., 1977) and at Brandeis (Mills and Bearman, 1975) by observing the Zeeman transition in a microwave spectroscopy experiment. The combined experimental value is given in Fig. 1. The theoretical value for the n=1 fine structure interval is (Kinoshita, 1979):

$$\Delta\nu\ ({}^3S_1 - {}^1S_0) = \alpha^2 c\ R_\infty\left[\frac{7}{6} - \frac{\alpha}{\pi}\left(\frac{16}{9} + \ell n 2\right) + \frac{5}{12}\alpha^2 \ell n \alpha^{-1}\right]$$

$$= 203\ 400.3\ \text{MHz} \tag{1}$$

Hence we find

$$\Delta\nu_{th} - \Delta\nu_{exp} = 14.6\ (1.0)\ \text{MHz} \tag{2}$$

Some terms of relative order α^2 have been evaluated but a complete

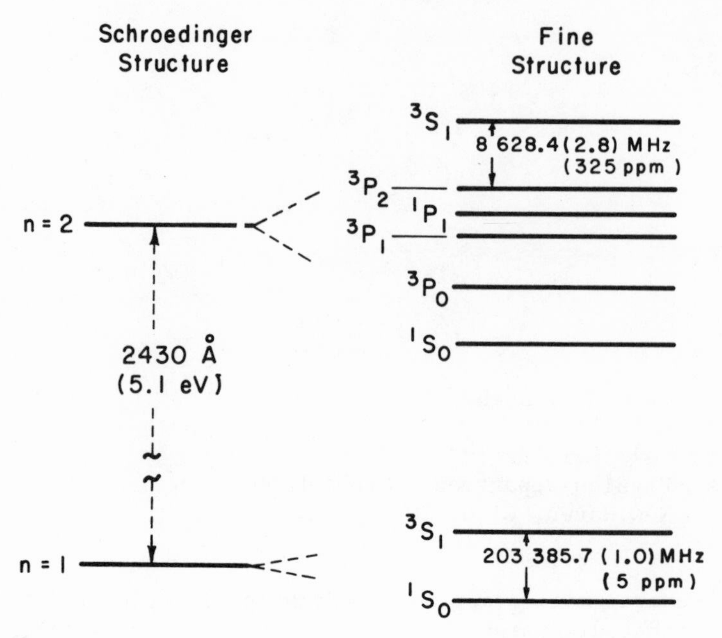

Fig. 1. Energy levels of the n=1 and n=2 states of positronium.

calculation to this order has not yet been done. A more meaningful comparison of theory and experiment requires a full calculation of all terms of relative order α^2.

The only other measured fine structure interval (Mills et al., 1975) is the $^3S_1$ to $^3P_2$ energy difference in the n=2 excited state. Its theoretical value is (Fulton and Martin, 1954)

$$\nu(2^3S_1-2^3P_2) = \frac{23}{480}\,\alpha^2 R_\infty(1 + 3.766\alpha) = 8625.14 \text{ MHz} \tag{3}$$

The agreement of theory and experiment is good.

$$\nu_{th}-\nu_{exp} = -\ 3.3(2.8) \text{ MHz} \tag{4}$$

In the past few years new experimental results on energy level measurements have not appeared. However we can expect soon new higher precision measurements on $\Delta\nu$ in the n=1 state (Yale) and on fine structure intervals in the n=2 state (Brandeis). Furthermore there is considerable interest in several laboratories in an experiment to measure the 1S→2S transition wavelength in positronium in order to determine the Lamb shift in the n=1 state. The technique being considered is two photon laser spectroscopy, similar to the corresponding hydrogen experiment (Hansch et al., 1975).

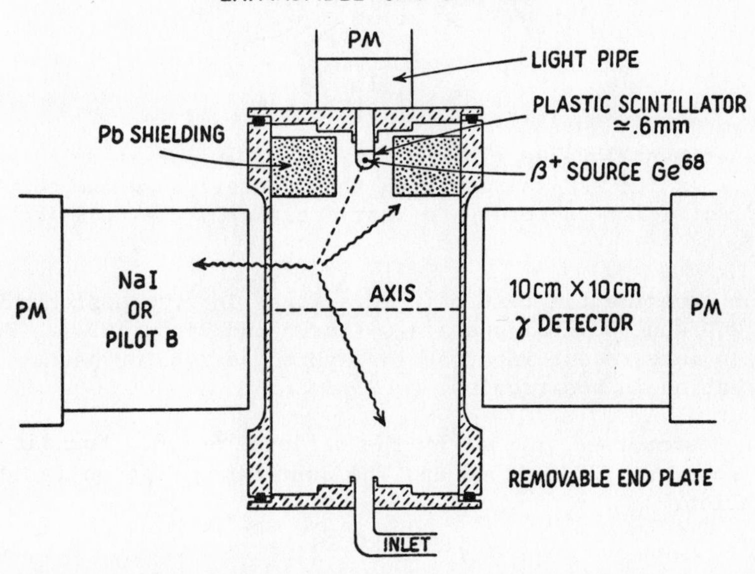

Fig. 2. Experimental apparatus used in the measurement at Michigan of the decay rate of the n=1 state of orthopositronium.

The principal advance in our basic knowledge of positronium recently has been of the n=1 state orthopositronium decay rate. The best current measurements involve positronium in a gas and give

$$\lambda(1^3S_1) = 7.056(7) \times 10^6 \ s^{-1} \qquad \text{(Gidley et al., 1978)}$$

$$\lambda(1^3S_1) = 7.045(6) \times 10^6 \ s^{-1} \qquad \text{(Griffith et al., 1978)} \tag{5}$$

A diagram of the Michigan experiment is shown in Fig. 2. A sizable extrapolation to zero gas pressure is made to obtain these values. The theoretical value is (Caswell et al., 1977; 1979)

$$\lambda = \frac{2}{9\pi} \alpha^6 (\frac{mc^2}{\hbar}) \ (\pi^2-9)[1-(10.348 \pm 0.070) \ \frac{\alpha}{\pi} - \frac{1}{3} \alpha^2 \ell n\alpha^{-1}] \tag{6}$$

$$= (7.0386 \pm 0.00016) \times 10^6 \ s^{-1}$$

in reasonable agreement with the experimental values.

The theoretical value of the singlet ground state decay rate is (Harris and Brown, 1957).

$$\lambda(^1S_0) = \frac{1}{2}\alpha^5 \frac{mc^2}{\hbar} \left[1 - \frac{\alpha}{\pi}\left(5 - \frac{\pi^2}{4}\right)\right]$$

$$= 0.7985 \times 10^{10}\ s^{-1} \tag{7}$$

The experimental value (Theriot et al., 1970) of $\lambda(^1S_0)$ = 0.799(11) x 10^{10} s^{-1} agrees with the theoretical value but is not of sufficient accuracy to test decisively the relative order α term.

An experimental test of C invariance in orthopositronium decay was done several years ago (Marko and Rich, 1974). There has been some recent interest in ways to search for parity non-conservation in positronium.

Research supported in part by the National Science Foundation under contract PHY78-15068 and The John Simon Guggenheim Memorial Foundation.

REFERENCES

Caswell, W.E., et al., (1979), Scheduled for Phys. Rev. A. July.
Caswell, W.E., et al., (1977), Phys. Rev. Lett. 38, 488.
Egan, P.O., et al., (1977), Phys. Rev. A15, 251.
Fulton, T., and Martin, P.C., (1954), Phys. Rev. 95, 811.
Gidley, D.W. et al., (1978), Phys. Rev. Lett. 40, 737.
Griffith, T.C., et al., (1978), J. Phys. B 11, L743.
Hansch, T.W., et al., (1975), Phys. Rev. Lett. 34, 307.
Harris, I., and Brown, L.M., (1957), Phys. Rev. 105, 1656.
Hughes, V.W., (1973), Physik 1973, Germany Physical Society Conf. (Physik Verlag Gamblt, Germany) p. 123; (1969), Physics of the One-and-Two Electron Atoms, eds. F. Bopp and H. Kleinpoppen (North-Holland Publishing Co,. Amsterdam) p. 407.
Kinoshita, T., (1979), Proceedings of the 19th Int'l. Conf. on High Energy Physics Toyko, 1978 (Physical Society of Japan), p. 571.
Marko, K., and Rich, A., (1974), Phys. Rev. Lett. 33, 980.
Mills, A.P., Jr., et al., (1975), Phys. Rev. Lett. 34, 1541.
Mills, A.P., Jr., and Bearman, G.H., (1975), Phys. Rev. Lett. 34, 246.
Mills, A.P., Jr., et al., (1977), Atomic Physics 5, eds. R. Marrus, M. Prior, and H. Shugart (Plenum Press, New York) p. 103.
Theriot, E.D., et al., (1970), Phys. Rev. A2, 707.

ANOMALOUS MUON CAPTURE AND LEPTON NUMBER CONSERVATION

Beat Hahn and Thomas Marti

Department of High Energy Physics
University of Berne
CH-3012 Berne, Switzerland

1. LEPTON CONSERVATION SCHEMES[1)]

Usually, lepton numbers L_e, L_μ are assigned as shown in the following table:

Particle	L_e	L_μ
e^-, ν_e	1	0
e^+, $\overline{\nu}_e$	-1	0
μ^-, ν_μ	0	1
μ^+, $\overline{\nu}_\mu$	0	-1
other particles	0	0

The lepton number L_τ, which can be assigned correspondingly to the $\tau^\pm$, ν_τ and $\overline{\nu}_\tau$ particles, will not be considered for the moment.

There are essentially three lepton conservation schemes which have been proposed to explain

i) the nonobservation of processes like $\mu \to e\gamma$, $\mu \to 3e$, $\mu^+ \to e^+\nu_\mu\overline{\nu}_e$ and of the neutrinoless muon capture process on nuclei, $\mu^- A \to Ae^-$ (1), and

ii) the possible observation of $\mu^- A \to A'e^+$ (2) and $K^+ \to \mu^+ e^+ \pi^-$ (2)'

Rates of processes (1) and (2) will be compared to the normal, charged current muon capture process: $\mu^- A \to \nu_\mu$ + anything (3).

The three lepton conservation schemes are the following:

i) Additive law: ΣL_e = const., and ΣL_μ = const.
ii) Konopinski-Mahmoud-law: $\Sigma (L_e - L_\mu)$ = const.
iii) Multiplicative law: $\Sigma (L_e + L_\mu)$ = const., and $(-1)^{\Sigma L_e}$ = const.

The two latter conservation schemes are both less stringent than the first one.

Of course, all of these schemes allow the normal muon capture process (3), as well as all observed leptonic or semileptonic processes. The doubly charged currents (2), (2)', which will not be discussed here any further (see Ref. 2), are not forbidden by the Konopinski-Mahmoud scheme but forbidden by the two others, while process (1) is forbidden by all three schemes. Recent data[3] seem to disfavour strongly the multiplicative law, but a definite decision between the three schemes cannot be taken yet.

In a unified gauge theory of electromagnetic and weak interactions, lepton number conservation might be only an approximate law, and rates are predicted[4-7] for the processes $\mu^- A \to A e^-$, $\mu \to e\gamma$, $\mu \to 3e$ which are not far from the experimental upper bounds.

2. QUARK MIXING AND LEPTON MIXING

In order to understand lepton-flavour conservation (or nonconservation) in weak interactions, let us assume an analogy between leptons and quarks, and look first to the quark sector. In the minimal SU(2) x U(1) model of Salam and Weinberg[8], the left-handed quarks are arranged in doublets of weak isospin, while the right-handed quarks are singlet states

$$\begin{pmatrix} u \\ d' \end{pmatrix}_L, \quad \begin{pmatrix} c \\ s' \end{pmatrix}_L, \quad u_R,\ d_R,\ c_R,\ s_R$$

(to start with, heavier quarks are ignored)

Here, the weak eigenstates d', s' differ from the mass eigenstates d, s ("physical" states) according to the Cabibbo mixing scheme:

$$d' = d \cos \tilde{\theta}_c + s \sin \tilde{\theta}_c$$

$$s' = -d \sin \tilde{\theta}_c + s \cos \tilde{\theta}_c$$

The value of the Cabibbo angle $\tilde{\theta}_c$ may be found[9] by comparing muon decay ($\mu^+ \to \bar{\nu}_\mu\ e^+\ \nu_e$) with nuclear beta decay ($u \to d\ e^+\ \nu_e$), the latter being suppressed by a factor $\cos^2 \tilde{\theta}_c$. The experimental data give[9] $\cos^2 \tilde{\theta}_c \simeq 0{,}95$ or $\cos \tilde{\theta}_c = 0{,}9737 \pm 0{,}0025$ and $\tilde{\theta}_c = (13{,}2 \pm 0{,}6)^\circ$.

u-s-transitions are suppressed, apart from phase space, by

a factor $\sin^2\tilde{\theta}_c \simeq 0{,}05$ (e.g. $K^+ \equiv u\bar{s} \to \mu^+\nu_\mu$). As the weak neutral current in the Cabibbo mixing scheme contains no strangeness changing piece[10], s-d-transitions cannot occur through a single Z^o exchange. A non-vanishing s-d-amplitude does in fact arise from two W exchange, e.g. in $K_L^o \to \mu^+\mu^-$, as explained below. The mass eigenstates K_L^o, K_S^o are linear combinations of the states K^o, $\overline{K^o}$ of definite strangeness:

$$K_L^o \simeq \frac{A}{\sqrt{2}}(K^o-\overline{K^o}) + \varepsilon_{CP}\cdot\frac{A}{\sqrt{2}}(K^o+\overline{K^o})$$

$$K_s^o \simeq \frac{A}{\sqrt{2}}(K^o+\overline{K^o}) + \varepsilon_{CP}\cdot\frac{A}{\sqrt{2}}(K^o-\overline{K^o}) \qquad \text{with } A=\left\{\sqrt{1+|\varepsilon_{CP}|^2}\right\}^{-1}$$

Neglecting the small CP-violating terms ($\varepsilon_{CP} \simeq 2\cdot10^{-3}$), we have

$$K_L^o \simeq \frac{1}{\sqrt{2}}(K^o-\overline{K^o}), \quad K_s^o \simeq \frac{1}{\sqrt{2}}(K^o+\overline{K^o})$$

The diagrams for $K_L^o \to \mu^+\mu^-$ then are the following:

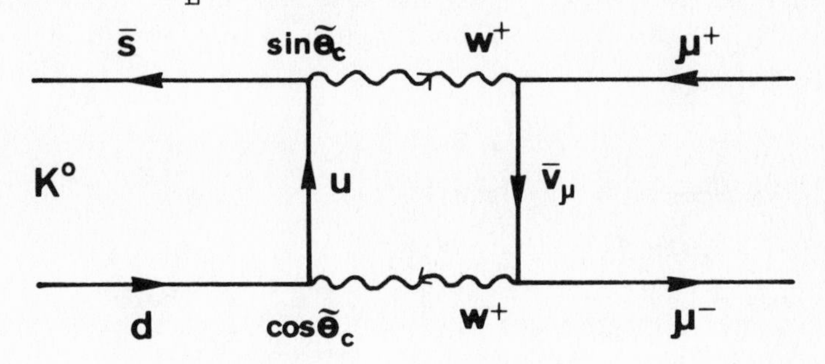

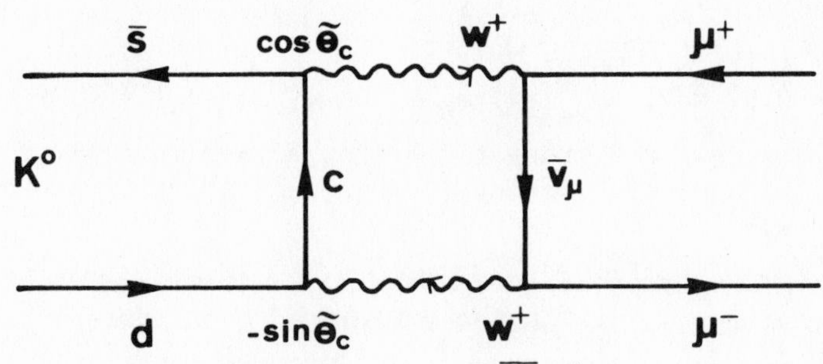

(plus the corresponding diagrams for $\overline{K^o}\to\mu^+\mu^-$).

In the limit $m_c=m_u$ the amplitudes of these two diagrams are $\sim \sin\tilde{\theta}_c \cos\tilde{\theta}_c$ and $\sim -\sin\tilde{\theta}_c \cos\tilde{\theta}_c$ respectively, i.e. they cancel out (GIM-mechanism[10]). In reality, without assuming $m_c=m_u$, s-d-transitions are suppressed by a factor

$$\cos^2\tilde{\theta}_c \sin^2\tilde{\theta}_c \left(\frac{m_c{}^2-m_u{}^2}{m_W{}^2}\right)^2 \simeq 6\cdot10^{-9} \text{ (with } m_c=1{,}5 \text{ GeV, } m_W=80 \text{ GeV).}$$

For instance, $K_L^o\to\mu^+\mu^-$ is expected weak at a level somewhat below the level at which it is expected electromagnetically[11]. The experimentally found value[12] $R = \frac{\Gamma(K_L^o \to \mu^+\mu^-)}{\Gamma(K_L^o \to \text{all})} \simeq 9 \cdot 10^{-9}$ is only slightly higher than the theoretical lower limit for $K_L^o \to \mu^+\mu^-$ through electromagnetic interactions (unitarity limit), $R \geq 6 \cdot 10^{-9}$. With this the GIM-suppression is clearly demonstrated, but not tested quantitatively. Such a test has been the calculation of the K_S^o - K_L^o mass difference, which through the GIM mechanism gave a good estimate of the charmed quark mass of $m_c \simeq 1{,}5$ GeV (see below).

Thus the fact $\widetilde{\theta}_c \neq 0$ together with $m_c \neq m_u$ and $m_d \neq m_s$ ensures that in weak processes there do not exist two separately conserved fractional baryon numbers, i.e. a u-d-number and a c-s-number.

Coming back to the leptonic sector, a violation of the separate conservation of L_e, L_μ could be analogously obtained by introducing a leptonic version θ_c^ℓ of the Cabibbo angle. With mass eigenstates ν_e, ν_μ and weak eigenstates ν_e', ν_μ' we would have

$$\begin{pmatrix} \nu_e' \\ e \end{pmatrix}_L , \begin{pmatrix} \nu_\mu' \\ \mu \end{pmatrix}_L , \quad e_R, \quad \mu_R, \nu_{eR}, \nu_{\mu R}$$

with $\nu_e' = \nu_e \cos\theta_c^\ell + \nu_\mu \sin\theta_c^\ell$

$\nu_\mu' = -\nu_e \sin\theta_c^\ell + \nu_\mu \cos\theta_c^\ell$

(such a scheme makes sense only if $m_{\nu_e} \neq m_{\nu_\mu}$; otherwise, the rotated, like the unrotated states, would be mass eigenstates).

Analogous to the s-d-transition in the quark sector, any μ-e-conversion will be suppressed by a factor

$$\cos^2\theta_c^\ell \sin^2\theta_c^\ell \left(\frac{m_{\nu\mu}^2 - m_{\nu e}^2}{m_W^2}\right)^2 \leq 6 \cdot 10^{-22}$$

(with $m_{\nu_\mu} \leq 0{,}57$ MeV, $m_{\nu_e} \simeq 0$, $m_W \simeq 80$ GeV and $\theta_c^\ell = 45^o$).

This shows that all μ-e-transitions such as $\mu \to e\gamma$, $\mu N \to eN$, etc. will be enormously suppressed, because of the small (or even vanishing) neutrino masses.

What about $\nu_e \leftrightarrows \nu_\mu$ transitions? Assuming $m_{\nu\mu} > m_{\nu_e}$ we might look for the decay $\nu_\mu \to \nu_e\,\gamma$. The decay width of this process[14)]

$$\Gamma_{\nu_\mu \to \nu_e \gamma} = \frac{9}{16}\,\frac{G^2}{128\pi^4}\,\alpha\, m_{\nu_\mu}^5 \cos^2\theta_c^\ell \sin^2\theta_c^\ell \left(\frac{m_\mu^2 - m_e^2}{m_W^2}\right)^2 \leq 2 \cdot 10^{-42}\,\text{MeV}$$

corresponds to a lifetime exceeding the age of the universe by many orders of magnitude ($\tau = \frac{\hbar}{\Gamma} \geq 10^{13}$ y).

Therefore, with only four leptons (e^-, ν_e, μ^-, ν_μ), a mixing between ν_e, ν_μ will produce no μ-e-transitions on a measurable level, and the decay of the heavier of the two neutrinos will not be detectable either.

Possibly the only way to detect a measurable effect of the ν_e-ν_μ-mixing in this case would be through neutrino oscillations[15)], a phenomenon similar to the well-known K^o-$\overline{K^o}$ oscillations: there, an initially pure K^o-beam will contain after some time a certain fraction of $\overline{K^o}$. We look now at the development in time of an initially pure ν_μ'-beam (e.g. from $\pi^+ \to \mu^+ \nu_\mu'$ decay), which is given by

$$\nu_\mu' \; (t) = \nu_\mu \cos \theta_c{}^\ell \; e^{-iE_{\nu_\mu} \cdot t} - \nu_e \sin \theta_c{}^\ell \; e^{-iE_{\nu_e} \cdot t} =$$

$$= a(t) \; \nu_\mu'(0) + b(t) \; \nu_e'(0)$$

A simple calculation (assuming $p \gg m_{\nu_e, \nu_\mu}$) leads to

$$P_{\nu_e'}(t) = |b(t)|^2 = (\sin 2\theta_c{}^\ell)^2 \; (\sin \pi \frac{R}{L})^2$$

showing that for $\theta_c{}^\ell \neq 0$ there will arise a certain fraction $P_{\nu_e'}(t)$ of ν_e' which may be found experimentally in $\nu_e' \to e^-$ transitions. $R \simeq ct$ is the distance from the ν_μ' production place, and L is the oscillation length:

$$L = \frac{4 \pi p}{|m_{\nu_e}{}^2 - m_{\nu_\mu}{}^2|} = 2{,}5 \; \frac{p \; [\mathrm{MeV}]}{|m_{\nu_e}{}^2 - m_{\nu_\mu}{}^2| \; [\mathrm{eV}^2]} \quad \text{meters}$$

A maximal ν_e'-admixture in the $\nu_\mu'(t)$-beam is reached at a distance $D = L/2$ from the ν_μ' production point: $P_{\nu_e}'(D) = (\sin 2\theta_c{}^\ell)^2$.

Thus the intensity of the $\nu_e' \rightleftarrows \nu_\mu'$ oscillation effect is not affected by the smallness of the neutrino masses, it depends only on the mixing angle $\theta_c{}^\ell$ and may be maximal as long as $\theta_c{}^\ell$ cannot be bounded. Note however that the oscillation length will be very large for very small neutrino masses.

In conclusion it can be said, that for the lepton case, apart from the possibility of observable neutrino oscillations, the smallness of the neutrino masses guarantees an at least approximate separate conservation of L_e, L_μ.

Now, there might be mixing with a heavier lepton doublet. Again, let's first have a look at the situation in the quark sector. Adding a new left-handed quark doublet (top and bottom quark) we arrive at:

$$\begin{pmatrix} u \\ d' \end{pmatrix} \quad \begin{pmatrix} c \\ s' \end{pmatrix} \quad \begin{pmatrix} t \\ b' \end{pmatrix}$$

(θ_c^ℓ between the first and second doublets, δ between the second and third, ε between the first and third)

with the weak eigenstates d', s', b' being related to the mass eigenstates d, s, b by a 3-dimensional unitary matrix U,

$$\begin{pmatrix} d' \\ s' \\ b' \end{pmatrix} = \mathbf{U} \begin{pmatrix} d \\ s \\ b \end{pmatrix}$$

which can be parametrized[16,17] by 3 mixing angles (θ_c, ε, δ) and one phase parameter (ϕ)

$$\begin{pmatrix} u \\ \cos\varepsilon\ d_c + \sin\varepsilon\ b \end{pmatrix}_L, \begin{pmatrix} c \\ \cos\delta\ e^{i\phi}\ s_c + \sin\delta\ (\cos\varepsilon\ b - \sin\varepsilon\ d_c) \end{pmatrix}_L$$

$$\begin{pmatrix} t \\ -\sin\delta\ e^{i\phi}\ s_c + \cos\delta\ (\cos\varepsilon\ b - \sin\varepsilon\ d_c) \end{pmatrix}_L$$

d_c, s_c are the Cabibbo rotated d and s quarks, respectively:

$$d_c = d \cos \theta_c + s \sin \theta_c$$
$$s_c = -d \sin \theta_c + s \cos \theta_c$$

The Cabibbo angle θ_c defined in this way is not exactly identical with the Cabibbo angle $\tilde{\theta}_c$ discussed before, with only four quark flavours present. Before, $\cos \tilde{\theta}_c$ described the $u \leftrightarrows d$ coupling, which is now given by $\cos \theta_c \cos\varepsilon$. Therefore we have now

$$\cos \tilde{\theta}_c \equiv \cos \theta_c \cos\varepsilon = 0{,}9737 \pm 0{,}0025 \qquad \text{(i)}$$

The strength $\sin \theta_c \cos \varepsilon$ of the u-s-coupling may be found from $s \to ue^- \bar{\nu}_e$ transitions in hyperon decays ($\Lambda \to p\, e^- \bar{\nu}_e$, $\Sigma^- \to n\, e^- \bar{\nu}_e$, etc.) or in K_{e3}-decays ($K \to \pi e \nu$), with the result[9]

$$\sin \theta_c \cos\varepsilon = 0{,}219 \pm 0{,}011 \qquad \text{(ii)}$$

From (i) and (ii) we get

$$\sin \varepsilon = 0{,}06 \pm 0{,}05$$

i.e.

$$(\sin \varepsilon)^2 \simeq \varepsilon^2 < 10^{-2} \quad \text{or} \quad \varepsilon < 6^\circ \quad \text{(one S.D.)}$$

Thus the "old" and the "new" Cabibbo angle, $\tilde{\theta}_c$ and θ_c, are practically identical: $11{,}9^\circ < \theta_c \leq \tilde{\theta}_c = 13{,}2^\circ$.

The smallness of ε shows that $\sin \delta$ essentially measures the coupling between the second and the third "generation" ($c \leftrightarrows b$), while $\sin \varepsilon$ measures the coupling between the first and the third generation ($u \leftrightarrows b$), and the Cabibbo angle θ_c measures the coupling between the first two generations ($u \leftrightarrows s$) (as indicated by the arrows between the doublets above). Assuming $\varepsilon \ll \theta_c$, the angle δ may be bounded by the $K_L^o - K_S^o$-mass difference[18,19]: Remember that the charmed quark mass m_c had been successfully estimated[11] from a calculation of the $K_L^o - K_S^o$ mass difference based on the diagram shown in Fig. (1a).

Now, with a new $Q = \frac{2}{3}$ -quark present (the t-quark), this additional quark has to be included into the calculation of the $K_L^o - K_S^o$-mass difference (Fig. 1b)).

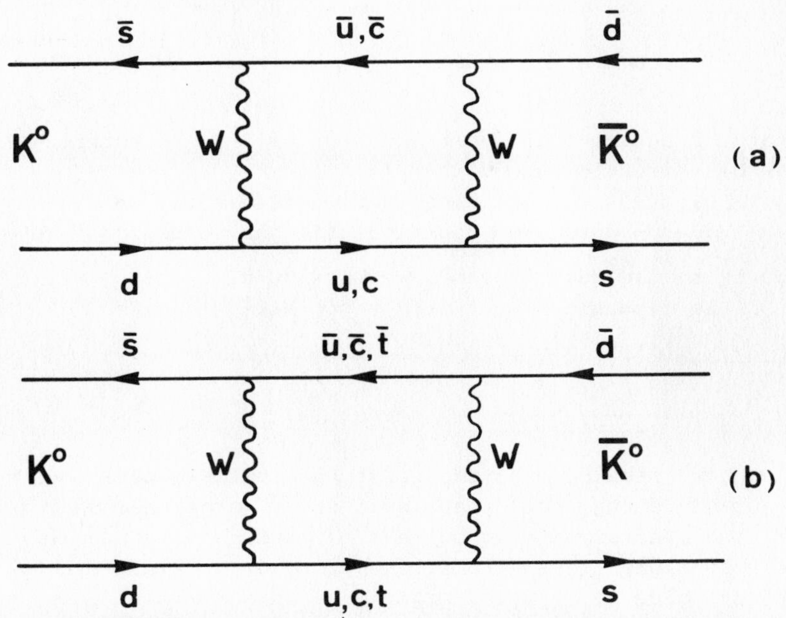

Fig. 1: Box diagrams used for calculating the $K_L^o - K_S^o$ mass difference.

In order that this additional t-quark doesn't influence too strongly the previous estimate of m_c, the t - d and t - s couplings have to be sufficiently suppressed, which, together with[20]
$m_t \geq 8$ GeV , yields the upper limit[21]
$\sin^2 \delta \leq 0{,}3$ i.e. $\delta \leq 30^o$.

The CP violating phase parameter ϕ may be bounded from below by the observed CP violation in the K^o-$\overline{K^o}$ system (which is described by the parameter $\varepsilon_{CP} \simeq 2 \cdot 10^{-3} \simeq \frac{\Gamma(K_L^o \rightarrow 2\pi)}{\Gamma(K_S^o \rightarrow 2\pi)}$, mentioned before), giving[18,19)]
$\phi \geq 5 \cdot 10^{-3}$ or $\phi \geq 0{,}3^o$

Taking all these experimental limits and putting them into our mixing scheme, we get (only absolute values are given):

$$\begin{pmatrix} & u & \\ 0{,}97\ d & 0{,}22\ s & \leq 0{,}10\ b \end{pmatrix}$$

$$\begin{pmatrix} & c & \\ {\leq 0{,}24 \atop \geq 0{,}13}\ d & {\leq 0{,}97 \atop \geq 0{,}80}\ s & \leq 0{,}50\ b \end{pmatrix}$$

$$\begin{pmatrix} & t & \\ \leq 0{,}20\ d & \leq 0{,}55\ s & \geq 0{,}83\ b \end{pmatrix}$$

Note that for instance an u-s-transition will be proportional to the u-s-coefficient squared, i.e. $(0{,}22)^2 \simeq 0{,}05$. The sum, e.g. of the c - s, c - s and c - b squared coefficients, or of the u - d, c - d, t - d squared coefficients, has to be equal to one.

Analogously to the quark sector, we may now construct a mixing scheme in the leptonic sector with three mixing angles θ_c^ℓ, β, γ:

$$\begin{pmatrix} \nu_e' \\ e \end{pmatrix} \quad \begin{pmatrix} \nu_\mu' \\ \mu \end{pmatrix} \quad \begin{pmatrix} \nu_\tau' \\ \tau \end{pmatrix}$$

θ_c^ℓ (e–μ), γ (μ–τ), β (e–τ)

As we have seen before, the first two lepton generations do not mix measurably among each other (even for θ_c^ℓ large) because of $m_{\nu_e}, m_{\nu_\mu} \simeq 0$, and any mixing between electron and muon would have to come from higher generations. Therefore we are left with only two mixing angles β, γ, which might give measurable effects, the first one (β) describing the mixing between e- and τ-doublet, the second one (γ) describing the mixing between μ- and τ-doublet, giving the scheme of Altarelli et. al.[4]:

$$\begin{pmatrix} (1 - \frac{1}{2}\beta^2)\ \nu_e - \beta\gamma\ \nu_\mu + \beta\ \nu_\tau \\ e^- \end{pmatrix}_L \quad \begin{pmatrix} (1 - \frac{1}{2}\gamma^2)\ \nu_\mu + \gamma\ \nu_\tau \\ \mu^- \end{pmatrix}_L$$

$$\begin{pmatrix} (1 - \frac{1}{2}\beta^2 - \frac{1}{2}\gamma^2)\ \nu_\tau - \beta\nu_e - \gamma\nu_\mu \\ \tau^- \end{pmatrix}_L$$

If m_{ν_τ} should turn out to be small or if even $m_{\nu_\tau} = 0$, there would again be no measurable $\mu \leftrightarrows e$ transitions and we would have to "hope for" a possible next generation[22-24] with a massive neutral lepton L^o. In this case β, γ would describe the $e - L^o$, $\mu - L^o$ couplings respectively.

Now, what do we know of β, γ? If m_{ν_τ} (or m_{L^o}) is assumed to be heavy ($\geq m_\pi$), the neutrinos produced e.g. in $\pi \to \mu\nu$ decay will be ν_μ, the ν_τ-degree of freedom will be frozen in. In contrast to ν_μ', this state ν_μ is in general not orthogonal to ν_e': $\langle \nu_e' | \nu_\mu \rangle = -\beta \cdot \gamma$. Consequently, there is a probability $P = \beta^2\gamma^2$ to produce an electron instead of a muon with this ν_μ-beam. Experimental limits give[4,13]

$$\beta^2\gamma^2 = \frac{\Gamma(\nu_\mu\ A \to e^- + ..)}{\Gamma(\nu_\mu\ A \to \mu^- + ..)} < 2 \cdot 10^{-3}$$

At high energy (Cnops et al.[13]).

$$\frac{\Gamma(\nu_\mu\ A \to e^- + ..)}{\Gamma(\nu_\mu\ A \to \mu^- + ..)} < 3 \cdot 10^{-3}$$

In this case, the e^- may either be produced directly or originate from τ^--decay ($\tau^- \rightarrow e^- + \ldots$). The τ^- production probability is $|\langle\nu_\tau'|\nu_\mu\rangle|^2 = \gamma^2$. Taking into consideration an additional phase-space suppression factor for the τ production relative to the μ-production of the order of 0,6 (see Ref. 13)) and with a measured $\tau^- \rightarrow e^- + \ldots$ branching ratio $\geq 16\%$ we get $\gamma^2 \leq 0{,}031$

The good validity of μ-e-universality as found in comparing $\pi \rightarrow e\nu$ and $\pi \rightarrow \mu\nu$ decay implies[21] $-0{,}008 \leq \gamma^2 - \beta^2 \leq 0{,}070$.

Taking all these limits together we arrive at the following scheme:

$$\begin{pmatrix} \geq 0{,}981\ \nu_e & \leq 0{,}035\ \nu_\mu & \leq 0{,}197\ \nu_\tau \\ & e^- & \end{pmatrix} \begin{pmatrix} \geq 0{,}985\nu_\mu & \leq 0{,}176\nu_\tau \\ \mu^- & \end{pmatrix}$$

$$\begin{pmatrix} \leq 0{,}197\ \nu_e & \leq 0{,}176\ \nu_\mu & \geq 0{,}966\ \nu_\tau \\ & \tau^- & \end{pmatrix}$$

3. EXPERIMENTS ON μ-e-CONVERSION PROCESSES AND THEORETICAL MODELS

Transitions between the electron and the muon doublet may be looked for in charged current processes like $\nu_\mu \rightarrow e^-$ or in neutral current processes like $\mu \rightarrow e\gamma$, $\mu \rightarrow 3e$, $\mu^- A \rightarrow Ae^-$.

The cross section ratio of the anomalous charged current process $\nu_\mu A \rightarrow A'e^-$ to the normal charged current reaction $\nu_\mu A \rightarrow A'\mu^-$ is given directly by the two mixing angles β, γ, and its experimental upper limit provides us therefore with an upper bound for these angles, as mentioned before:

$$\frac{\Gamma(\nu_\mu\ A \rightarrow e^- + ..)}{\Gamma(\nu_\mu\ A \rightarrow \mu^- + ..)} = \beta^2\gamma^2 < 2 \cdot 10^{-3}$$

We should mention that this ratio is difficult to measure because of the ν_e-contamination of the ν_μ-beam ($K^{\pm}_{e3}$ and $D^{\pm}_{e3}$ decays).

Now we discuss the neutral current processes. In muon-electron-conversion (anomalous μ-capture) a negative muon is brought to rest and an outgoing electron of $\sim$ 100 MeV kinetic energy is searched for. It is therefore, like $\mu \rightarrow e\gamma$ and $\mu \rightarrow 3e$, a low energy experiment in contrast to the high energy experiment $\nu_\mu \rightarrow e^-$.

The ratio of the anomalous neutral current $\mu^- A \rightarrow A\ e^-$ to the charged current $\mu^- A \rightarrow A''\nu_\mu$ is roughly given by

$$R_{\mu e} = \frac{\Gamma(\mu^- A \rightarrow A\ e^-)}{\Gamma(\mu^- A \rightarrow A''\nu_\mu)} \simeq \beta^2\gamma^2 \left(\frac{\alpha}{m_W^2} \frac{m^2}{\sin^2\theta_W} \ln \frac{m_W^2}{m^2}\right)^2$$

(here, m is the mass of the heavy "neutrino"). Thus we have a rather substantial additional suppression factor beside $\beta^2\gamma^2$. In the case of the ν_τ, the limit[25] $m_{\nu_\tau} \leq 250$ MeV gives for this additional suppression factor a value of $\leq 10^{-11}$, and consequently μ-e-conversion will be down to a level of $R_{\mu e} \leq 10^{-14}$, a level which is far beyond present experimental possibilities. In order to get a measurable μ-e-conversion rate, a heavy neutrino is needed. For instance, with m = 2 GeV, the additional suppression factor is $\simeq 2 \cdot 10^{-8}$, and from this one obtains $R_{\mu e} \leq 4 \cdot 10^{-11}$, a value close to our experimental upper bound[26] of

$R_{\mu e} < 7 \cdot 10^{-11}$ (90% C.L.)

The experimental technique used to measure this bound is described in Ref. 26. The apparatus is shown in fig. 2. A sulfur target, where the muons were stopped, was placed in the center of a streamer chamber, which was used to measure the momentum

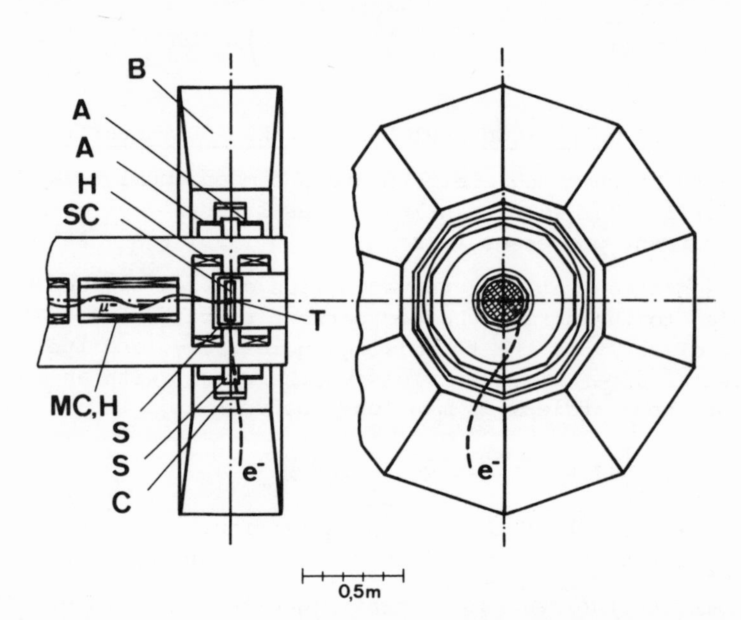

Fig. 2: Schematic of the apparatus for the muon-electron conversion experiment. The side-view on the left shows a typical $\mu^- \to e^-$-event. A muon coming down the superconducting muon channel (MC) enters the streamer chamber (SC) and stops in the target (T). The outcoming electron goes through scintillation (S)- and Čerenkov (C)-counters and stops in a scintillation calorimeter block (B). Helmholtz coils (H) provide the necessary magnetic fields. Anticounters (A) and iron shielding (not shown) were used to reduce cosmic ray background. On the right, a front view of the apparatus is shown.

of e^- coming from the target. Only electrons leaving the target with an angle near 90° relative to the muon beam axis were measured. A trigger system selected those events which had a trigger pattern corresponding roughly to the curvature of e^- from $\mu^- \to e^-$ conversion. Low momentum particles were trapped by the applied magnetic field of 3,5 T. Anticounters and calorimeter counters were used to further reduce background. Intrinsic background processes were muon decay in orbit and radiative muon capture. The fast background from radiative pion capture was suppressed by pulsing the primary proton beam and measuring only during the "beam-off" periods. The number of captured muons was measured with a Ge(Li) detector using the 1,27 MeV nuclear gamma rays from ${}^{31}P^* \to {}^{31}P + \gamma$ following $\mu^- + {}^{32}S \to {}^{31}P^* + n + \nu_\mu$.
The experimental limits for the other two μ-e-transition processes mentioned before are[27)]

$R_{\mu \to e\gamma} < 1,9 \cdot 10^{-10}$ (90% C.L.)

and[28)]

$R_{\mu \to 3e} < 1,9 \cdot 10^{-9}$ (90% C.L.)

Let's compare now these experimental values with theoretical models. In the model with left-handed neutrino mixing,

$$\begin{pmatrix} \nu_e' \\ e^- \end{pmatrix}_L \quad \begin{pmatrix} \nu_\mu' \\ \mu^- \end{pmatrix}_L \quad \begin{pmatrix} \nu_\tau' \\ \tau^- \end{pmatrix}_L \quad \begin{pmatrix} L_o' \\ L^- \end{pmatrix}_L$$

with β, γ describing the e-L, μ-L couplings respectively and L^o being a heavy neutral lepton (mass m), the rate for $\mu^- A \to Ae^-$ is calculated from the diagrams shown in fig. 3, with the result[4)]

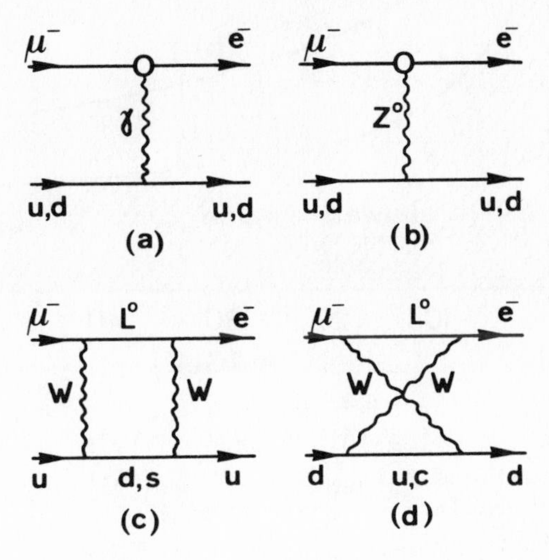

Fig. 3: Diagrams contributing to μ-e-conversion in the left-handed model of Altarelli et al.; L^o is a heavy neutrino.

$$R_{\mu e} \simeq \frac{Z\, F_N{}^2}{c(1+3g_A{}^2)} \left\{ \frac{9\alpha}{8\pi} \; \frac{\beta\ \gamma\ m^2}{m_W{}^2 \sin^2\theta_W} \right\}^2 \cdot \left\{ \ell n \frac{m_W{}^2}{m^2} - 1 + \frac{4}{9} \sin^2\theta_W (\ell n \frac{m_W{}^2}{m^2} - 2) \right\}^2$$

(This formula is an approximation valid for nuclei with A = 2Z and m $\leq$ 10 GeV).
Here, F_N is the elastic charge form factor[29], c is a statistical facotr, $g_A \simeq 1{,}25$ is the axial coupling of the nucleon, and θ_W is the Weinberg angle.

The experimental upper bound $R_{\mu e} < 7 \cdot 10^{-11}$ confines the region of allowed masses m and mixing angles $\beta \cdot \gamma$ as shown in fig. 4. The corresponding curves for $\mu \to e\gamma$, $\mu \to 3e$ are also shown.

In this case, the experimental bound for $\mu^- A \to Ae^-$ is the most restrictive of the three experimental results. Furthermore, in this model $R_{\mu N \to eN} < 7 \cdot 10^{-11}$ implies $R_{\mu \to 3e} < 2 \cdot 10^{-13}$ and $R_{\mu \to e\gamma} < 10^{-12}$ for m $\leq$ 10 GeV. The figure shows furthermore that a finite value of $\beta^2\gamma^2$ would imply an upper limit on the heavy "neutrino" mass. For instance, with $\beta^2\gamma^2 = 2 \cdot 10^{-3}$ we get $m_{Lo} \leq 2{,}5$ GeV, or with $\beta^2\gamma^2 = 2 \cdot 10^{-4}$, $m_{Lo} < 5$ GeV. If somebody finds a neutral heavy lepton with e.g. m = 8 GeV, we could give a new limit for the corresponding mixing angles: $\beta^2\gamma^2 \leq 5 \cdot 10^{-5}$.

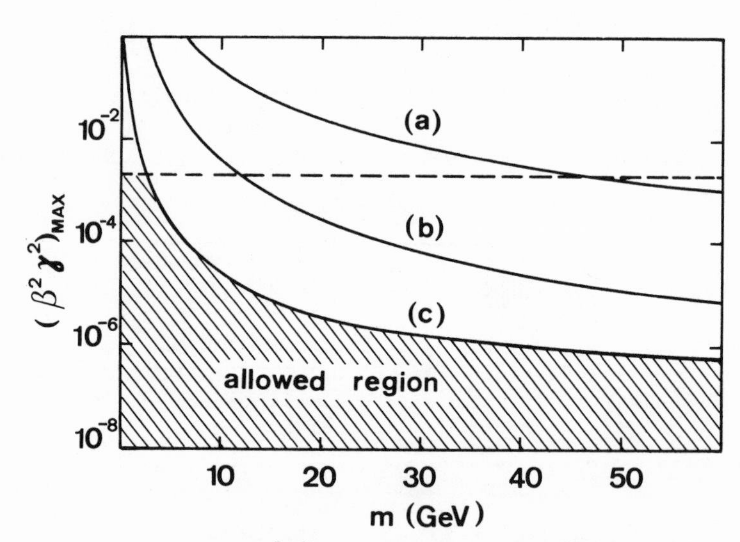

Fig. 4: Upper bounds for $\beta^2\gamma^2$ as implied by the experimental limits $R_{\mu \to 3e} < 1{,}9 \cdot 10^{-9}$ (curve (a)), $R_{\mu \to e\gamma} < 1{,}9 \cdot 10^{-10}$ (curve (b)) and $R_{\mu N \to eN} < 7 \cdot 10^{-11}$ (curve (c)). The dashed curve is the upper bound $\beta^2\gamma^2 < 2 \cdot 10^{-3}$ from $\nu_\mu A \to A'e^-$.
$\beta^2\gamma^2|_{Max}$ is shown as a function of the neutral heavy lepton mass m.

There is another theoretical model which until recently earned a lot of interest, which is the model of Cheng and Li[5]. In this model there are lefthanded as well as righthanded doublets:

$$\begin{pmatrix} \nu_e \\ e^- \end{pmatrix}_L \begin{pmatrix} \nu_\mu \\ \mu^- \end{pmatrix}_L \begin{pmatrix} L_e^o \\ e^- \end{pmatrix}_R \begin{pmatrix} L_\mu^o \\ \mu^- \end{pmatrix}_R$$

ϕ

with

$L_e^o = N_1 \cos\phi + N_2 \sin\phi$

$L_\mu^o = -N_1 \sin\phi + N_2 \cos\phi$

(N_1, N_2 being mass eigenstates).

In this model we have

$$R_{\mu e} = \frac{Z\,F_N^2}{c(1+3g_A^2)} \cdot \left\{ \frac{9\alpha}{8\pi} \cdot \frac{\sin\phi \cos\phi \; (m_1^2 - m_2^2)}{m_W^2 \sin^2\theta_W} \right\}^2 \cdot \left\{ \ell n \frac{m_W^2}{m^2} - 2 - \frac{4}{9} \sin^2\theta_W \cdot (\ell n \frac{m_W^2}{m^2} - 3) \right\}^2$$

where $m^2 = \frac{1}{2}(m_1^2 + m_2^2)$ and the righthanded up and down quarks are taken as singlet states. For example, taking $\phi = 45^o$ and $m_2 = 0$, $m_1 = 2$ GeV one obtains $R_{\mu e} \simeq 10^{-9}$.

Unfortunately for this model, the recent SLAC experiment with polarized electrons[30] shows that the coupling of e^-_R to a righthanded neutral lepton is strongly disfavoured; this experiment, which investigated the dependence of the deep inelastic scattering cross section $e^- + N \rightarrow e^- + X$ on the longitudinal polarization of the initial electron gave an experimental value of

$$\frac{A}{Q^2} \equiv \frac{\sigma_R - \sigma_L}{\sigma_R + \sigma_L} \cdot \frac{1}{Q^2} = (-9{,}5 \pm 1{,}6) \cdot 10^{-5} \; (\mathrm{GeV/c})^{-2}$$

for the parity violating asymmetry. Fig. 5 shows the predictions for A in various models compared with experimentally allowed values of A and $\sin^2\theta_W$.

Finally, μ-e-conversion could occur through Higgs exchange if there exists more than one Higgs doublet. Bjorken and Weinberg[7] roughly estimate the rates of $\mu \rightarrow e\gamma$ and $\mu N \rightarrow eN$ from diagrams like the one shown in fig. 6. Together with the experimental upper bounds of the $\mu \rightarrow e\gamma$- and $\mu N \rightarrow eN$-rates, these estimates may be used to put a lower bound on the Higgs mass. A more accurate theoretical calculation might improve the reliability of such a bound.

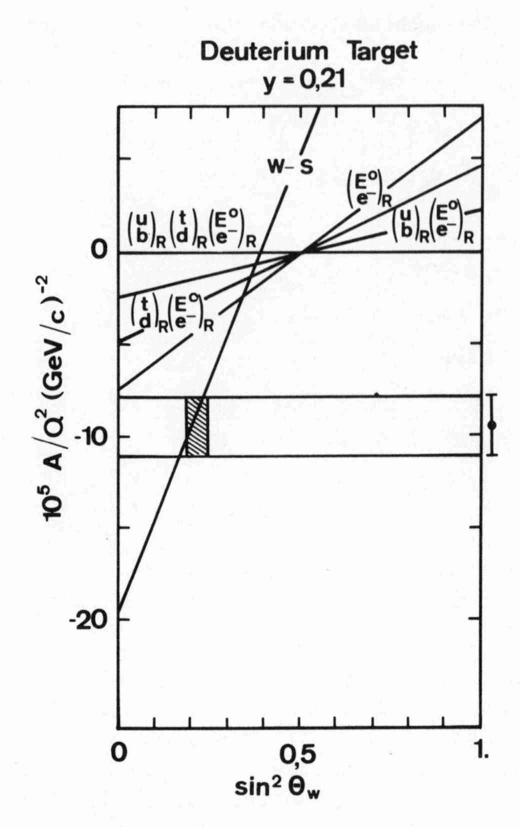

Fig. 5: Asymmetry in the Weinberg-Salam-model and in models with a righthanded leptondoublet $\begin{pmatrix} E^0 \\ e^- \end{pmatrix}_R$ as a function of $\sin\theta_W$. Experimentally allowed values of A/Q^2 and $\sin^2\theta_W$ (shaded region) strongly disfavour the latter models. y is the inelasticity, i.e. the fractional energy transfer to the hadrons.

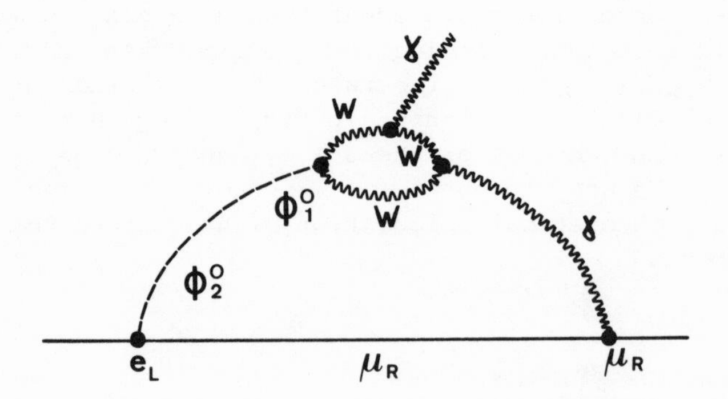

Fig. 6: Two-loop diagram for $\mu \to e\gamma$ through Higgs exchange.

We thank our colleagues of the μ-e-conversion experiment[2,26) for many valuable discussions.

Referenzen

1) S. Frankel, Rare and Ultrarare Muon Decays, in: Muon Physics vol. II, eds. V. Hughes and C.S. Wu (Academic Press, New York 1975).
2) A. Badertscher et al., Phys. Lett. 79B, 371 (1978), and H. Hänni, Thesis, Bern (1978), to be published.
3) J. Blietschau et al., Nuclear Physics B133, 205 (1978) P. Nemethy, Bulletin of the APS 24/4, 671 (1979).
4) G. Altarelli, L. Baulieu, N. Cabibbo, L. Maiani, and R. Petronzio, Nuclear Physics B125, 285 (1977).
5) T.P. Cheng and L.F. Li, Phys. Rev. Lett. 38, 381 (1977).
6) W.J. Marciano and A.I. Sanda, Phys. Rev. Lett. 38, 1512 (1977).
7) J.D. Bjorken and S. Weinberg, Phys. Rev. Lett. 38, 622 (1977).
8) For recent reviews see H. Quinn, Proceedings of the 1978 Stanford Summer Institute on Particle Physics (Weak Interactions - Present and Future) or S. Weinberg, Proceedings of the Tokyo conference 1978.
9) R.E. Shrock and L. Wang, Phys. Rev. Lett. 41, 1692 (1978); see also M. Ross, as quoted by K. Kleinknecht at the London Conference 1974.
10) See, for example, L.H. Ryder, Elementary Particles and Symmetries (Gordon & Breach, 1975), chapter 13.
11) M.K. Gaillard and B.W. Lee, Phys. Rev. D10, 897 (1974).
12) M.J. Shochet et al., Phys. Rev. Lett. 39, 59 (1977).
13) E. Bellotti et al., Nuovo Cimento Lett. 17, 553 (1976) A.M. Cnops et al., Phys. Rev. Lett. 40, 144 (1978). See also ref. 4).
14) S.T. Petcov, Yad. Fiz. 25, 641 (1977).
15) S.M. Bilenky and B. Pontecorvo, Phys. Lett. 41C, 225 (1978).
16) M. Kobayashi and K. Maskawa, Progr. Theoret. Phys. 49, 652 (1973); H. Harari, Phys. Lett. 42C, 235 (1978).
17) The parametrization given here is different, but of course equivalent to that of ref. (16).
18) J. Ellis, M.K. Gaillard, D.V. Nanopoulos, S. Rudaz, Nuclear Physics B131, 285 (1977).
19) J. Ellis, M.K. Gaillard, and D.V. Nanopoulos, Nucl. Phys. B109, 213 (1976).
20) L. Lederman, Proceedings of the Int. Conference on High Energy Physics Tokyo 1978.
21) G. Altarelli, Rapporteur Talk presented at the XIX International Conference on High Energy Physics, Tokyo 24-30 August 1978.

22) In the standard big-bang cosmology, the number of neutrino flavours is limited by the primordial abundance of ^{4}He. The present cosmic ^{4}He-abundance of $\simeq$ 29% implies that there cannot be more than about 5 to 7 types of light neutrinos. There is strong evidence that the primordial abundance was $\leq$ 25%, limiting the number of neutrino flavours to $\leq$ 3 or 4. See also G. Steigman, D.N. Schramm, J.E. Gunn, Phys. Lett. 66B, 202 (1977).
23) A laboratory experiment to measure directly the number of neutrino types has been proposed by E. Ma and J. Okada, Phys. Rev. Lett. 41, 287 (1978).
24) There have been speculations on a possible (charged-) lepton mass formula. For the mass of the next charged lepton L^-, if it exists, these speculations lead to values around 10 GeV/c^2. See A.O. Barut, Phys. Rev. Lett. 42, 1251 (1979) and S. Weinberg, SLAC-PUB-2195 (1978) (to be published in Phys. Rev. D).
25) W. Bacino et al., Phys. Rev. Lett. 42, 749 (1979).
26) A. Badertscher et al., Phys. Rev. Lett. 39, 1385 (1977) and A. Badertscher et al., to be published.
27) J.D. Bowman et al., Phys. Rev. Lett. 42, 556 (1979).
28) S.M. Korenchenko et al. JETP (Sov.Phys.) 43, 1 (1976).
29) S. Weinberg and G. Feinberg, Phys.Rev.Lett. 3, 111 (1959); (E) 3, 244 (1959).
30) C.Y. Prescott et al., Phys. Lett. 77B, 347 (1978); S.T. Petcov, Phys. Lett. 80B 83 (1978).

A NEW MEASUREMENT OF THE MUON CAPTURE RATE IN LIQUID HYDROGEN

J. Duclos

DPh-N/HE, Centre d'Etudes Nucléaires de Saclay

BP 2, 91190 Gif-sur-Yvette, France

The muon capture rate in hydrogen is a fundamental weak process, particularly sensitive to the pseudo-scalar coupling constant: F_p. Its measurement is very difficult for various reasons[1] :

i) The capture rate is very small compared to the decay rate.

ii) The capture rate depends strongly on the initial hyperfine state which has to be known exactly in order to interpret the result.

iii) The hydrogen must be extremely pure because of the high cross section for muons to be transferred to deuterium or nuclei.

iv) In all the previous experiments the 5 MeV neutron, produced by the reaction, had to be detected amongst an intense background of electrons and γ rays due to muon decays. The main limitation comes from the neutron detector efficiency which has to be computed by sophisticated Monte-Carlo programmes.

Several experiments have been performed measuring the capture rate in gas[2,3] and liquid[4,5]. They are interpreted as being respectively the capture rate in the singlet μp state and in the ortho-molecular $p\mu p$ state. The best results are

$$\Lambda_S = 651 \pm 57 \quad s^{-1} \qquad \text{(Ref. 2)}$$

$$\Lambda_{OM} = 464 \pm 42 \quad s^{-1} \qquad \text{(Ref. 5).}$$

These measurements should be improved in order to get more significant values of the coupling constants involved[6].

A new experiment is in progress at Saclay[7] which avoids the neutron problem by comparing accurately the μ^- and μ^+ lifetimes in liquid hydrogen. In addition by comparing the lifetime of neutrons from μ capture and electrons from μ decay, uncertainties about the initial state of the pμp molecule can be resolved. In this lecture I will describe the principle of the experiment and give some preliminary results.

The method, has been used already for measuring the μ^+ lifetime[8] and the total μ^- capture rate in ^{6}Li and ^{7}Li [9]. It is based on the particular beam structure of the Saclay Linac. The muons are stopped in the target during the 3 μs beam burst and the decay electrons from the surviving muons are detected after the burst by telescopes of scintillators. Their time distribution gives an accurate measurement of the muon lifetime because of two favorable effects: the low background level and the absence of any regeneration of the muon source during the decay study. In such conditions, systematic errors are smaller than 3.10^{-5} and the capture rate can be measured with a 3 % precision.

The liquid hydrogen is contained in a spherical copper vessel (Fig. 1). The high purity is obtained in a, now, traditional way by using a protium gas (1 ppm deuterium) passing through a palladium filter and then a heat exchanger. The muons enter the target through a lead collimator and a copper degrader. About 90 % of the muons are stopped in hydrogen. Those stopped in copper or lead are rapidly captured and do not contribute inside the measuring gate which starts about 1 μs after the burst.

The decay electrons are detected in 6 telescopes of 3 plastic counters surrounding the target. The total energy threshold (6 MeV) and the threefold coincidence insures a sufficient rejection of the neutral delayed events (mostly γ rays from thermal neutron capture). The measuring gate opens 6 digitrons supplied by a common quartz oscillator of 500 MHz (stability better than 10^{-5}). The first event occurring in a telescope closes the corresponding digitron. If two

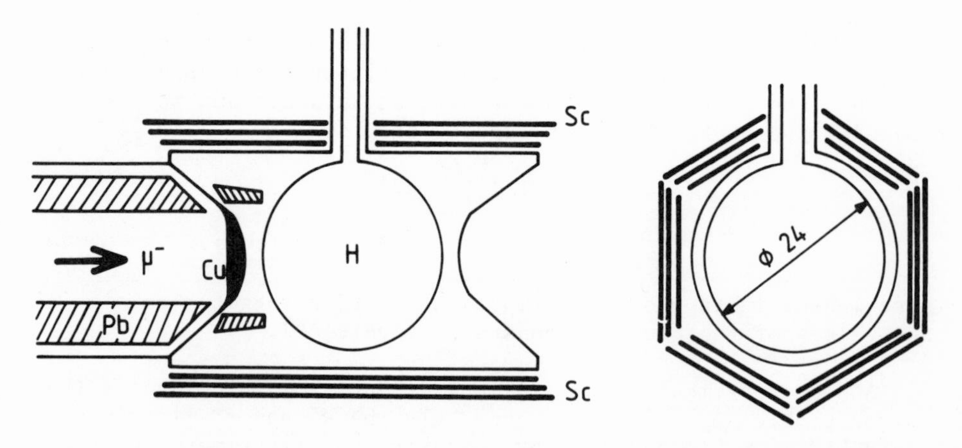

Fig. 1 Experimental set up : target and electron detectors.

or more electrons occur in the same gate, this "multiple" event is rejected.

The rate of "single" events is expected to follow the exponentional law :

$$R(t) = R_o \exp(- t/\tau\mu).$$

As showed in Ref.[8] the observed law is rather :

$$R'(t) = R(t) \left[1 + A.R(t)/R_o\right] + B R_o.$$

The distortion term A is due to the finite resolution of the electronics circuits. The effect is proportional to the rate and is measured by increasing the nominal rate by a factor 4, using a π^+ beam. For the nominal rate between 0.06 and 0.1 event per gate and telescope, A is smaller than 10^{-4}. At the level 10^{-5} there is a time dependent component (period : about 150 μs) due to thermal neutrons. These backgrounds are measured from the distribution of the delayed events between two machine burst (see Fig. 2). By stopping μ^- in the copper walls of the target (target empty) one checks the behaviour of the background in the region of μ decays (Fig. 3).

The results obtained until now concerning τ_{μ^+} and τ_{μ^-} are displayed on Fig. 4. They are compared to the world value of τ_{μ^+} and the corresponding expected value for τ_{μ^-} assuming the theoretical capture rate : 500 s^{-1}. The present preliminary numbers obtained are

$$\tau_{\mu^+} = (2197.19 \pm .14)\text{ns}$$

$$\tau_{\mu^-} = (2194.97 \pm .15)\text{ns}$$

$$\Lambda_C = (467 \pm 43)s^{-1}.$$

The value of Λ_C takes into account the effect of the μ^- binding which, according to Überall[10], is $\alpha^2/2$.

In order to interpret Λ_C we have to know the initial state of the pμp molecule. It is well established now[1] that the molecule is formed in an "ortho" state for which the capture rate can be expressed in terms of the singlet and triplet rates by :

$$\Lambda_{OM} = 2\,\gamma_o\,(\tfrac{3}{4}\,\Lambda_S + \tfrac{1}{4}\,\Lambda_t)\ ; \qquad 2\,\gamma_o = 1.01 \pm .01\ .$$

But the transition rate to the "para" ground state, though expected to be weak[11], could have a non negligible effect. Indeed, the para-molecular capture rate is much smaller than Λ_{OM} :

$$\Lambda_{PM} = 2\,\gamma_P \left[\tfrac{3}{4}\,\Lambda_T + \tfrac{1}{4}\,\Lambda_S\right] \qquad 2\,\gamma_P + 1.15 \pm .01\ .$$

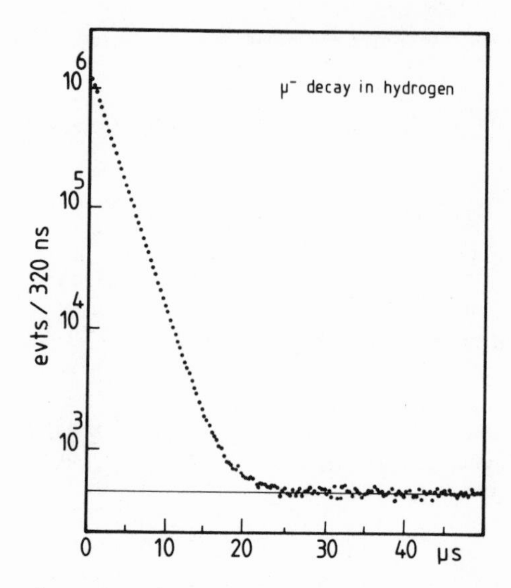

Fig. 2 Time distribution of the threefold coincidence events inside the gate.

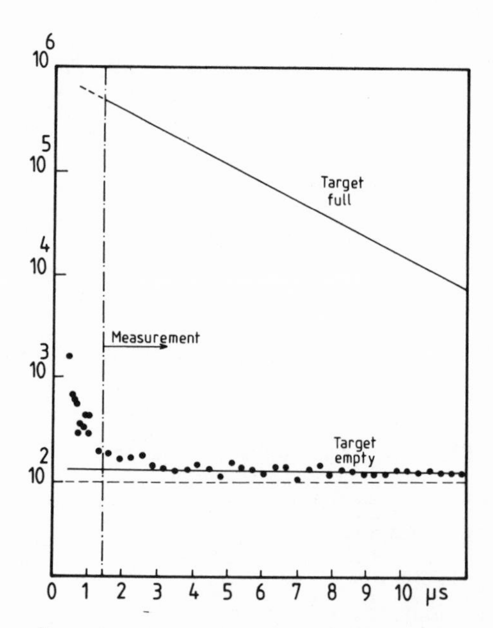

Fig. 3 Time distribution of events, target empty, starting at the end of the beam burst. The muons are stopped in the copper walls. In the first µs we see the contribution of the copper atoms. The dashed line shows the level of cosmic rays ; the full line includes the time dependent background analyzed from the delayed events with the procedure used for "target full" data.

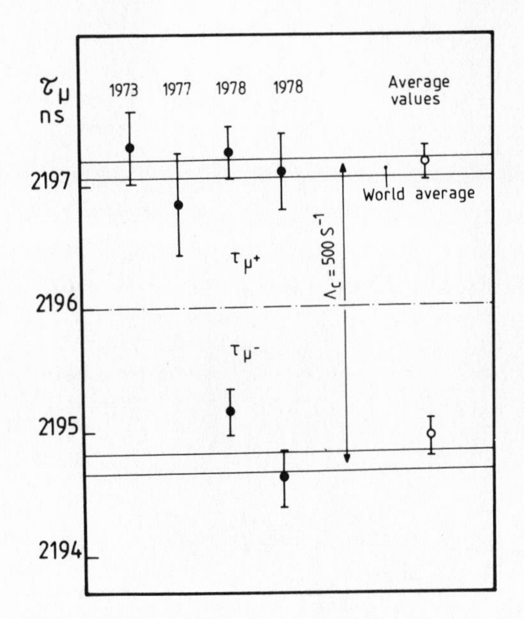

Fig. 4 Values of τ_{μ^+} and τ_{μ^-} obtained from the different runs.

The last theoretical prediction[12] gives :

$$\Lambda_{OM} = 502\ s^{-1}\ ; \qquad \Lambda_{PM} = 199\ s^{-1}.$$

A classical method to investigate the effect of a transition in the initial state in muon capture is to study the time distribution of the capture products[13]. If $N_\mu(t)$ is the number of living muons at the time t the rates of electrons from decay and neutrons from capture are

$$e(t) = \Lambda_o . N_\mu(t)$$

$$n(t) = \Lambda_c . N_\mu(t)\ .$$

Λ_o is the decay rate. An ortho-para transition results in a time dependence of the capture rate Λ_c. Particularly the mean time of the neutrons compared to the electrons is :

$$\tau_n = \tau_e \left(1 - \frac{\Lambda_{OP}}{\Lambda_o} \frac{\Lambda_{OP} - \Lambda_{PM}}{\Lambda_{OM}}\right)$$

if the ortho-para transition rate Λ_{OP} is small compared to Λ_o. A first measurement by the Colombia group gives[5] :

$$\Lambda_{OP}/\Lambda_o \lesssim .11\ .$$

Such a measurement is also done at Saclay by using 6 liquid scintillators (NE213) a huge background of γ rays produced by the decay electrons is removed by pulse shape discrimination and other backgrounds like photo-neutrons and cosmic neutrons are subtracted Very preliminary data give

$$\tau_n/\tau_e = .91 \pm .07 \; ; \; \frac{\Lambda_{OP}}{\Lambda_o} \leqslant .2 \; .$$

A small amount of ortho-para transition would explain the systematic (though non significant) deviation from the theoretical predictions of all the experimental rates in liquid hydrogen. In terms of the pseudo-scalar form factor F_p these values contribute to give $(F_p/F_V)_\mu = 12 \pm 2$ slightly higher than the theoretical prediction (about 9)[6].

REFERENCES

1) See for example E. Zavattini, Muon physics II (Academic Press. Inc., NY, 1975) p. 219.
2) A. Alberigi Quaranta et al., Phys. Rev. 177, 2118 (1969).
3) Bystristky et al., Dubna Preprint-D1-7300 (1973).
4) E. Bleser et al., Phys. Rev. Lett. 8, 288 (1962).
5) J.E. Rothberg et al., Phys. Rev. 132, 2664 (1963).
6) See E. Zavattini, Proc. First Course of the Intern. School of Physics of Exotic Atom, Erice 24-30 April, 1977.
7) Collaboration : Saclay (G. Bardin, J. Duclos, A. Magnon, J. Martino, D. Measday, A. Richter), CERN (E. Zavattini), Bologne (A. Bertin, M. Piccinini, A. Vitale).
8) J. Duclos et al., Phys. Lett. 47B, 491 (1973).
9) G. Bardin et al., Phys. Lett. 79B, 52 (1978).
10) H. Uberall, Phys. Rev. 119, 365 (1960).
11) S. Weinberg, Phys. Rev. Lett. 4, 585 (1960).
12) A. Santisteban and R. Pascual, Nucl. Phys. A260, 392 (1976).
13) R. Winston, Phys. Rev. 129, 2766 (1963).

POLARIZATION EXPERIMENTS IN THE GODFREY-CYCLE

Laszlo Grenacs

Université Catholique de Louvain
Institut de Physique Corpusculaire
B - 1348 Louvain-la-Neuve, Belgium

ABSTRACT

From the "new" experiments performed in the A = 12 mass multiplet, viz. i) beta-decay asymmetry of aligned $^{12}B/^{12}N$ mirror nuclei and ii) the average and longitudinal polarizations of ^{12}B (g.s.) produced by μ-capture in ^{12}C, we learned about the nature of weak interactions :
- *there are no second class axial-vector currents,*
- *the weak magnetism is now verified in μ-capture,*
- *the helicity of the "muonic" neutrino is measured with good precision (≃ 10 %) (the result agrees with the standard V-A scheme).*

INTRODUCTION

The first quantitative test of the μ-e universality in weak interactions has been achieved by N.K. Godfrey (1952 to 1954), he measured the rate Γ_μ of the partial capture reaction $\mu^- + {}^{12}C \rightarrow \nu_\mu + {}^{12}B$ (g.s.) - with the aid of cosmic-ray muons ! - and compared it with the rate Γ_β of the inverse process ${}^{12}B \rightarrow {}^{12}C$ (g.s.) + $e^- + \tilde{\nu}_e$ (fig. 1). Admitting that relativistic effects have a negligable influence on these rates, the measured ratio Γ_μ/Γ_β implies that the "muonic" Gamow-Teller coupling constant vs. the "electronic" one is 0.97 ± 0.13 [1]. The involved weak reactions ${}^{12}C \rightleftharpoons {}^{12}B$, together with the electromagnetic processes ${}^{12}C \rightleftharpoons {}^{12}C^\star$ (analogue, 15.1 MeV), are often referred to as the "Godfrey-cycle". The work of Godfrey was followed by a series of critical weak interaction experiments in the A = 12 multiplet, including the ${}^{12}N \rightarrow {}^{12}C$ β^+-decay (extended Godfrey-cycle), which were designed to study specific relativistic effects in β-decay *and μ-capture. In one of these experiments - the verification of weak magnetism - decay of unoriented $^{12}B/^{12}N$ nuclei is investigated, in the others one deals with oriented nuclei. This discussion is devoted to the latter type of experiments.*

The status of weak magnetism (WM) as determined from beta-ray spectrum-shape measurements in the decay of $^{12}B/^{12}N$ mirrors is recalled in Ch. I.1. The result of the experiment which was designed to determine whether 2nd class axial-currents exist or not is presented in Ch. I.2.

Ch. II. is devoted to the measurements of the "average" and "longitudinal" polarizations of ^{12}B (g.s.) produced in the Godfrey-reaction. From these polarization effects we deduce the polarization state of leptons (in $\pi \to \mu, \nu$ decay and μ-capture) and obtain dynamical information in μ-capture, e.g. the verification of weak magnetism.

I.1. WEAK MAGNETISM FROM MIRROR β SPECTRA OF $^{12}B/^{12}N$

Following the Conserved Vector Current (C.V.C.) theory the isovector M1 transition-amplitude ($\sim\tau_3$) in $^{12}C^\star$ (15.1 MeV) $\to$ ^{12}C (g.s.) has "weak" ($\sim \tau^\mp$) analogs - the weak magnetism invented by M. Gell-Mann - which contribute to the corresponding $\beta^\mp$ mirror transitions $^{12}B/^{12}N \to {}^{12}C$. The weak magnetism - a relativistic effect - produces, through its interference with the "allowed" Gamow-Teller amplitude, distortion of the β-spectra. The departures from the allowed Fermi-shape of $\beta^\pm$-spectra in the g.s. decay of $^{12}B/^{12}N$ are predicted as : $a_\mp = \pm\, 8/3\ aE$ (retaining only the linear effect in the β-ray energy E). The Gell-Mann's "a" parameter is calculated from the M1-width of $^{12}C^\star \to {}^{12}C$ (g.s.) transition (form factor F_M) and from the ft-value of the considered $\beta^\pm$-transitions (form factor F_A) : $a = F_M/F_A$. The experiment shows the presence of WM in β-decay.

Table 1. Effect of weak magnetism on the sape of β-spectra in the g.s. decay of $^{12}B/^{12}N$

	Measured average slopes % per MeV	CVC-prediction [2]
a_-	0.38 (0.10) C	0.36
	0.68 (0.11) H	0.44
a_+	- 0.48 (0.09) C	- 0.45
	- 0.41 (0.09) H	- 0.53
$a_- - a_+$	0.86 (0.24) C	0.81
	1.09 (0.09) H	0.97

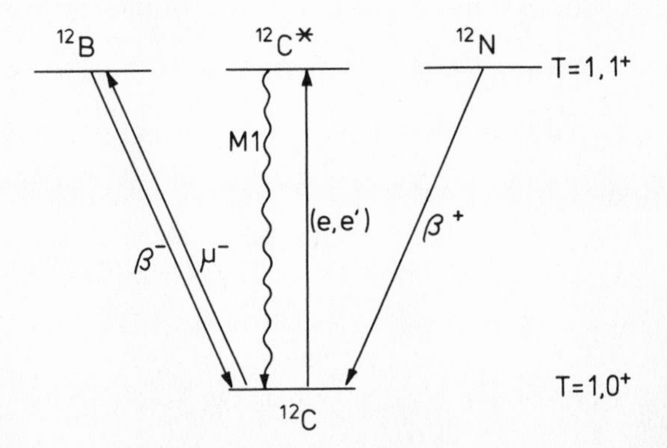

Figure 1. The relevent reactions in the A=12 multiplet.

The experimentally determined slopes a_- and a_+ are given in Table 1, these results are taken from the recent analysis by Koshigiri [2]. *Data "C" are those given by Wu, Lee and Mo and revised by them in 1977* [3] *using the latest result on end-point energies, Coulomb-functions, etc. Data "H" are those obtained recently by the Heidelberg Group, Soergel et al.* [4], *using a different experimental approach.*
There are small discrepancies in the individual values of the slope-parameters. However, the difference of slopes, $a_- - a_+$, agrees well with the C.V.C. prediction.

I.2. ABSENCE OF 2ND CLASS AXIAL-VECTOR CURRENTS IN β-DECAY

The "axial" analog ($\sim \sigma_{\mu\nu}q_\nu\gamma_5$) of the weak magnetism ($\sim\sigma_{\mu\nu}q_\nu$) could in principle exist. This coupling (called sometimes as weak electricity, WE) - as it was pointed out by S. Weinberg - has G-parity which is opposite to that of the primary Gamow-Teller ($\sim\gamma_\mu\gamma_5$) coupling. In order to test whether this 2nd class gradient-type current exist one has to look for an interference between this and the Gamow-Teller term. Such an interference is proportional to the momentum of the lepton field and changes sign from (+) to (-) β-transition. So that the experimental problem is similar to the one encountered in the investigation of WM. It turns out that the β-ray angular distribution $W(\theta,E)$ is particularly sensitive to WE. This distribution, specialized to $1 \to 0$ (no) transitions, is given [5] *:*

$$W(\theta,E)_\mp \sim F\,[\,1 \mp P(1+\alpha_\mp E)P_1(\theta) + A\alpha_\mp EP_2(\theta)]$$

with P(A) nuclear polarization (alignment) of the parent nuclei. The β-ray energy dependent corrections $\alpha_\mp E$ are contributed by gradient couplings [5,6] *: WM (Gell-Mann's "a" we met in I.1.) and WE. The We contribution can have 2nd class piece ($F_E^{(2)}$) due to the $\sigma_{\mu\nu}q_\nu\gamma_5$ coupling and 1st class piece ($F_E^{(1)}$) coming from the* *time-component of the axial vector current.* $F_E^{(1)}$ *does not lead to sign change from (+) to (-) transition. The parameters* $\alpha_{\mp}$ *are given as :*

$$\alpha_{\mp} = \pm \frac{2}{3}\frac{F_M}{F_A} - \frac{2}{3}\frac{F_E^{(1)} \pm F_E^{(2)}}{F_A} \equiv \pm \frac{2}{3} a - \frac{2}{3} b_{\mp} .$$

The parameter b is sometimes called as Gell-Mann's "b" [6]. *If b is charge-dependent (independent), there is (there is no) 2nd class current.*

Fig. 2 shows the experimental result obtained by the Louvain-ETHZ collaboration [7]*, including data-points of the Osaka group.*

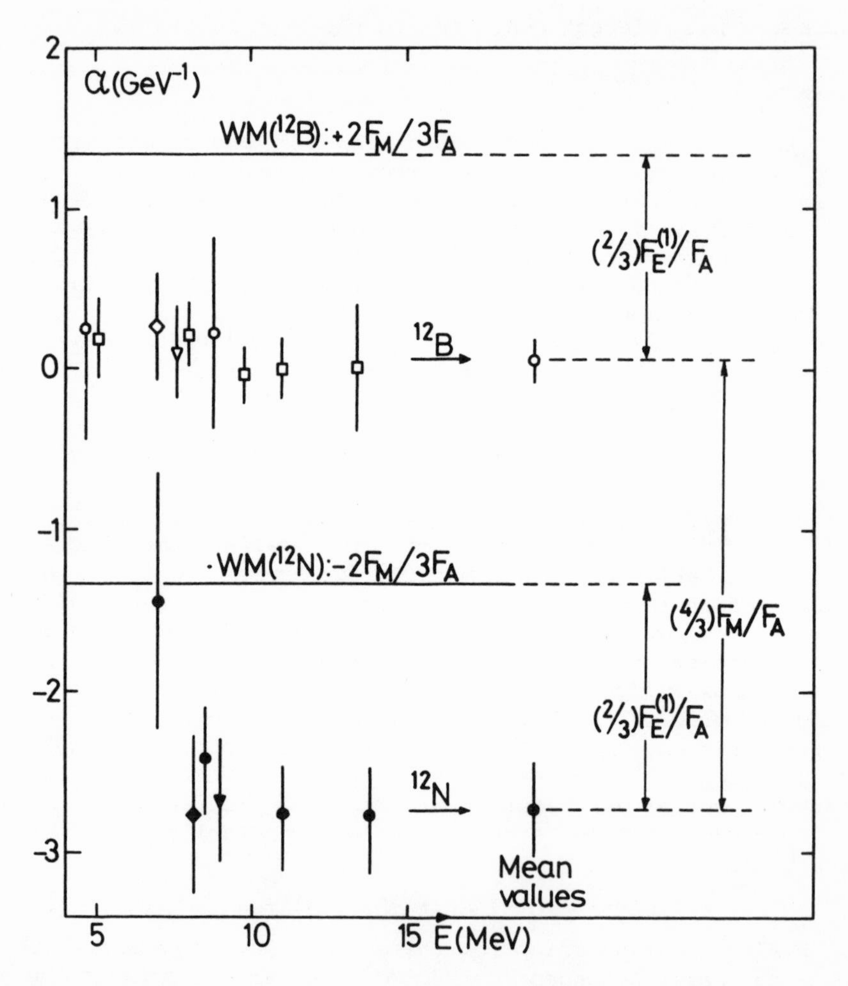

Figure 2. Asymmetry parameters $\alpha_-(^{12}B)$ and $\alpha_+(^{12}N)$ vs. the β-ray energy E. The WM contribution alone is indicated by the lines.

The principle of the experiment yielding these results is reminiscent of direct measurement of the g-factor anomaly of leptons, one measures the third term in $W(\theta,E)$ contributed entirely by gradient couplings.

i) *in a first step the nucleus (^{12}B, ^{12}N) is produced in purely polarized state, that is $P^o \neq 0$ and $A^o \equiv 0$*

ii) *by suitable magnetic substate population equalization the initial polarization is partly converted into alignment ($P^o \to A^{induced}$) with (+) or (-) sign for A^{ind} at will*

iii) *the difference of β-ray countings with (+) and (-) induced alignment allows to isolate the A^{ind} αE term in $W(\theta,E)$*

iv) *the conversion $P^o \to A^{ind}$ leads also to a new (diminished) polarization P^{ind}. From P^o-P^{ind}, through the second term in $W(\theta,E)$, we determine A^{ind}.*
This procedure leads to calibrated results for α.

The conclusion is : there is no 2nd class axial-vector current in β-decay since $b_+ \simeq b_-$ (fig. 2), $F_E^{(2)}/F_A$ is not larger than one tenth of F_M/F_A.
On the other hand, the sum $\alpha_- + \alpha_+$ yields $F_E^{(1)}/F_A = 3.79\ (0.48)/2M$. We will use this determination of $F_E^{(1)}$ in the interpretation of μ-capture measurements (Ch. II).

II. POLARIZATION MEASUREMENTS IN μ-CAPTURE

II.1. AVERAGE POLARIZATION (P_{av}) OF ^{12}B (g.s.) [8]

The average polarization, $P_{av} \equiv \langle J_B \cdot s_\mu \rangle$, of ^{12}B in this allowed ($0^+ \to 1^+$) reaction $\mu^- + {}^{12}C \to \nu_\mu + {}^{12}B$ (g.s.) can be exploited in two ways :

(a) To determine the helicity of the μ^- from π-decay, following an old suggestion of Jackson, Treiman and Wyld. This is analogous to the ν_e-helicity experiment performed in the allowed ($0^- \to 1^-$)$e^- + {}^{152}Eu^m \to \nu_e + {}^{152}Sm^\star$ e^--capture. There the recoil polarization is measured with respect to the emitted ν_e, while in the P_{av} experiment the recoil polarization is determined, via its known β-decay asymmetry, with respect to the incoming direction of μ^-,

(b) To gain information about weak induced currents.

The main practical problem in such an experiment is the preservation of the polarization of ^{12}B recoils (τ = 30 ms) in the (carbon) capture target. The road to this experiment was opened by Madansky and his co-workers who observed that the polarization of ^{12}B recoils (produced in the $^{11}B(d,p)^{12}B$ reaction) implanted in graphite can at least partly be preserved by a longitudinal magnetic field B_z. Following this observation, we systematically investigated the decoupling of ^{12}B implanted in various materials.

In the frame z' of a (left-handed [9]) neutrino emitted at an angle θ with respect to the μ^- spin direction (z), the recoil nucleus is described as a superposition of M = 1 and M = 0 substates (see fig. 2) as $\psi(z') = \cos(\theta/2)\sqrt{2}A|1\rangle + \sin(\theta/2)(A-B)|0\rangle$. Here $\sqrt{2}A$ and (A - B) obviously represent the transverse and longitudinal transition-amplitudes, respectively. The term B arises mainly from relativistic contributions. Defining $X \equiv [\mathrm{Re}\,(A-B)/A]$, P_{av} is given in the muon frame by $P_{av} = +(2/3)(1+2X)/(2+X^2)$; for the "allowed" G.-T. case (B = 0, i.e. X = 1) $P_{av} = +2/3$. The positive sign means that the spin of ^{12}B (J_B) is along the spin of the captured μ^- (s_μ). Fig. 3, a plot of $P_{av}(X)$, shows that $-1/3 \le P_{av} \le 2/3$. Thus the relativistic contributions cannot only induce a departure from $P_{av} = +2/3$, but even lead to a sign reversal of P_{av}. Therefore an unambiguous determination of the sign of the muon polarization from P_{av} without a priori knowledge of X is possible only if $|P_{av}| > 1/3$ (see fig. 3). Conversely, a determination of X from P_{av} yields information about the nature of the induced (relativistic) couplings.

The set-up is shown schematically in the insert of fig. 4, where FC(BC) is the forward (backward) beta-ray scintillator telescope.

The β-decay asymmetry $A(^{12}B)$, measured (ALS, Saclay) with "forward" and "backward" muons, is defined as : $A(^{12}B) = [BC/FC]_{B_z}/[BC/FC]_{B_z=0} - 1$. From $A(^{12}B)$ we deduce the magnitude of the effective polarization of ^{12}B and its sign relative to the incident beam direction. The results are shown in fig. 4 (sign of data with backward muons reversed). The calibration data obtained with (d,p)-produced ^{12}B nuclei implanted into <u>Au</u>, <u>Pd</u> and <u>graphite</u> are also plotted ; the scale is so chosen for these points as to make the (d,p) and μ-capture graphite data coincide at low field. Thus the asymptotic polarization in Au and Pd (the unity corresponds to 0.0820 ± 0.0064) represents the polarization of the ^{12}B recoil in μ-capture in the limit of complete spin decoupling (Paschen-Back limit). This polarization has still to be corrected for the incomplete polarization of μ^- in ^{12}C. The final result is $P_{av}(\mathrm{obs.}) = +0.452 \pm 0.042$, that is with J_B directed along the incident (forward" μ^-) beam.

For the "forward" muons (for which the decay kinematics cannot reverse the polarization in the π-frame) $P_{av}(\mathrm{obs.}) = +0.452 \pm 0.042$ would immediately imply poisitive μ-helicity, were it not for the fact that some recoils originate from captures to excited ^{12}B states, for which the $|P_{av}| > 1/3$ uniqueness argument might be vitiated. Fortunately this is not so. The major part (88 %) of recoils results from the direct (g.s. to g.s.) capture and the rest from a branch feeding almost exclusively an excited 1^- bound state (2.62 MeV). Since $P_{av}(X)$ depends only on the spin but not on the parity of the states, P_{av} for the excited (1^-) sta-

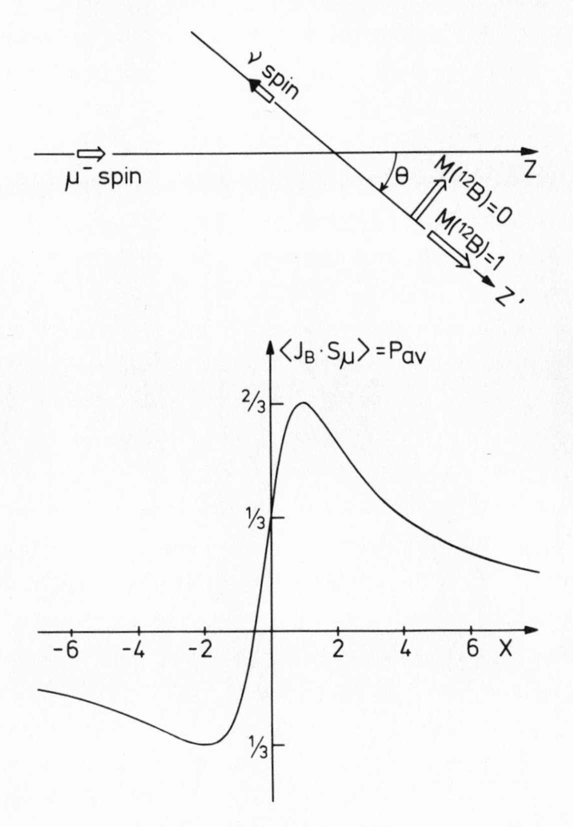

Figure 3. Average polarization P_{av} of ^{12}B (g.s.) vs. the "dynamical" parameter X.

te has the same limits as for the ground state (1^+), and thus the $|P_{av}| > 1/3$ criterion still holds. In actual practice, the P_{av} contributed by the 1^- capture is further attenuated (by $\simeq + 0.72$) through the γ-cascades leading to the ground state, and the strict limit is thus $|P_{av}(obs.)| > 0.32$. The conclusion is : the sign of the μ-helicity is ***positive***.

To determine the μ-helicity quantitatively one has to correct $P_{av}(obs.)$ for the effect of the 12 % 1^- branch. A calculation (see Ref. 8) yields for this state $P_{av} = - 0.24$. The corrected g.s. result is thus $P_{av} = + 0.537 \pm 0.049$. The theoretical value of P_{av} predicted on the basis of the "canonical" couplings (no 2nd class current) is + 0.53 to + 0.55. The ratio $P_{av}/P_{av}(th) = + 1.0 \pm 0.1$ corresponds thus to $h^- \gtrsim + 0.9$, that is the muon anti-neutrino is right-handed, with a left-handed component less or equal to 5 %.

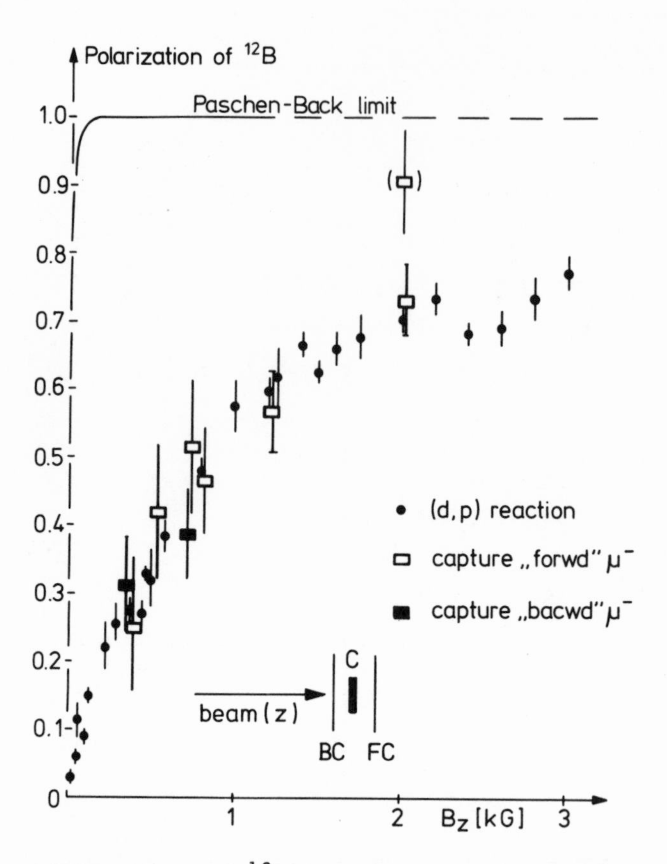

Figure 4. Polarization of ^{12}B vs. the decoupling field B_z. Polarization data with Au and Pd (not shown) fall on the Paschen-Back curve (Ref. 8).

We discuss now the higher order contributions to A and A-B. These contributions come from the gradient type couplings (q/M order effects), viz. the weak magnetism WM, the induced pseudoscalar IP and weak electricity WE (1st and eventually 2nd class) and from "forbidden" matrix elements. In terms of "effective" couplings G_A(axial) and G_P(pseudoscalar) one has : $A = G_A$ and $A-B = G_A - G_P$. To first order in q/M, the WM contributes to $G_A - G_P$ only.

In the following calculations we use two inputs, viz. the experimental rate $\Gamma_\mu = 6100 \pm 270$ sec^{-1} and our "dynamical" parameter $X = (G_A-G_P)/G_A$ determined from P_{av} as $X = 0.36 \pm 0.11$.

Comparing Γ_μ to the theoretical rate $\Gamma = 3.53\ (10^3\ s^{-1})\,|F_A(q^2)/F_A(0)|^2\ G_A^2(2+X^2)$, we calculate $|G_A|$. We

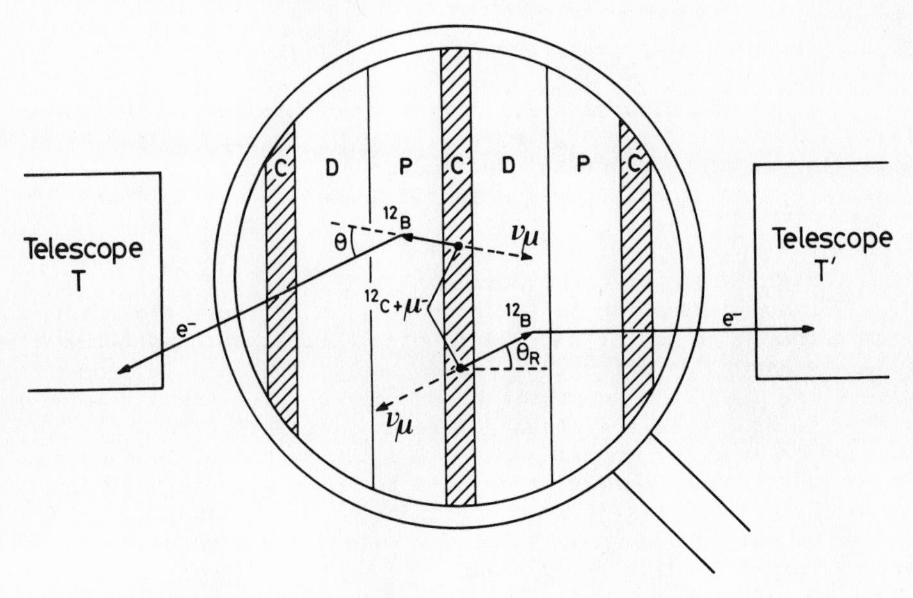

Figure 5. The stacked target for the measurement of P_L = P = polarization retaining, D= polarization relaxing material, C = carbon layer, θ_R = recoil angle and θ = electron emission angle.

have $G_A = 1.198 \pm 0.043$. *From* $G_A - 1 \simeq -(q/2Mg_A)\mu g_V$, *the WM coupling is obtained (with* $g_A = -1.25\ g_V$) *as :*

$$\mu = 5.3 \pm 1.1 \quad \text{(versus the CVC value 4.7).}$$

Next we deduce from $G_AX-1 \equiv G_A-G_p-1 \simeq -(q/2Mg_A)(g_p+g_T+yg_A) = -(0.57 \pm 0.13)$, *the ratio* $R = (g_p + g_T)/g_A$ *as :* $R = 8.1 \pm 2.7$. *The parameters* g_p *and* g_T *are the coupling constants of IP and 2nd class WE, y is the first class WE contribution* ($F_E^{(1)}$) *determined experimentally in I.2.*

The "approximate" ≃ *symbols indicate the neglect of small second forbidden terms ; with the inclusion of these, one obtains* $\mu = 4.5 \pm 1.1$ *and* $(g_p+g_T)/g_A = 7.1 \pm 2.7$. *The first of these results supports strong CVC, second agrees well with Partially Conserved Axial-Current hypothesis (PCAC) if there is no 2nd class coupling. Conversely, with the "canonical" value* $g_p/g_A = 6$ *one gets* $g_T/g_A = +1.0 \pm 2.7$ *that is compatible with vanishing 2nd class current* [10].

II.2. LONGITUDINAL POLARIZATION (P_L) OF ^{12}B (G.S.) [11]

The polarization of ^{12}B(g.s.) in the neutrino (left-handed) frame is clearly (fig. 3) : $P_L = 2/(2+X^2)$. This polarization is a parity-violating quantity, related to ν_μ-helicity : a measurement of P_L thus institutes the muonic analog of the Goldhaber-experiment in β-decay.

The principle of the experiment is illustrated in fig. 5. Consider muons stopping in a carbon foil, C sufficiently thin to permit the existence of the ^{12}B recoils, and sandwiched between two layers D and P, in which the recoils stop. The materials of these layers are ideally so chosen that one (P) retains the ^{12}B polarization fully, while the other (D) relaxes it instantly. P_L is then determined by comparing the ^{12}B decay rates on the two sides of the sandwich. One way of doing this is to alternately turn the D and P sides towards a fixed electron detector. Another is to choose materials such that P becomes polarization retaining only with a (weak) holding field $\vec{B}$ and to compare field on/field off counting rates.

A mechanically chopped muon beam (S.I.N.) (33 msec on, 56 msec off) is partially stopped in a target T consisting of a stack of sandwiches of the type illustrated in fig. 5. The decay electrons are detected with two telescopes, consisting of MWPC's and thin scintillators respectively ; this setup is designed to

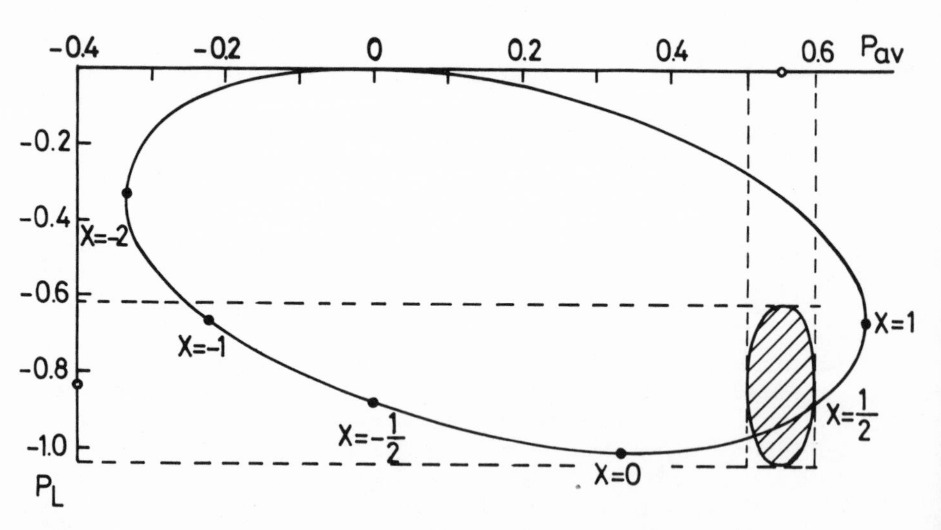

Figure 6. Average vs. longitudinal polarization of ^{12}B together wi[th] the experimental values of P_{av} and P_L. The point x = 0.[.] corresponds to canonical couplings (no 2nd class current[,] CVC and PCAC hold).

maximize solid angle and to minimize unwanted carbon signals, while providing reasonably fast time resolution. A holding field $B(\lesssim 100\ G)$ can be applied with the Helmholtz coils. Two targets were used : (A) 1500 sandwiches Ag(800 $\mu g/cm^2$/C(20 $\mu g/cm^2$)/ LiF(250 $\mu g/cm^2$), and (B) 200 sandwiches Au(1500 $\mu g/cm^2$)/C(50 $\mu g/cm^2$)/Al(400 $\mu g/cm^2$). LiF(Al) plays the role of D, while Ag(Au) plays that of P, provided that $B \geq 50G$ ($\geq 25G$). In zero field, both Ag and Au are depolarizing. From the net decay electron rates in the two telescopes, denoted as N(B) and N'(B), one derives the asymmetry $A = [N(B)/N(O)]/[N'(B)/N'(O)]-1$ for a given orientation of target. Indicating with A and A' the asymmetries measured with the two target orientations, i.e. the layers P facing the unprimed and primed telescopes respectively, one has to a crude approximation $A - A' = P_L$. This result corrected for attenuation effects yields

$$P_L = -0.83(21).$$

From this result we conclude :

(1) the neutrino emitted in muon capture is lefthanded (corroborating the result of II.1) ;
(2) the measured P_{av} and P_L values intersect the ellipse implied by $P_{av} = (2/3)(1+2x)/(2+x^2)$ and $P_L = 2/(2+x^2)$ in one point, i.e. are consistent (see fig. 6) ;
(3) they are furthermore consistent with the "canonical" values of the form factors, i.e. $X = 0.44$.

REFERENCES

[1] T.N.K. Godfrey, Thesis, Princeton University, 1954 (unpublished)
[2] K. Koshigiri, Osaka preprint (1979, to be published in Nucl. Phys.)
[3] C.S. Wu, Y.K. Lee and L.W. Mo, Phys. Rev. Letters 39 (1977) 72
[4] W. Kaina, V. Soergel, H. Thies and W. Trost, Phys. Lett. 70B (1977) 411
[5] M. Morita et al., Progr. Theor. Phys., Suppl. 60 (1976) 1
[6] S. Weinberg, Phys. Rev. 112 (1958) 1375
[7] H. Brändle et al., Phys. Rev. Letters 41 (1978) 299 (and references quoted therein) ; the fig. 1 is taken from the "extended" version of this reference (unpublished preprint)
[8] A. Possoz et al., Phys. Lett. 70B (1977) 265 and A. Possoz, Thesis, Louvain 1978 (unpublished)
[9] Since P_{av} is a scalar, this is merely a convenient choice
[10] More recent calculations [M. Kobayashi et al., Nucl. Phys. A312 (1978) 377 and W.-Y. Hwang (preprint 1979)] yield $P_{av}(1^-) > 0$. This affects the value of WM (μ) only very slightly, the value of $(g_P+g_T)/g_A$ becomes 10 ± 2.7 which is still compatible with the PCAC prediction
[11] P. Truttmann et al., Phys. Lett. 83B (1979) 48

PARITY NON-CONSERVATION IN ATOMS AND MOLECULES

P.G.H. Sandars

Clarendon Laboratory

OXFORD, U.K.

INTRODUCTION

We discuss in these two lectures a rapidly expanding field of particle physics: the search for PNC in atoms and molecules.[1] In the first part we briefly review the present situation, discussing what can be measured and how it is that atomic experiments can be made sensitive enough to detect weak interaction effects and then listing the various schemes which have been proposed and/or are under way. A more detailed description of the Oxford, Washington and Novosibirsk optical rotation experiments and their implications is deferred to section 8.

1 NEUTRAL CURRENT INTERACTIONS

The Weinberg-Salam and most other current models of the weak interactions predict the existence of PNC interactions between the constituents of an atom. Direct evidence for the existence of the PNC electron nucleon interaction has recently been obtained in the beautiful polarized electron scattering experiment at SLAC.[2] The polarization assymmetry is in excellent agreement with the Weinberg-Salam model.

Assuming non-derivative coupling and CP invariance there are just five interactions which can be measured in atomic physics (fig. 1). And it is possible in principle to measure each of these five interactions H^p_{AV}, H^n_{AV}, H^p_{VA}, H^n_{VA}, and H^e_{AV} separately. Thus experiments on hydrogen will be sensitive only to H^p_{AV} and H^p_{VA} whereas H^n_{AV} and H^n_{VA} will also contribute in deuterium. H_{AV} and H_{VA} can be readily separated because the former is nuclear spin-dependent whereas the latter is not. Bouchiat and Bouchiat [1] have pointed out that for heavy atoms the PNC produced by H_{VA} will be very much greater than that from H_{AV}. H^{e}_{AV} can in principle be distinguished because, unlike the other interactions, it is non-zero for electroni states with zero overlap with the nucleus. Unfortunately there are reasons to believe that its contribution to atomic PNC will be rather small and that it will be extremely difficult to detect in practice.

An important subsidiary point is that the success with which lepton hadron scattering can be related to quark models leads us to expect that measurement of the appropriate coupling constants will yield direct information on the fundamental structure of the weak interactions.

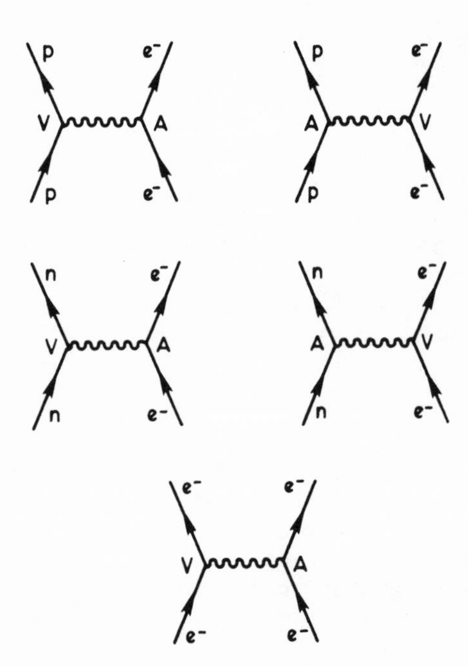

Fig. 1. Neutral current PNC interactions in an atom. The diagrams represent the interactions H^{p}_{VA}, H^{p}_{AV}, H^{n}_{VA}, H^{n}_{AV}, H^{e}_{AV} where the subscript represents the vector or axial vector structure of the currents involved; the hadronic current being listed first in the hadron electron interactions. The strengths of the five interactions are expressed in terms of the Fermi constant G_F through coupling constants C^{p}_{VA}, C^{p}_{AV}, C^{n}_{VA}, C^{n}_{AV}, C^{e}_{AV}.

2 PNC ENHANCEMENT

The most immediate difficulty in carrying out this programme is the small size of the weak interactions effects. They are expected to be of order G_F which in typical atomic units has the magnitude 10^{-14}; the small value reflects the low energies of the particles in atomic systems. In other words, the crucial PNC fraction

$$F = \frac{\text{PNC effect}}{\text{competing EM effect}}$$

will in general be of order 10^{-14}. Clearly this is unusuably small and enhancement is necessary.

Three lines of attack have been developed:

(i) Near Degeneracy in Hydrogen[3]

The near equality of the energies of the $ns_{1/2}$ and $np_{1/2}$ levels of opposite parity means that the mixing effect of the PNC Hamiltonian is greater than in other atoms by a factor of order 10^4.

(ii) Z^3 Law

Bouchiat and Bouchiat [1] were the first to point out that the PNC atomic matrix elements of H^n_{AV} and H^p_{AV} scale as Z^2 and H^n_{VA} and H^p_{VA} as Z^3 so that the use of very heavy atoms can lead to an enhancement of order 10^4 and 10^6 respectively.

(iii) Forbidden Transitions[1]

The use of M1 or forbidden M1 transitions greatly reduces the competing EM effect and increases F by α and α^3 respectively.

In many experiments more than one enhancement process is utilised leading to PNC fractions $\sim 10^{-3}$. But, of course, a large PNC fraction is not the only criterion and questions of observability and signal to noise are equally significant.

3 HYDROGENIC EXPERIMENTS (ELECTRONIC AND MUONIC)

$2s \to 1s$

One of the earliest experiments proposed [4] was the search for circular polarization in the highly forbidden $2s \to 1s$ single photon decay in hydrogen. The PNC fraction is quite high ($\sim 10^{-3}$) but the difficulty of the experiment is indicated by the fact that this single photon decay has never been observed in hydrogen. As far as the author is aware no experiments of this type are under way.

A similar situation exists for the proposed muonic experiments which all appear to have been abandoned.

2s → 3s Laser Absorption

Lewis et al [5] at Michigan have proposed a laser absorption experiment on a beam of $2s_{\frac{1}{2}}$ metastable hydrogen atoms. The expected PNC fraction is reasonably large but background problems are likely to be very severe.

2s → 2s rf Experiments

At least three groups [6] at Michigan, Washington and Yale are setting up radiofrequency experiments on hydrogen of the general type in which one observes a PNC E1 matrix element in interference with a P allowed matrix element, either M1, or E1 induced by an external electric field. The interference is observed through the change in intensity on 180° reversal of the relative phase of the two rf fields which interact with the PNC and the P allowed matrix elements. In principle one can make the PNC fraction large by reducing the magnitude of the P allowed transition field, though this is at the expense of transition probability and a limit is set by background effects. Signal to noise calculations suggest that such experiments are feasible, though clearly they are not easy.

A basic feature of all experiments is to work at a magnetic field (560 Gauss or 1180 Gauss) for which the $2p_{\frac{1}{2}}$ and $2s_{\frac{1}{2}}$ levels cross. The PNC admixture of $2p_{\frac{1}{2}}$ into $2s_{\frac{1}{2}}$ then becomes:

$$\delta = \frac{\langle 2s_{\frac{1}{2}}\; m_j\; m_I | H^{PNC} | 2p_{\frac{1}{2}}\; m'_j\; m'_I \rangle}{W2s_{\frac{1}{2}}\; m_j\; m_I - W2p_{\frac{1}{2}}\; m'_j\; m'_I + i\Gamma_{2p_{\frac{1}{2}}}}\,, \qquad \begin{array}{l} m_j + m_I \\ = m'_j + m'_I \end{array} \tag{1}$$

where we have included the nuclear spin quantum number m_I which is necessary because the H_{AV} part of H^{PNC} is nuclear spin dependent.

It is useful to note that because H_{VA} is spin independent ($\Delta m_I = 0$) its contribution to δ vanishes at the low field crossing point for which $\Delta m_I = 1$. The main advantage of the crossing point is not, as might be thought at first sight, the enhanced size of δ because of the small energy denominator; this is counter balanced by increased electric field quenching and a smaller electric field must be used with the same ultimate sensitivity. The two main advantages of the crossing points are: (i) there is a (relatively) rapid variation in PNC transition probability with magnetic field which helps in background discrimination and (ii) δ can be real or imaginary depending on the value of the magnetic field, allowing flexibility in the types of P conserving matrix element which can be used in interference with the PNC E1 element.

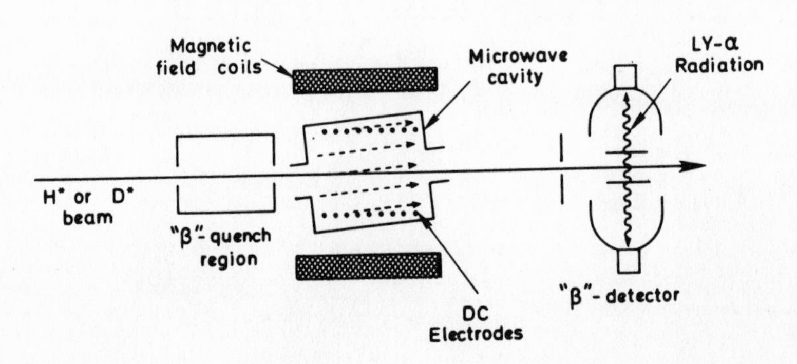

Fig. 2. Schematic of the Michigan experiment

We now outline briefly the most significant features of the various experiments.

Michigan

A schematic of the Michigan apparatus is shown in fig. 2. A beam of order 10^{14} hydrogen metastables per sec. is produced in a duoplasmatron source followed by a cesium charge exchange cell. The atoms are prepared in the $m_J = +\frac{1}{2}$ $m_I = -\frac{1}{2}$ state by familiar quenching plus transition methods. The 1600 MHz transition $m_J = \frac{1}{2}$ $m_I = -\frac{1}{2} \rightarrow m_J = -\frac{1}{2}$ $m_I = \frac{1}{2}$ is induced at the high field crossing point by a combination of transverse DC electric field E_x and transverse rf electric field ε_x giving a matrix element α E_x ε_x for the $\Delta M = 0$ transition. Atoms which have undergone the transition are detected by quenching in a Lyman α detector.

The PNC E1 matrix element interacts with the Z component of the rf field ε_z. The PNC fraction is then given by $F = \frac{E1^{PNC}\ \varepsilon_z}{\alpha E_x\ \varepsilon_x}$. This can be 'improved' by making $\frac{\varepsilon_z}{\varepsilon_x}$ large which is achieved by slightly tilting the cavity so that $\frac{\varepsilon_z}{\varepsilon_x} \sim 10$.

Williams and his group at Michigan estimate that an integration time of order 30 minutes will be required to observe the PNC signal against background. Clearly, very great care will be needed to remove spurious background effects but the experiment has the great advantage that the signal of interest must change sign with E_x, B, $\varepsilon x/\varepsilon z$ as well as changing in a known manner as a function of magnetic field.

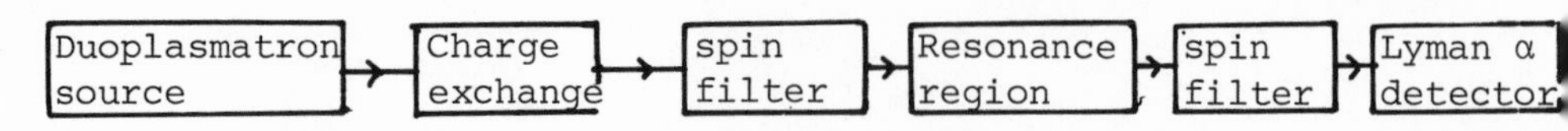

Fig. 2b. Schematic of the Yale experiment

Washington

No details of this experiment have been published but it is similar in principle to the Michigan scheme described above.

Yale

A schematic of this experiment is shown in Fig. 2b. It differs in a number of respects from the others:

(i) The low field crossing point is used. (ii) An 80 MHZ nuclear spin transition ($\Delta m_I = 1$) is used. (iii) The P-conserving and PNC amplitudes are stimulated in different rf regions. (iv) The state selection is achieved by means of spin filters of the type designed for use in polarized hydrogen sources for accelerators. The experiment closely resembles a classic atomic beam set-up with source, polarizer, resonance region, analyser and detector.

4 HIGHLY FORBIDDEN M1 TRANSITIONS

(i) Circular Dichroism

In their pioneering 1974 paper [1] the Bouchiats proposed a very elegant experiment to search for circular polarization in the highly forbidden 6s → 7s M1 transition of Cs. The fractional circular polarization is given by

$$PC = 2 \text{ Imag. } \frac{E_1^{PNC}}{M1}$$

$E1^{PNC}$ has been calculated both by the Bouchiats [7] and by the author [8] with satisfactory agreement on a value $-i \times 1.7 \times 10^{-11}|e|a_o$. The M1 matrix element was initially unknown but has been measured by the Bouchiat group [9] who find $M1 = -4.24 \times 10 \mu B$. These numbers give a theoretical prediction $PC = 2.2 \times 10^{-4}$ for the circular polarization.

The PNC fraction is satisfactorily large but this is achieved at the expense of a transition probability $\sim 10^{-12}$ of that for a normal atomic transition. Nonetheless calculations indicate that present laser intensities are in principle sufficient to give adequate signal to noise. Unfortunately, the Paris group has found in practice that at useable Cs pressures the background from collision induced transitions seriously exceeds the M1 transition rate and precludes straightforward application of the originally proposed method. It may be possible to avoid the difficulty while retaining sufficient signal to noise by using reduced Cs vapour pressure or an atomic beam [10], but the Bouchiat team have adopted an alternative approach.

(ii) E1 Field Induced Polarization [11]

They look at the same 6s → 7s transition, but induced by the presence of a static electric field E_o. Interference between $E1^{PNC}$ and $E1^{ind}$ then leads to an atomic polarization

$$\underline{P}_e = \frac{8\,F\,(F+1)}{3(2I+1)^2}\ \text{Imag.}\ \frac{E1^{PNC}}{E1^{ind}}\ (\underline{\sigma}_{in} \cdot \underline{k}_{in})\ \underline{k}_{in} \times \underline{E}_o$$

which is proportional to the circular polarization $\underline{\sigma}_{in}.\underline{k}_{in}$ of the incoming photon and is directed perpendicular to both $\underline{E}_o$ and $\underline{k}_{in}$. The presence of $\underline{P}_e$ leads to circular polarization in fluorescence observed along $\underline{E}_o \times \underline{k}_{in}$. Thus the experiment looks for the pseudo-scalar quantity

$$(\underline{\sigma}_{in} \cdot \underline{k}_{in})\ (\underline{k}_{in} \times \underline{E}_o \cdot \underline{k}_{out})\ (\underline{\sigma}_{out} \cdot \underline{k}_{out}).$$

This new scheme illustrated in fig. 3 has the two great merits (i) that the magnitude of the transition can be increased with E_o, though this is at the expense of fractional polarization and there is a practical limit set by electric field breakdown.
(ii) The effect reverses with both $\underline{E}_o$ field direction and with the

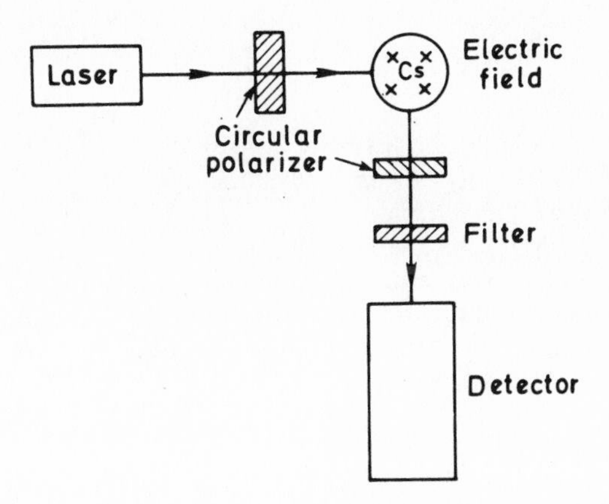

Fig. 3. Schematic diagram of the Bouchiat experiment on the highly forbidden 6s → 7s transition in Cs.

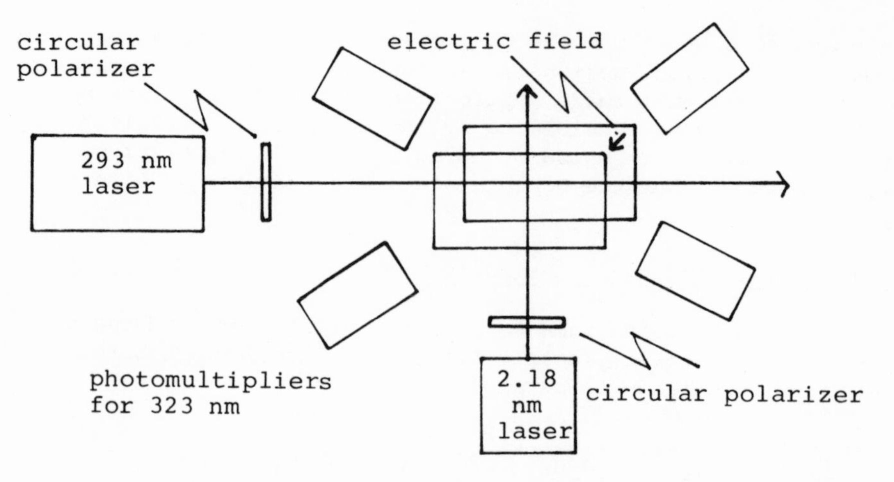

Fig. 4. Schematic diagram of Tℓ experiment

circular polarization of both incoming and outgoing photons. Calibration is also straightforward since a magnetic field parallel to E_o produces a polarization $\underline{P}_e^1$ along $\underline{k} \times \underline{B}$ in a well understood manner.

The apparatus has been constructed and performs well, the only limitation being signal to noise. This has been improved by using multiple passes for the incident light, and further work along these lines is in progress. The present result [12] is

$$\left|\frac{E1^{PNC}}{M1}\right| = (0.56 \pm 1.8) \times 10^{-3} \qquad 90\% \text{ confidence level}$$

c.f. theory (1.1×10^{-4})

(iii) $6p_{\frac{1}{2}} \rightarrow 7p_{\frac{1}{2}}$ in Tℓ [13]

A somewhat similar experiment is being carried out by Commins and his group at Berkeley on the 293 nm $6p_{\frac{1}{2}} \rightarrow 7p_{\frac{1}{2}}$ transition in Tℓ. Here the circular polarization is expected to be large PC ≈ 2.5×10^{-3}. The M1 element has been determined experimentally:[14] $M1 = (-2.11 + 0.30) \times 10^{-5}\mu_o$. As in the Bouchiat experiment this M1 turns out too small to use directly as the parity conserving transition element and an auxiliary electric field is applied. This again produces a polarization

$$\underline{P}_z \;\alpha\; (\underline{\sigma}_{in} \cdot \underline{k}_{in})(\underline{k}_{in} \times \hat{\underline{E}}_o)$$

proportional to the helicity of the 293 nm photon and directed perpendicular to both the photon direction and the electric field.

The Commins (figs.4,5) experiment differs from the Bouchiat experiment in the way that the polarization is detected. Commins applies to the cell a strong circularly polarized beam at 2.18μ, the $7p_{1/2} \rightarrow 8s_{1/2}$ transition frequency. The polarization of the $7p_{1/2}$ state therefore determines the intensity of the 2.18μm transition and hence the intensity of the resulting 323 nm $8s_{1/2} \rightarrow 6p_{3/2}$ fluorescence. If I_+ and I_- are the fluorescence intensities for infra red photons with $J_z = \pm 1$ then there is an symmetry $\Delta_o = (I_- - I_+)/(I_- + I_+) = 0.7\ P_z$ where the factor of 0.7 is an experimentally determined factor. This value of the asymmetry is to be compared with the small circular polarization $\approx 0.08\ P_z$ of the direct fluorescence from the $7p_{1/2}$ state, showing clearly the advantage of the two laser method.

Details of the experimental technique are found in Conti et al.[13] A number of precautions were taken against spurious systematic effects. The result obtained from some two hundred hours of running is (δ = CP).

$$\delta_{exp} = +(5.2 \pm 2.4) \times 10^{-3}$$

compared to [15]

$$\delta_{theor} = (2.3 \pm 0.9) \times 10^{-3}$$

The result constitutes persuasive evidence in favour of parity non-conservation in atoms. The Berkeley group are confident that improvements now under way should permit a more precise determination of δ.

(iv) J=0→J=0 Transitions

As Bouchiat and Bouchiat [7] pointed out J=0→J=0 transitions have the important feature that only the nuclear spin dependent part of

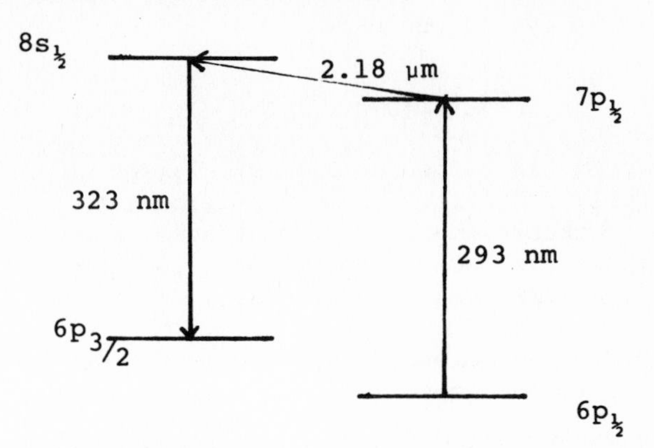

Fig. 5. Energy levels involved in Tℓ experiment

H^{PNC} can induce circular polarization. Thus even though this interaction in heavy atoms is ~100 smaller than the spin-independent part it is in principle separately measurable. Nor is the PNC fraction particularly small; Khriplovich and co-workers[16] predict PC = -6.5×10^{-5} for the $6p^2 J = 0 \to J = 0$ transition in lead. Unfortunately, the M1 transition element is very small ~0.18×10^{-6} and the transition wavelength would require second harmonic generation. Even so, as techniques develop, one can envisage this experiment being carried out.

(v) Diatomic Molecules

In a recent paper Labzovskii [17] has suggested that very high PNC circular polarization fractions may be obtained in diatomic molecules. He has pointed out the electron-electron interaction can mix the opposite parity states split by the Λ doubling. In certain cases where this doubling energy is very low (~10 Hz) one could expect CP fractions of order unity if one could resolve the doublet. This is not likely to be possible for such a small splitting but experiments on molecules with doublings in the kHz region may yield viable experiments in the future.

5 OPTICAL ROTATION [18]

A PNC term in the Hamiltonian can lead to difference between the refractive index for right and left handed photons and hence to the phenomenon of optical rotation. Close to an M1 transition the rotation angle has a dispersive dependence on wavelength:

$$\phi^{PNC} = \frac{-4\pi(n-1)L}{\lambda}\{\mathrm{Imag.}\frac{E1^{PNC}}{M1}\}$$

where n is the normal refractive index, L is the length and λ is the wavelength. Absorption places a limit on attainable rotation angle; at one absorption length one has

$$\phi^{PNC}_{1A.L.} = \frac{(\omega - \omega_o)}{\Gamma/2}\{\mathrm{Imag.}\frac{E1^{PNC}}{M1}\}$$

For an allowed M1 transition in a heavy element $\phi^{PNC}_{1A.L.}$ ~10^{-7} radians. It is even higher for a forbidden transition but in practice it is not possible to obtain one absorption length - and for a given density of gas the attainable PNC is proportional to M1 and consequently much reduced for forbidden transitions.

The equation above suggests two possible strategies: one can work close to a resonance so that $(\omega - \omega_o) \approx \Gamma/2$ and $\phi^{PNC}_{1A.L.} \approx \frac{E1^{PNC}}{M1}$ and one has rapid wavelength dependence which can be used for discrimination against background; or one can move far away from resonance where

$$\phi^{PNC}_{1A.L.} \approx \frac{(\omega - \omega_0)}{\Gamma/2} \times \frac{E1}{M1}^{PNC} >> \frac{E1}{M1}^{PNC}$$

but the wavelength dependence is much reduced.

There are four experiments of the first type under way at Oxford [19], Washington [20], Novosibirsk [21], and Moscow [22] - detailed description is deferred to the next section. One experiment of the second class is in progress in Moscow [23]; the absence of wavelength discrimination makes this a very hard project. Khriplovich and Novikov [24] have suggested an optical rotation experiment using microwaves tuned to an atomic hyperfine transition. However the effect, which depends on H_{AV}, is of order 10^{-9} and is likely to be very difficult to detect experimentally.

An interesting extension of the optical rotation experiments to molecules has been suggested by Sushkov and Flambaum.[25] They argue that H_{AV} can mix Λ doublet states of diatomic molecules and produce fractional effects of order 10^{-4}, provided the hyperfine structure can be resolved. The possibility of sensitive experiments of this type is clearly of considerable interest.

6 ENERGY SHIFTS IN CHIRAL MOLECULES

For a non-degenerate atomic system P conservation rules out any first order energy shift due to H^{PNC}. Even in an odd parity external electric field where the parity argument fails T invariance implies $\Delta W^{PNC} = 0$. This is a generalization of the well-known theorem that an electric dipole moment would imply T violation.

None of these theorems hold for degenerate chiral molecules, and one may therefore expect a first order energy shift ΔW^{PNC}. The magnitude of such a shift is of considerable interest [26,27,28,29,30,31] because initial experiments are already under way [31]. It seems likely that ΔW^{PNC} will be appreciably smaller than a simple order of magnitude calculation might suggest. Gajzago and Marx [26] pointed out the spin-dependence of H^{PNC} makes the effect zero unless one has spin-orbit coupling present. This together with the Z^3 enhancement makes the use of a heavy atom imperative. Zeldovich et al [29] have carried out a ligand field discussion in which the potential on the heavy atom was expanded in a series of spherical harmonics $V = \sum_{k,q} A^k_q Y^k_q(\theta,\phi)$. They showed that V is required in at least third order in perturbation theory and even then ΔW^{PNC} is the result of interference between multipole k = 2, k = 3 and k = 4 terms in the expansion of V. Zeldovich et al [29] give an order of magnitude estimate which gives $\Delta W^{PNC} \approx 10$ kHz for the heaviest atoms.

The magnitude of such energy differences has been investigated further by Hegstrom, Rein and the author [32]. We find that there are a number of theorems which reduce the size of ΔW^{PNC} below the value expected from an order of magnitude estimate. In order to pursue the question further we have carried out calculations on twisted ethylene and a dialkyl sulphide both of which have appreciable optical activity. We find $\Delta W^{PNC} \approx 2 \times 10^{-4}$ Hz and 2×10^{-6} Hz respectively. These very small values suggest that even in heavier molecules the energy difference is unlikely to be large enough to measure.

7 OTHER PROPOSED EXPERIMENTS

(i) E,B Resonance in Heavy Atoms

Loving and Sandars [33] have suggested that one look for E1,M1 interference at low frequencies in an atomic beam resonance experiment. This method would be sensitive only to the nuclear spin dependent terms H_{AV}. Preliminary design considerations suggest that such an experiment would be difficult but feasible with present techniques.

(ii) Two Photon Transitions in Hydrogen Atoms

Drukarev and Moskalev [34] have argued that PNC effects in the two photon $1S_{\frac{1}{2}} \rightarrow 2P_{\frac{1}{2}}$ can reach $10^{-4} - 10^{-6}$ in a suitable experimental arrangment.

(iii) Internal Conversion

Koonin [35] has pointed out that internal conversion associated with a magnetic dipole nuclear transition can lead to PNC effects of order 10^{-6} in favourable circumstances.

(iv) Two Electron Atoms

Gorshkov et al [36] have noted that in two electron atoms there are close coincidences between opposite parity states for certain values of Z. They estimate that PNC effects such as circular polarization of forbidden decay photons and auto-ionization could be as high as $10^{-6} - 10^{-8}$.

(v) Parity Violation in Liquid He

Legget [37] has argued that H^{PNC} should give rise to a thermodynamically stable electric dipole moment density directed along the characteristic rotation axis $\underline{w}$ of superfluid ^{3}He-B and that this may be measurable. If feasible, this would constitute the first observation of PNC on a macroscopic scale.

(vi) Josephson Effect

Vainshtein and Khriplovich [38] have pointed out that the nuclear spin dependent term H_{AV} will modify the properties of the Josephson effect and that these are in principle observable. But the practical difficulties are likely to be severe, not least because the effect depends on a high degree of nuclear polarization.

8 OPTICAL ROTATION IN BISMUTH

(1) The Transitions

There are four experiments currently in progress to look for the optical rotation in atomic bismuth at Oxford [19], Washington [20], Novosibirsk [21], and Moscow [22]. The two transitions under investigation are shown below. The 648 nm line is in a region covered by continuous wave dye lasers which have excellent intensity and sufficiently narrow line-width that individual hyperfine components can readily be resolved. But it has the disadvantage that the region of the spectrum is overlaid with molecular Bi_2 lines which reduce the usable intensity. The 876 nm line is free from molecules but only recently have continuous wave lasers been produced for this wavelength and the original Washington experiment was carried out with a pulsed parametric laser of wide linewidth.

A number of calculations of the expected optical rotation in bismuth have been made on the basis of the Weinberg-Salam model.

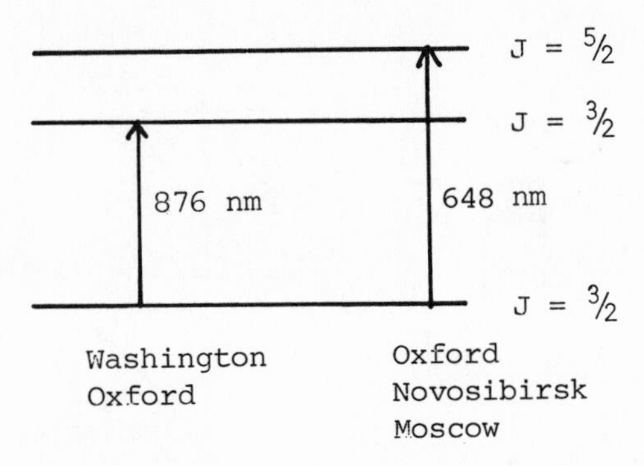

There is general agreement that the value of R is likely to be of order 1×10^{-7} though the precise value will depend on the magnitude of the 'atomic shielding' [41]. Best current estimates are given in table 1 though these values may be modified when the more complete calculations which are presently under way have been completed.

(2) Experiment

While the important details differ significantly, the basic method in all three experiments relies on the fact that the intensity of plane polarized light transmitted through crossed polarizers is proportional to the square of the angle of rotation of the plane of polarization in the region between them: $I = I_o\phi^2$. In the experiment $\phi = \phi^{PNC} + \phi^M + \phi^R$ where ϕ^{PNC} is the optical rotation angle of interest, ϕ^M is a controlled angle of order 10^{-3} radians applied by means of a Faraday cell in the Oxford and Washington experiments or by rotating one polarizer in the Novosibirsk case; ϕ^R is a residual angle of order 10^{-5} radians due for example to misalignment of the polarizers, optical rotation in the optical components etc. By looking for the change in signal on reversing ϕ^M one can obtain a signal which is linear in ϕ^{PNC}: $\delta I = 2I_o\ \phi^M(\phi^{PNC} + \phi^R)$. Because of the high intensity of the lasers available this method has excellent sensitivity - thus the measured figure for the Oxford apparatus is a signal to noise of 1×10^{-8} radians ($^1/_{10}$ the Weinberg-Salam angle) in two minutes. But in order to make use of this we have to get rid of the much larger ϕ_R. This is done by changing the laser wavelength and looking for the rapid wavelength dependence of the true PNC signal at the various hyperfine transition frequencies.

Schematic diagrams of the three experiments are shown in figures 6, 7, and 8. The most significant features of each are discussed below.

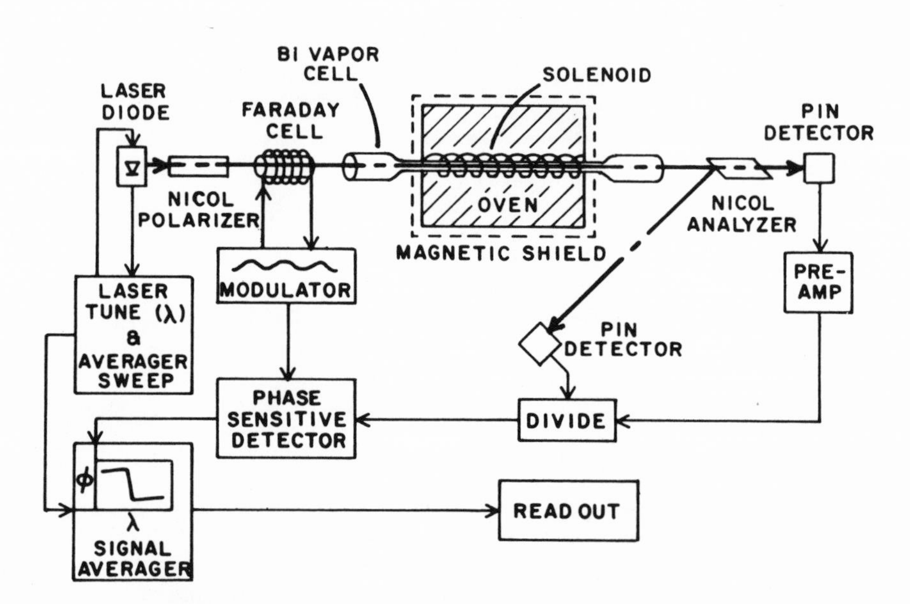

Fig. 6. Schematic of the Washington bismuth optical rotation experiment

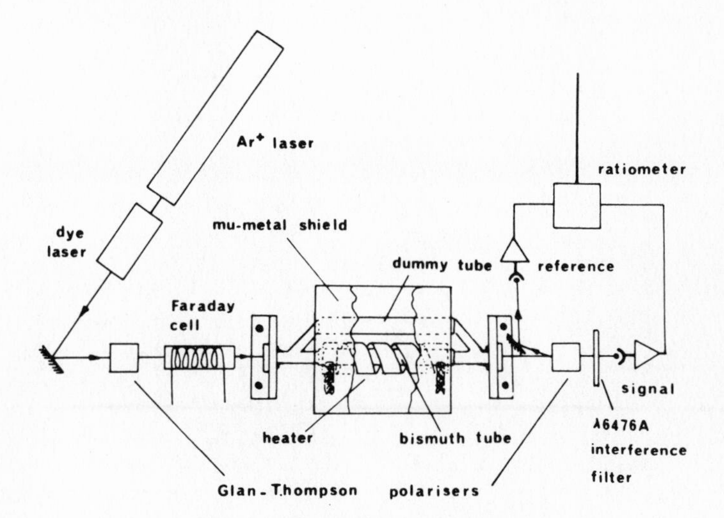

Fig. 7. The Oxford experiment

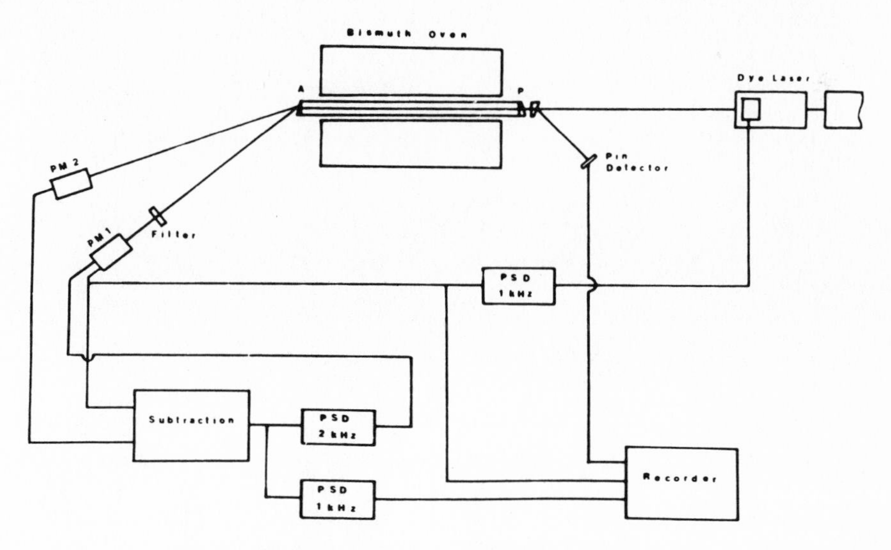

Fig. 8. The Novosibirsk experiment

(i) The Washington Experiment (876 nm) [39]

Except for the use of different lasers the set-up for the two Washington experiments is essentially the same (fig. 6).

In the early experiment the group used a pulsed parametric laser. This had such a wide spectral output that it was not possible to resolve the hyperfine structure. Normalization and modulation were also difficult with the pulse system. The pulsed laser was therefore replaced by a diode laser when the latter became commercially available. The different hyperfine components can be readily resolved and light intensity and optical characteristics are improved.

In both experiments a Faraday modulator is used to produce ϕ_M and the angle is measured via the ratioed intensity $\delta I/I$ which is directly proportioned to $(\phi^{PNC} + \phi^{R})$. The two are separated by sweeping the wavelength through the hyperfine pattern and analysing the data for a component with the wavelength dependence expected for ϕ^{PNC}.

(ii) The 648 nm Oxford Experiment [40]

The apparatus is illustrated in fig. 7. Angle measurement used a Faraday cell modulated at ~350 Hz, the intensity signal being ratioed as shown and analysed by analogue phase sensitive electors. The wavelength discrimination was achieved by moving from side to side of the $F = 6 \rightarrow 7$ hyperfine component at a modulation rate ~1 Hz. Residual wavelength dependent angles were subtracted by means of the double oven systems which allowed interchange of the bismuth tube for an empty one every few minutes without any movement of an optical component. In the latest experiments the analogue detection system has been replaced by a computer analysis scheme based on a square wave Faraday modulation pattern with angles ϕ_M, O, O, $-\phi_M$. ϕ the rotation angle to be measured is given by the ratio

$$\phi = (I_1 - l_4)/(I_1 + I_4 - I_3 - I_2),$$

independent of overall beam intensity. In both experiments the Bi Faraday effect is removed by (a) a magnetic shield surrounding the oven and (b) a feedback using a small applied magnetic field which locks the laser so that the two wavelengths used in the 1 Hz modulation have equal (and nearly zero) Faraday effects. Calibration of the Bi optical depth is made through the measurement of the well understood Bi Faraday effect in an auxiliary field.

The main difficulty in these experiments is a small ~10^{-7} wavelength dependent angle of optical origin. While it can be removed by use of the double oven, it varies in time and produces 'optical noise' which is greatly in excess of the intrinsic laser noise.

(iii) The 876 nm Oxford Experiment

This second experimental set-up at Oxford is basically similar to the first one illustrated in fig. 7, except that only a single oven is used. The main differences from the first experiment (apart from the different transition) are (i) the wavelength dependence of the PNC angle is utilized by recording the measured angle as a function of wavelength as the laser scans the whole hyperfine spectrum. The data is then computer analysed to extract that part of the signal which the wavelength dependence expected for the PNC angle. (ii) Because there is no molecular absorption it is possible to use greater Bi density than in the 648 nm experiment with correspondingly larger PNC angles to be measured. The experiment has reached design sensitivity and results can be expected shortly.

(iv) The Novosibirsk Experiment [21]

The two main features of the experiment are (i) no Faraday modulator is used, the linearizing angle ϕ_M is produced by small rotations of the polarizer. (ii) The dye laser used has no facility for continuous wavelength tuning; the wavelength can only be changed by 400 MHz mode jumps. The procedure in the experiment is to modulate this laser between two modes at a 1 kHz repetition rate; the polarizer angle is changed at a rate of order 1 Hz. A series of analogue feedback circuits are so arranged that a signal results which is proportional to $\Delta\phi$ the wavelength dependent angle. Intrinsic wavelength dependent angles are claimed to be small and this is checked by carrying out the experiment on a number of hyperfine lines and also at points in the hyperfine spectrum where no PNC effect is expected. The lack of fine wavelength tuning precludes the use of wavelength points with equal Faraday but opposite PNC angles as in the first Oxford experiment and Bi Faraday suppression relies on a very high degree of magnetic shielding.

(v) The Moscow Experiment [22]

The main features of this new experiment are: (i) the use of an improved heat pipe oven for the Bi, (ii) the manufacture of carefully designed single prism polarizers and (iii) a completely enclosed optical system avoiding unnecessary windows.

3 Results

Table 1 Theoretical and experimental values for $R = \text{Imag.}\frac{E1}{M1}$

		$R \times 10^8$ 648 nm	876 nm	
Theory				
(a) Empirical central field including shielding (ref. 41)		-14	-12	
(b) Semiempirical matrix-elements (ref. 42)		-18	-14	
Experiment				
	Wash. old	-0.7 ± 3.2	2.7 ± 4.7	Ox.
	Wash. new	-1.1 ± 1.9	-20 ± 5.5	Nov.

The published Bi results are set out in the table. It is clear that a significant discrepancy exists between the Washington and Oxford groups on the one hand and the Novosibirsk group on the other. It is expected that the improved experiments now under way will resolve the uncertainty within a matter of months.

9 CONCLUSION

The Bi and Tℓ experiments have demonstrated that atomic physics experiments can be performed at the level of sensitivity required to see the PNC effect of the weak interactions. While the present experimental position is not completely clear it seems likely that PNC in atoms has been seen. If this proves to be the case, the next step is to obtain quantitative information about the weak interactions. This will require both the new generation of experiments many of which have been mentioned in this review, and good atomic theory. The latter is no problem in hydrogen and hopefully will ultimately not prove a lasting problem for the heavy atoms. Indeed the calculation of PNC effect in many electron atoms is one of the most fascinating theoretical problems in theoretical atomic physics.

REFERENCES

1. M.A. Bouchiat and C. Bouchiat, Phys. Lett. 48B, 111 (1974). P.G.H. Sandars, Atomic Physics 4 Plenum Press 1975
2. C.Y. Prescott, W.B. Atwood, R.L.A. Cottrell, H. DeStaebler, E.L. Garwin, A. Gonidec, R.H. Miller, L.S. Rochester, T. Sato, D.J. Sherden, C.K. Sinclair, S. Stein and R.E. Taylor, Phys. Lett. 77B 347 (1978)
3. Ya B. Zel'dovich and A.M. Perelomov, J.R.T.P. 12 777 (1961) G. Feinberg and M.Y. Chan, Phys. Rev. D10 3789 (1974)

4. J. Bernabeu, T.E-O. Ericson, C. Jarlskog, Phys. Lett. 50B 467 (1974)
   A.N. Moskalev, J.E.T.P. Lett. 19 141 (1974)
5. R.R. Lewis and W.L. Williams, Phys. Letters 59B 70 (1975)
6. R.W. Dunford, R.R. Lewis and W.L. Williams, preprint (1976)
   E.N. Fortson, Private Communication (1977)
   E.A. Hinds and V.W. Hughes, Phys. Lett. 67B 487 (1977)
7. M.A. Bouchiat and C.C. Bouchiat, J. Physique 35 899 (1974), 36 493 (1975)
8. C. Loving and P.G.H. Sandars, J. Phys. B. 8 336 (1976)
9. M.A. Bouchiat and L. Pottier, K. Physiques Lettres 37 79L (1976)
10. S.D. Bloom, Preprint Neutral currents and parity breakdown in atomic transitions: three proposed experiments, UCRL-51750 (1955)
11. C.C. Bouchiat, M.A. Bouchiat, L.Pottier, Atomic Phys. 5 ed. R. Marrus, M. Prior, H. Shugart, (Plenum Press 1977) p.L.
12. M.A. Bouchiat and L. Pottier, in Unification of Elementary Forces and Gauge Theories ed. D.B. Cline and F.E. Mills (1977)
13. R. Conti, P. Bucksbaum, S. Chu, E. Commins and L. Hunter, Phys. Rev. Letters 42 343 (1979)
14. S. Chu, E. Commins and R. Conti, Phys. Lett. 60A 96 (1977)
15. D.V. Neuffer and E. Commins, Phys. Rev. A. 16 16 (1977)
16. O.P. Sushkov, V.V. Flambaum and I.B. Khriplovich, J.E.T.P. 24 461 (1977)
17. L.N. Labzovskii, J.E.T.P. 46 853 (1977)
18. Ya B. Zel'dovich, J.E.T.P. 36 682 (1959)
19. P.E.G. Baird, M.W.S.M. Brimicombe, G.J. Roberts, P.G.H. Sandars and D.N. Stacey, Atomic Physics V, ed. R. Marrus, Plenum Press (1977)
20. E.N. Fortson, Atomic Physics 5 Plenum Press p.27 (1977)
21. L.M. Barkov, M.S. Zolotorev, J.E.T.P. Lett. 26 379 (1977); L.M. Barkov, I.B. Khriplovich and M.S. Zolotorev, comments Atom. Molec. Physics 8 79 (1979)
22. I. Sobelman and V. Sorokin, private communication
23. V.A. Alexeev et al Proceedings of the V. Vavilov Conference on Atomic Spectroscopy (1977)
24. I.B. Khriplovich and V.N. Novikov, J.E.T.P. Letters 22 162 (1977)
25. O.P. Sushkov and V.V. Flambaum, Zh Exsp. Teor. Fiz. 75 1208 (1978)
26. E.E. Gajzago and G. Marx, Proceedings of the V GIFT Seminar in Theoretical Physics (1974)
27. D.W. Rein, J. Mol. Evol. 4 15 (1974)
28. V.S. Letokhov, Physics Letters 53A 275 (1975)
29. Y.B. Zel'dovich et al, J.W.T.P. Letters 25 94 (1977)
30. O.N. Kompenets, A.R. Kikudshanov, V.S. Letokhov, Optics Communications 19 414 (1976)
31. E. Arimondo, P. Glorieux and T. Oka, Optics Communications 23 369 (1977)
32. D.W. Rein, R.A. Hegstrom and P.G.H. Sandars, Phys. Lett. 71A 499 (1979)
33. C.E. Loving and P.G.H. Sandars, J. Phys. B. 10 2755 (1977)

34. E.G. Drukarev and A.N. Moskalev, J.E.T.P. 46 1078 (1977)
35. S.E. Koonin (1979) Cal. Tech. Print.
36. V.G. Gorshkov et al J.E.T.P. 45 666 (1978)
37. A.J. Leggett, Phys. Lett. 39 587 (1977)
38. A.I. Vainshtein and I.B. Khriplovich, J.E.T.P. Lett. 20 34 (1974)
39. L.L. Lewis, J.H. Hollister, D.C. Soreide, E.G. Lindhal and E.N. Fortson, Phys. Rev. Letters 39 795 (1977). E.N. Fortson, Proceedings of Neutrino -78 ed. E.C. Fowler p.417 (1978)
40. P.E.G. Baird, M.W.S.M. Brimicombe, R.G. Hunt, G.J. Roberts, P.G.H. Sandars and D.N. Stacey, Phys. Rev. Letters 39 798 (1977)
41. M.J. Harris, C.E. Loving and P.G.H. Sandars, J. Phys. B. 11 L749 (1978). See also P.G.H. Sandars,Physica Scripta to be published
42. I.B. Khriplovich, V.V. Novikov and O.P. Sushkov, J.E.T.P. 71 1665 (1976)

Part II
Quark Atoms

QUARK ATOMS

Thomas Appelquist

Yale University
Physics Department
New Haven, Connecticut 06520

ABSTRACT

Some of the important theoretical issues and recent experimental developments in the physics of heavy quark-antiquark systems are reviewed. On the theoretical side, special emphasis is given to unresolved questions in the use of quantum chromodynamics (QCD) to describe the lifetime of $Q\bar{Q}$ states. The possible impact of recent experimental work on the pseudoscalar charmonium states and on the upsilon states is discussed and a list of critical experimental questions which may be answered soon is offered.

INTRODUCTORY NOTE

In preparing the lectures, I relied considerably on a recent review, Charm and Beyond, written by R. M. Barnett, K. Lane and myself. It has just appeared in the 1978 edition of the Annual Review of Nuclear and Particle Science(1). A number of figures and tables were taken directly from sections II and III of that paper which dealt respectively with quantum chromodynamics and charmonium.

This contribution to the proceedings will not reproduce all the material of the lectures. Instead, I have tried to write something that is a kind of pedagogical companion to Charm and Beyond and also an updating of its experimental data review. The most important issues are stressed without reproducing all the detail of Charm and Beyond. Hopefully this will make it easier for nonspecialists to get some appreciation for the main theoretical ideas and for the experimental questions being asked. There have been a number of important theoretical and experimental developments since the spring of 1978 when Charm and Beyond was completed. They bear directly on several fundamental questions and I will discuss them and their implications as completely as possible.

I. INTRODUCTION

There is a great deal of evidence from experiments which probe the structure of hadrons at short distances that the strong interactions become progressively weaker as the distance scale is reduced. From deep inelastic lepton hadron scattering and e^+e^- annihilations into hadrons, one can conclude that below distance scales of perhaps 1/10 fermi, the forces which bind quarks have become weak enough to treat the quarks as noninteracting to a good zeroth order approximation. The interactions can then be taken into account perturbatively.

This important fact does not, however, allow us to compute the properties of mesons and baryons composed of the usual u, d, and s quarks. These hadrons have a typical size of one fermi and at this distance scale, the quark binding forces have risen to their full strength. Perturbation theory cannot be used and in addition (and partly as a consequence), another complication arises. Since the quarks are confined to a one fermi radius, they must, according to the uncertainty principle, have momentum components of at lease 200-300 MeV. But since the effective u and d quark masses are no more than 300 MeV, it seems very likely that ordinary hadrons are relativistic systems. One is perhaps somewhat better off with the s quark ($m_s \simeq 500$ MeV) but certainly for any hadron containing even one u or d quark, the combination of strong forces and relativistic motion has prevented any computation of its properties starting from an underlying theory.

We have had, for several years now, a leading candidate for an underlying quantum field theory of strong interactions. Quantum chromodynamics, which I will briefly review in section III, has many of the necessary properties to do the job. It becomes more and more weakly coupled at short distances and its predictions based on perturbation theory are so far consistent with the relevant high momentum transfer experiments. Whether this theory will eventually explain the masses, magnetic moments and other properties of strongly interacting, relativistic hadrons is one of the most important (and difficult) questions in elementary particle physics.

The discovery of the new heavy quarks, c (mass ≃ 1.5 GeV) and b (mass ≃ 4.5 GeV), has had a considerable impact on this state of affairs. The evidence strongly suggests that both the $c\bar{c}$ (charmonium) and $b\bar{b}$ bound states are nonrelativistic and if yet heavier quarks are discovered, this will be even more the case. The ground state of charmonium, for example, has a mean radius of about 0.4 fermi (1/500MeV) and the uncertainty principle allows a 1.5 GeV quark to be comfortably nonrelativistic at these distance scales. This eliminates one of the complicating features of ordinary hadrons and as a result, these new hadrons can be thought of as quark atoms.

Nevertheless, the strong coupling problem remains with the $c\bar{c}$ and $b\bar{b}$ systems. It is only at distances of 1/10 fermi or less that quantum chromodynamics can be treated perturbatively and so a phenomenological (but nonrelativistic) approach must be taken to discuss the binding. A rather successful nonrelativistic phenomenology based loosely on anticipated strong coupling features of QCD has been developed. This work will be summarized in section IV with special emphasis on the inclusion of spin-dependent effects. As the mass of the heavy quark is increased, the $Q\bar{Q}$ system becomes smaller. Below 1/10 fermi, the potential is expected to be nearly Coulombic according to QCD. It corresponds to the exchange of a single massless quantum called a gluon. If nature has been good enough to provide us with a sequence of heavier and heavier quarks, then perhaps we shall eventually be able to see these $Q\bar{Q}$ systems becoming more and more Coulombic.

Even with the existing $c\bar{c}$ and $b\bar{b}$ systems, there is the possibility of testing perturbative QCD. In the decay of these states into light hadrons, the $Q\bar{Q}$ annihilation might take place at distance scales on the order of the quark compton wave length. This is just getting into the weak coupling regime for the c quark and is well within it for the b quark. The application of perturbation theory to $Q\bar{Q}$ annihilation has been described many times in the literature and is reviewed in Charm and Beyond. It leads to a number of striking experimental predictions, most of which have not yet been well tested. The fragmentary evidence that exists so far, however, seems to be in accord with the basic theoretical ideas. The discussion here will concentrate on the experimental developments within the last year and on some of the theoretical underpinnings.

Whether this use of QCD perturbation theory can be established as a sound theoretical procedure is still unclear. There has been some interesting work on the problem during the last year and there is some ongoing work that I am aware of. In section V, I will try to summarize this work and underscore some of the important unresolved problems.

II. AN EXPERIMENTAL OVERVIEW

Section I of Charm and Beyond (C&B) provides a brief review of quark properties including charm. The properti of the $c\bar{c}$ system are shown in C&B Table 1 and a level dia gram with indicated electromagnetic transitions is shown in C&B Fig. 2. Section 3.1 of C&B contains some commenta on these data.

The most important experimental development since spring 1978 concerns the experimental candidates for the pseudoscalar states $\eta_c(1^1S_o)$ and $\eta_c'(2^1S_o)$. An η_c candidate, the X(2830), has been reported for some time by experimenters at DESY[2]. They claim to have seen it in the sequential decay $\psi \to \eta_c\gamma$ followed by $\eta_c \to \gamma\gamma$ with a combined branching ratio of $(1.3 \pm 0.4) \times 10^{-4}$. During the last year, the crystal ball collaboration[3] at SPEAR has searched for the η_c unsuccessfully. They have recent presented some preliminary data in which an upper bound below the DESY value is reported. They claim, with 90% confidence, that[3]

$$BR(\psi \to \eta_c\gamma) \times BR(\eta_c \to \gamma\gamma) < .8 \times 10^{-4}. \tag{2.1}$$

Nothing could be more important in heavy quark physi than to refine this search for the η_c. If it does exist at 2830 and if the DESY value of $(1.3 \pm 0.4) \times 10^{-4}$ for the above combined ratio is correct, all its properties seem very difficult to understand theoretically. Some discussion of these questions is included in sections IV and V.

The existence of the X(3455) state, the candidate

for the excited pseudoscalar η_c', is also now in doubt. It had been reported at SLAC in the cascade decay shown in C&B Fig. 1:

$$\psi'(3684) \rightarrow \gamma + X(3455)$$
$$\gamma + \psi(3095).$$

This state has been searched for using the new Mark II detector without success. An upper limit for the combined (cascade) branching ration of about 0.1% has been reported,[4] compared to an earlier report of 0.8 ± 0.4%. An upper limit of 0.6% for the M1 transition $\psi' \rightarrow \gamma + X(3455)$ has also been reported. The X(3455) is just as troublesome a candidate theoretically for the η_c' as is X(2830) for the η_c.

The intermediate P states seem well established and they provide a testing ground for some theoretical ideas on decay widths. These ideas and their experimental status will be discussed in section V. Finally I should mention that in the fall of 1978, some evidence for another intermediate state at 3600 was reported at DESY[5]. This state too, has been searched for with the Mark II detector without success. An upper limit of 0.4% for the M1 branching ratio $\psi' \rightarrow \gamma + (3600)$ has been reported[4]. For the cascade down to the ψ, an upper limit of 0.06% is reported for the product of branching ratios.

The Υ seems now to be well established as a $b\bar{b}$ bound state with $m_b \simeq 4.5$ GeV and $|Q_b| = 1/3$. After the original discovery at Fermilab[6] in the reaction $p + Be \rightarrow \Upsilon + \ldots \rightarrow \mu^+\mu^- + \ldots$, the DORIS e^+e^- colliding beam ring at DESY was increased enough in energy to be able to see two narrow states at 9.46 and 10.02 GeV[7]. The Fermilab experiment suggests that there are three narrow levels and if the first two are constrained to the DESY values, the third is found to be at 10.38 GeV. These could then be the three lowest lying $^3S_1$ states of a $b\bar{b}$ system.

There is, as yet, no information on the existence of intermediate P states or ^1S_o hyperfine partners of the T $^3S_1$ states. The leptonic width of the T(9.46) can be extracted directly from the total cross section since the full width is small compared to the experimental resoluti[on] of about 10 MeV. The result is [8]

$$\Gamma(T(9.46) \rightarrow e^+e^-) = 1.26 \pm 0.21 \text{ KeV} \qquad (2.2)$$

From the reaction $e^+e^- \rightarrow \mu^+\mu^-$ on resonance, the leptonic branching ratio has been increased to be[8]

$$\frac{\Gamma(T \rightarrow \mu^+\mu^-)}{\Gamma(T \rightarrow \text{all})} = 2.6 \pm 1.4\% \qquad (2.3)$$

These two numbers suggest a total T(9.46) width of around 50 KeV but with large errors. For the T(10.02), the leptonic width is found to be [8]

$$\Gamma(T(10.02) \rightarrow e^+e^-) = 0.36 \pm 0.09 \text{ KeV} \qquad (2.4)$$

These values for the leptonic widths suggest very strongly[9] that the electric charge of the b quark is smaller in magnitude than 2/3. A value of 1/3 is consistent with the data.

III. THEORETICAL FOUNDATIONS

An introductory description of quantum chromodynamics (QCD) along with references to the original literature, can be found in Charm and Beyond. What I will do here is only to summarize the essential features of the theory in outline form.

1. Quantum chromodynamics is the local gauge theory of quarks and gluons (gauge fields) based on the color group SU(3). Forces between quarks arise from gluon exchange.

2. Each quark u, d, s, c, b,..., comes in three colors and the SU(3) group operates on the color labels for each of the above quark types[10]. The mesons and baryons are color singlets. The role of color in quark statistics and in the computation of certain measured transition probabilities is explained in C&B.

3. The theory possesses the remarkable property of asymptotic freedom[11]. Higher order corrections in perturbation theory lead to the replacement of the coupling constant $\alpha = g^2/4\pi$ in the Coulomb potential α/r by an effective, r-dependent coupling constant $\alpha(r)$. This is also the case in quantum electrodynamics (QED) but in QCD, explicit computation shows that unlike QED, $\alpha(r) \propto (\log r)^{-1}$ as $r \to 0$. This is the property of asymptotic freedom and it means that the coupling between color charges such as quarks becomes weaker and weaker as the separation decreases. The physical origin of asymptotic freedom is discussed in C&B.

4. The scale at which asymptotic freedom sets in must be determined experimentally. From high momentum transfer experiments such as deep inelastic electron scattering[12], the data suggest that $\alpha(r) \lesssim 0.3$ at distances of 1/10 fermi

or less. The expansion parameter is essentially $\alpha(r)/\pi$ so that perturbation theory should converge and become better and better as $r \to 0$.

5. As r is increased from 1/10 fermi (correspondin to momentum transfers below 2 GeV), the coupling grows and at distance scales of about 1/4 to 1/3 fermi and beyond, the perturbation expansion completely breaks down. It is conjectured that such strong attractive forces develop in color single channels at these distance that colored quarks and gluons are permanently confined to the interior of hadrons.

6. This has not been established theoretically and a review of the large body of work on this problem is beyond the scope of this outline[13]. Nevertheless, one expected qualitative feature of confinement dynamics pla an important role in $Q\bar{Q}$ physics and I want to mention it briefly. A variety of computations and physical analogie suggests that at distances beyond 1/4 - 1/3 fermi, the ch moelectric flux lines connecting, say, a quark and anti-quark become more and more collimated into a one dimensional configuration. This "string-like" structure is quite different from the Coulombic configuration associat with the approximate 1/r potential at shorter distances a leads naturally to a potential V(r) which grows linearly with the separation distance. The use of a linear potential has been quite successful in dealing with the $c\bar{c}$ system which has a ground state mean radius of about 1/2 fermi.

IV. $Q\bar{Q}$ LEVEL SPECTROSCOPY

The first round of work on the level structure of charmonium made use of the "coulinear" potential

$$V_o(r) = -\frac{\kappa}{r} + \frac{r}{a^2} \tag{4.1}$$

This incorporates the anticipated linear confining potential at large distances and the Coulomb potential (with asymptotic freedom corrections neglected) at short distances. The $c\bar{c}$ fits and predictions following from this potential are described in C&B. Suffice it to say here that a Schroedinger formalism using V_o leads to a good qualitative understanding of the spin-triplet level splittings of charmonium including the P states. Estimates of the mean square c quark velocity then show that a nonrelativistic formalism is consistent. The potential V_o is shown in C&B Fig. 14 and the ground state $c\bar{c}$ wave function is shown in C&B Fig. 15. The mean radius is about 1/2 fermi and it can be seen that charmonium is dominantly sensitive to the linear part of V_o.

With the discovery of the upsilon, some of the shortcomings of the potential V_o have become clear. It was anticipated[14] using the V_o potential, that with $m_b \approx 4.5$ GeV, the splitting between the ground state and first excited state of the upsilon should be about 400 MeV. Both these states would also move predominantly in the linear part of V_o (see C&B Figs. 13 and 14). Instead, it has turned out that this splitting is 560 MeV. What this perhaps means is that the transition from linearity to an approximate Coulomb shape begins at somewhat larger distances than 0.1 - 0.2 fermi so that the upsilon ground state at least, senses this more steeply sloping part of the potential. This could then account for the larger splitting than had been anticipated.

A variety of potentials have been suggested which incorporate this feature and some of these are described in Charm and Beyond. More recently, a simple and elegant potential has been proposed by Richardson[15] which I wa to mention. He suggests the following single parameter potential which is easiest to write down in momentum space:

$$\tilde{V}(\vec{q}^2) = \left(\frac{-16\pi}{33 - 2N_f}\right)\left(\frac{1}{\vec{q}^2}\right)\left(\frac{1}{\ln(1 + \vec{q}^2/\Lambda^2)}\right) \tag{4.2}$$

For $q \gg \Lambda$, this is the Coulomb potential corrected by asymptotic freedom and for $q \ll \Lambda$, it behaves like $1/\vec{q}^4$, the Fourier transform of the linear potential. A remarkably good fit to the known spin triplet $c\bar{c}$ and $b\bar{b}$ levels is obtained with $\Lambda \simeq 400$ MeV, $m_c \simeq 1.5$ Gev and $m_b \simeq 4.9$ GeV. The 1^3P_J $b\bar{b}$ states are predicted to be centered at about 9.9 GeV.

While I don't want to oversell any one potential, i does look as though something like Richardson's might provide a smooth connection between linear confinement a asymptotic freedom. The ground state of the upsilon is apparently moving largely in this transitional region an if yet heavier quarks are discovered, they may help us map out the potential right down into the region where perturbation theory is applicable. The physics of the transitional region is, however, almost surely very complicated. One effect stressed by Poggio and Schnitzer[1 is the vacuum polarization of light $q\bar{q}$ pairs. They point out that at these intermediate distances (1/4 - 1/3 fermi) and beyond, the rising field energy can begin to strongly activate and polarize the $u\bar{u}$, $d\bar{d}$ and $s\bar{s}$ pairs in the vacuum. The linear potential sensed by charmonium is then a "shielded" potential with this effect included. The setting in of this effect is one of the complicating features of the transitional region.

Since both the charmonium and upsilon systems are too large to deal with perturbatively, it isn't at all clear a priori how to include spin dependence. What has been done essentially is to try a generalization of the Breit-Fermi Hamiltonian for a positronium. The Coulomb part of the $Q\bar{Q}$ potential is expected to be the zeroth component of a Lorentz four vector and that leads immediately to the well-known Coulombic spin orbit, spin-spin and tensor interactions. These are shown in C&B, section 3.4.

Since it isn't clear whether the longer range part of the potential transforms as a Lorentz scalar or four vector, one approach is to allow it to be a mixture of both with a parameter $\eta (0 \leq \eta \leq 1)$, giving the relative strength of each[17]. This is blatantly phenomenological but at least one then has a way of interpreting the data on spin splittings.

To be even more general, a color anomolous moment λ can be given to the quark and then an attempt can be made to fit the $\psi - \eta_c$ and $\psi' - \eta_c'$ splittings and the splittings among the J = 0,1, and 2 P-states.

The result of doing this for the V_o potential (4.1) is shown in C&B table 6 and can be summarized as follows If the η_c and η_c' are really located at 2800 MeV and 3455 MeV respectively, an extremely large anomolous moment ($\lambda \simeq 5$) is required. If it is assumed, following recent indications(3,4), that the η_c and η_c' have not yet been found, then a reasonable fit to the P-state splittings can be obtained with $\lambda = 0$, and with $\eta \simeq 1$. This would correspond to a dominantly Lorentz scalar confining potential. The $\psi - \eta_c$ splitting is predicted to be between 70 and 100 MeV and the $\psi' - \eta_c'$ splitting somewhat less. It would then be quite reasonable not to have seen them yet since the M1 transitions from the ψ and ψ' respectively would be very small. I shall return to this point in Section V.

It would be quite useful to build in spin effects in this way with something like the Richardson potential (4.2). A fit to the charmonium P-state splittings could then be used to make predictions for spin splittings in the upsilon.

To summarize, with the possible exception of η_c and η_c', it seems that the heavy quark bound states are behaving in a sensible way. Although we don't know how to proceed from first principles at these "large" distance scales, nothing terribly anomalous seems to be going on. If the η_c and η_c' are ultimately found within 100 MeV of the ψ and ψ' respectively, then this statement can be made with much more confidence. The exciting prospect is the existence of yet heavier quarks which will sink down into the Coulomb tip of the potential where firmer theoretical predictions can be made.

V. TRANSITIONS AND DECAYS

One of the earliest theoretical suggestions about charmonium was that the narrow widths of the states below charm threshold could be simply explained by quantum chromodynamics[18]. The idea is basically that $c\bar{c}$ annihilation into light hadrons must proceed through gluons (and at short distances on the order of $1/m_c$) and since the effective gluon coupling constant should be small at these distances, the decay will be inhibited. The dominant contribution will dome from the minimum number of intermediate gluons which depends upon the quantum numbers of the charmonium state. Some rather striking experimental predictions can be made and many of these are reviewed in Charm and Beyond. I will concentrate here on new experimental and theoretical developments.

Consider the decay of the $\psi\,{}^3S_1$ state. Its dominant electromagnetic decay is shown in C&B Fig. 3. The $c\bar{c}$ pair must come together to annihilate into the virtual photon and if the bound state is nonrelativistic, then, to first approximation, the decay width will be given by

$$\Gamma(\psi \to \ell^+\ell^-) = \frac{4\alpha^2(2/3)^2}{M^2} |\psi(0)|^2 \tag{5.1}$$

The charge of the charmed quark is $2/3\,|e|$ and M is the charmonium mass. $\psi(0)$ is the nonrelativistic wave function at the origin and one cannot expect to be able to compute it in perturbation theory. The reason for this is that the mean radius of charmonium is on the order of 1/2 fermi, a distance scale at which the effective coupling strength for the binding has become large. Thus $\psi(0)$ will be determined primarily by the nonperturbative, long range part of the potential. The hadronic decay of the ψ must proceed through a minimum of three gluons. If this is indeed the dominant contribution, that is to say, if perturbation theory is truly relevant to this problem, th decay will proceed as shown in C&B Fig. 4. The $c\bar{c}$ annihilation will be essentially local (on the order of $1/m_c << <n>$). The computation of the decay matrix elemen is then done in analogy to the parton model computation of σ_{tot} ($e^+e^- \rightarrow$ hadrons) as if the final state consisted of three on mass shell gluons. This amounts to the state ment that the transition from the three gluon state into physical hadrons takes place with unit probability. If this mechanism is correct, then the total hadronic width of the ψ is given by

$$\Gamma(\psi \rightarrow \text{hadrons}) = \frac{40}{81\pi} (\pi^2-9) \frac{\alpha_s^3(M)}{M^2} |\psi(0)|^2 \tag{5.2}$$

The strong coupling constant is defined at the ψ mass and as before, $\psi(0)$ is the nonrelativistic wave function at the origin.

The ratio $\Gamma(\psi \rightarrow \text{hadrons})/\Gamma(\psi \rightarrow e^+e^-)$ is experimental known to be about 10 (48 kev/4.8 kev). From this, a valu of $\alpha_s(M)$ can be extracted and one finds that

$$\alpha_s(M) \approx 0.2 \tag{5.3}$$

This is a very small value of $\alpha_s(M)$ and whether it is con sistent with deep inelastic lepton hadron scattering and $\sigma_{tot}(e^+e^- \rightarrow$ hadrons) is not yet clear.

The $\eta_c(^1S_0)$ decay can proceed via two gluons and wit the above value of α_s, it's width can be predicted. One finds

$$\Gamma(\eta_c \to \gamma\gamma) = \frac{4}{3}\,\Gamma(\psi \to e^+e^-) = 6 - 8\ \text{MeV}$$

$$(\eta_c \to \text{hadrons}) = \frac{9}{8}\,(\frac{\alpha_s}{\alpha})^2\Gamma(\eta_c \to \gamma\gamma) \qquad (5.4)$$

$$= 738\ \Gamma(\eta_c \to \gamma\gamma) = 4\text{-}5\ \text{MeV}.$$

Thus the $\gamma\gamma$ branching ratio is predicted to be

$$BR(\eta_c \to \gamma\gamma) = 1.3 \times 10^{-3}. \qquad (5.5)$$

With this prediction, one encounters another problem with the identification of the possible state at 2830 with the η_c. The DESY experiment reported a product of branching ratios

$$BR(\psi \to \eta_c\gamma) \times BR(\eta_c \to \gamma\gamma)$$

$$= 1.3 \pm 0.4 \times 10^{-4} \qquad (5.6)$$

while, from the inability to detect monochromatic photons in ψ decay, it is known that

$$BR(\psi \to \eta_c\gamma) < 1.7\% \,. \qquad (5.7)$$

One would then conclude that

$$BR(\eta_c \to \gamma\gamma) \gtrsim 7 \times 10^{-3} \qquad (5.8)$$

and the discrepancy is at least a factor of four or five. In addition to the inability of the crystal ball collaboration to find this state[3], I can mention one other development which suggests that the η_c is yet to be found. A recent re-examination of the three photon Dalitz plot in the DESY experiment by Meshkov and Samios[19] suggests that the evidence for its existence is perhaps less compelling than was claimed. Essentially all of the above problems are encountered for the η_c' if the state at 3455 exists and is so identified. However, the experimental developments mentioned in Section II suggest that the η_c' is even less likely to have been discovered than the η_c.

The minimal gluon mechanism leads to interesting pre dictions such as the large η_c width (5.4) but can one rea trust QCD perturbation theory used in this way? A necess condition is that no large dynamical factors enter in hig er orders to make the perturbation expansion break down. These would possibly come in the form of "large logarithm whose arguments would involve a small dimensional parameter such as the binding energy. This problem is discus in Charm and Beyond and I will only summarize the situation as follows. For ψ and η_c decay, there is no indication that this happens as long as one works only to lowes order in a nonrelativistic expansion. A proof of this to all orders in α_s is, however, still lacking and it involv some subtle questions having to do with the separation of binding effects from corrections to the decay matrix element.

Even if there are no infrared problems of this sort, it is still possible that large numerical coefficients co enter in higher orders. This had not been anticipated but recently the next order corrections to η_c decay have been computed with a rather surprising result[20]. The branching ratio (5.5) is found to be

$$BR(\eta_c \to \gamma\gamma) = \frac{8}{9}\left(\frac{\alpha}{\alpha_s(M)}\right)^2 \left\{1 - 22.14\ \frac{\alpha_s(M)}{\pi} + \ldots\right\} \tag{5.9}$$

and with $\alpha_s(M) \simeq 0.2$, perturbation theory looks very bad indeed. It is not yet clear whether a similar thing happens in ψ decay or what (5.9) itself really means. This is clearly a very important problem which should be carefully studied.

Finally, let me summarize the situation with respect to electromagnetic transitions. They don't probe short distances in the way that direct decays may well do, but they do provide another way of testing our qualitative understanding of the structure of the bound states. The E_1 transitions ($\psi' \to {}^3P_J$) seem to be in reasonable accord with expectation as discussed in Charm and Beyond. If the η_c and η_c' are yet to be found, then no M1 transitions have yet been seen. If these states are within 100 MeV of the ψ and ψ' respectively then the M1 widths should be very small ($\simeq$ 1/3 KeV), well below the current upper bounds. For the $\psi \to \eta_c\gamma$ transition, this width implies a branching ratio of less than 1/2%. Given these small expected widths it may be better to try and produce the η_c and η_c' directly in hadronic collisions or by photoproduction. The Einhorn-Ellis mechanism of gluon fusion[21] suggests that the η_c production cross section in hadron collisions should be reasonably large, especially at Fermilab energies and above.

VI. SUMMARY AND OUTLOOK

A critical issue for the near future is the existen of the η_c and η_c'. What are the hyperfine splittings and what are the total hadronic widths? The computation of Barbieri et al[20] has made the minimal gluon mechanism for the decay look somewhat dubious. The question of th convergence of QCD for these processes will have to be examined carefully and at the same time, it is important to collect as much experimental information as possible.

The Υ provides an even better system for addressing all of these questions. The radius is smaller and the direct decay coupling constant is presumably logarithmic smaller due to asymptotic freedom. Even more importantl perhaps, the mass is high enough to check QCD prediction on details of the final state in hadronic decay.

If the $^3S_1$ Υ states decay predominantly into three gluons, then more than three GeV will go into each gluon and that should be enough energy to produce three identi- fiable jets of final state hadrons. The expected angula distributions and other properties of these jets have been studied by a number of authors[22] but there is not yet enough data from Υ decay to clearly test the idea. There is, however, already one indication from Υ decay that the three gluon mechanism may be correct. Three gluons coming out of a point must lie in a plane by momen tum conservation. Thus the final hadrons in the decay should form a flattened disc as opposed to something lik a spherical distribution. An analysis[23] of events with four or more charged hadrons in the final state suggests that this is indeed the case. The question of whether some of these disc events will be resolved into three jets remains open. Preliminary results such as this and more established results such as the P state widths suggest that there is something right about the minimal gluon mechanism. On the other hand, the result of Barbieri et al is cause for concern and a great deal of important work remains to be done.

FOOTNOTES AND REFERENCES

1. T. Appelquist, R. M. Barnett and K. Lane, Ann. Rev. Nucl. Part. Sci. 28, 387 (1978).
2. W. Braunschweig et al, Phys. Lett. 67B, 249 (1977). For a recent review see B. H. Wiik and G. Wolf, DESY preprint, DESY 78/23 (1978).
3. R. Schwitters, private communication.
4. R. M. Barnett, private communication.
5. G. Kramer, Invited talk at the DESY workshop on New Quarks and Leptons, Hamburg, Sept., 1978.
6. S. W. Herb et al, Phys. Rev. Lett. 39, 252 (1977); T. Yamanouchi, parallel session A10, Tokyo Conference.
7. Ch. Berger et al, Phys. Lett. 76B, 243 (1978); C. W. Darden et al, Phys. Lett. 76B, 246 (1978).
8. J. K. Bienlien et al, Phys. Lett. 78B, 360 (1978); C. W. Darden et al, Phys. Lett. 78B, 364 (1978).
9. C. Quigg, Fermilab preprint, FERMILAB-Conf-78/82-THY, (1978).
10. O. W. Greenberg, Ann. Rev. Nucl. Part. Sci. 28, 329 (1978).
11. H. D. Politzer, Phys. Rev. Lett. 26, 1346 (1973); D. Gross and F. Wilczek, Phys. Rev. Lett. 26 1343 (1973).

12. For a recent critical review with references to earlier literature, see L. F. Abbot and R. M. Barnett, SLAC-PUB-2325, May 1979.
13. For an extensive review of QCD, see W. J. Marciano a H. Pagels, Phys. Rev. C36, 137 (1978).
14. E. Eichten and K. Gottfried, Phys. Lett. B66, 286 (1
15. J. L. Richardson, SLAC-PUB-2229, December 1978.
16. E. Poggio and H. J. Schnitzer, Phys. Rev. Lett. 41, 1344 (1978); Brandeis University preprint, Sept. 1978.
17. This work, done by several different authors, is reviewed in Charm and Beyond (Ref. 1).
18. T. Appelquist and H. D. Politzer, Phys. Rev. Lett. 34, 43 (1975); Phys. Rev. D12, 1404 (1975).
19. S. Meshkov and N. P. Samios, Brookhaven preprint BNL-25926, April 1979.
20. R. Barbieri et al, CERN preprint TH-2622-CERN, Janua 1979.
21. M. B. Einhorn and S. D. Ellis, Phys. Rev. D12, 2007 (1975).
22. For a recent review, see K. Koller and T. Walsh, DESY preprint DESY 78/16, March 1978.
23. PLUTO Collaboration, Ch. Berger et al, DESY preprint DESY 78/71.

TEST OF QCD WITH HEAVY QUARK BOUND STATES

Johann Rafelski and Raoul D. Viollier

Theoretical Physics Division
CERN
1211 Genève 23

With the forthcoming results from the electron-positron storage rings PETRA and SPEAR, new possibilities open for the study of new exotic atoms - the bound states of heavy quarks and antiquark pairs. At present, two families of such states have been discovered, but many theoretical studies indicate that at least one more family which consists of bound states of still heavier "top" quark and antiquark pairs may be discovered.

In view of these facts we have studied in more detail the experimental spectra in the framework of the currently accepted theory of strong interactions, QCD. While the suggestion that the ψ states are bound states of a charmed quark-antiquark pair has been made very early, the theoretical study of the spectrum with Coulomb + linear potential [1)] did not adequately reflect the strong polarizability of the vacuum as described by the q dependence of the running coupling constant in first order :

$$\alpha_\Lambda(q^2) = \frac{1}{B_n \log(-q^2/\Lambda^2)} \tag{1}$$

$$B_n = \frac{1}{4\pi}\left(11 - \frac{2}{3}n\right) = \begin{cases} .77 & n=2 \\ .72 & n=3 \end{cases}$$

n is the number of light quarks, Λ determines the strength o α and is known to be within the interval 300 to 600 MeV from the scaling violations in deep inelastic scattering. Its most recent value is $\Lambda = 470 \pm 200$ MeV [2].

We assume that heavy quarks interact via a static potentia $V_{QCD}(r)$, which is the space Fourier transform of the Born one polarized gluon exchange matrix element $(q^2 = q_0^2 - \vec{q}^{\,2})$:

$$\tilde{V}_{QCD}(-\vec{q}^{\,2}) := \frac{4}{3}\,\frac{\alpha_\Lambda(-\vec{q}^{\,2})}{-\vec{q}^{\,2}} = \frac{1}{4\pi}\int V_{QCD}(r)\, e^{i\vec{q}\cdot\vec{r}}\, d^3r \tag{2}$$

The approximate form (1) for $\alpha_\Lambda(q^2)$ can only be trusted up to distances $r \lesssim 1/\Lambda \approx 0.4$ fm. Thus in order to describe the excited states of "onium", we must introduce further corrections to $V_{QCD}(r)$ which shall be discussed below.

In order to compute $V_{QCD}(r)$ we will go through an intermediate step ; we first find its spectral representation with the help of the Cauchy's theorem :

$$\tilde{V}_{QCD}(q^2) = \frac{1}{2\pi i}\oint_C \frac{\tilde{V}_{QCD}(M^2)}{M^2 - q^2}\, dM^2 \tag{3}$$

Here the path c encloses only the singularity at $M^2 = q^2$. We note that $\tilde{V}_{QCD}(q^2)$ as given by Eqs. (1), (2), has a first order pole at $q^2 = -\Lambda^2$ and a cut on the real axis connecting zero with infinity - there is no pole at $q^2 = 0$. The usual methods of complex analysis lead us in a straightforward fashio to

$$\tilde{V}_{QCD}(q^2) = \frac{4}{3} B_n^{-1}\left[\frac{1}{q^2 + \Lambda^2} - \int_0^\infty \frac{dM^2}{M^2}\,\frac{1}{M^2 - q^2}\,\frac{1}{\ln^2 M^2/\Lambda^2 + \pi^2}\right] \tag{4}$$

The inversion of the Fourier transform in Eq. (2) is now easy and we find

$$V_{QCD}(r) = \frac{4}{3} B_n^{-1} \left[\frac{1-\cos \Lambda r}{r} - \int_0^{\infty} \frac{dM^2}{M^2} \frac{1-e^{-Mr}}{r} \frac{1}{\ln^2 M^4/\Lambda^2 + \pi^2} \right] \tag{5}$$

Before considering the quantitative results, let us generalize the spectral method to arbitrary forms of the running coupling constant. Denoting by $\alpha\, q^2\, \Pi_\Lambda(q^2)$ the one-gluon polarization function, renormalized at $-\Lambda^2$, we have :

$$\alpha_\Lambda(q^2) = \frac{\alpha}{1 + \alpha\, \Pi_\Lambda(q^2)} \tag{6}$$

Depending on the pole and cut structure of $\alpha_\Lambda(q^2)$ we have several possible generalizations of Eq. (4). Denoting by C the cut contribution, the generalization of the second part in Eq. (4) :

$$C = \int_{\lambda_0^2}^{\infty} \frac{dM^2}{M^2} \left\{ \frac{-\frac{1}{\pi}\alpha\, \mathrm{Im}\, \Pi_\Lambda(M^2+i\varepsilon)}{(1+\alpha \mathrm{Re}\, \Pi_\Lambda)^2 + (\alpha\, \mathrm{Im}\, \Pi_\Lambda)^2} \right\} \frac{1}{M^2 - q^2}$$

we have :

(a)

$$\tilde{V}(q^2) = \frac{4}{3} \left(\frac{a_1}{q^4} + \frac{a_2}{q^2} + C \right) \tag{7}$$

when $\alpha(q^2) \underset{q^2 \to 0}{\to} (a_1/q^2) + a_2$ and α has a cut from λ_0^2 to infinity along the real axis (recall $1/q^4$ corresponds to $V(r) \sim r$).

(b)

$$\tilde{V}(q^2) = \frac{4}{3} \left(\frac{a_1}{q^2 - t} + C \right) \tag{8}$$

when $\alpha(q^2) \underset{q^2 \to 0}{\to} 0$ and α has a pole at $q^2 = t$ with residuum a_1, and a cut as above. This is the case described by Eq. (4).

(c)

$$\tilde{V}(q^2) = \frac{4}{3} \left(\frac{\alpha_0}{q^2} + C \right) \tag{9}$$

where $\alpha(q^2) \underset{q^2 \to 0}{\to} \alpha(0) = \text{const.}$, cut as above and no further poles are found. This is the case of usual QED.

We should emphasize that knowledge of $\mathrm{Im}\,\Pi(M^2 + i\epsilon)$ is sufficient to determine $\mathrm{Re}\,\Pi$ via a dispersion relation. $\mathrm{Im}\,\Pi$ is in turn related to the decay amplitude of the gluon into other coloured objects inside the hadronic volume. In particular we recall here the contribution of the massive quarks

$$\frac{1}{\pi}\mathrm{Im}\,\Pi_{g\to q\bar{q}} = -\frac{1}{6\pi}\sqrt{1-\frac{4m^2}{q^2}}\left(1+\frac{2m^2}{q^2}\right), \quad q^2 > 4m^2, \tag{10}$$

and that of massless gluons

$$\frac{1}{\pi}\mathrm{Im}\,\Pi_{g\to gg} = \frac{11}{4\pi}, \quad q^2 > 0. \tag{11}$$

Of course, Eqs. (10), (11) and (8) lead straightforwardly to (4a).

In actual calculations below we have used Eq. (8) with two massless quarks and the strange quark of the mass $m_s =$ $= 300$ MeV. This differs somewhat from three massless quarks, since the p-state sees only a strongly reduced influence of the polarization induced by strange quarks, which is of shorter range.

We return now to the discussion of potential (5). We note that the first part is always positive, while the contribution of the integral is negative. The first part originates in the singular behaviour of $\alpha_\Lambda(q^2)$ at $q^2 = 0$ and $q^2 = \Lambda^2$. Thus this term is little reliable and shall be modified. (It is, however, interesting to notice that for a range of $0 < r \lesssim 0.4$ fm it corresponds to a linear potential $a^{-2}r$ with the slope $a^{-2} = \frac{1}{2}\Lambda^2$). We consider this part of the potential as ill-defined and for that reason extrapolate it linearly from a point $r_0 \sim 1/\Lambda$. r_0 is to be fitted so that best agreement between theory and experiment is achieved.

The second part of the potential (5), which is the attractive part, is much better defined, since it originates mostly from $q^2 > \Lambda^2$. It is less singular than the Coulomb-like $1/r$ potential, however, for large distances $r > 1/\Lambda$ it approaches the Coulomb-like potential of strength B_n^{-1} ($\sim$1.4).

In Fig. 1 we show the first (---) and second (-.-.-) con-

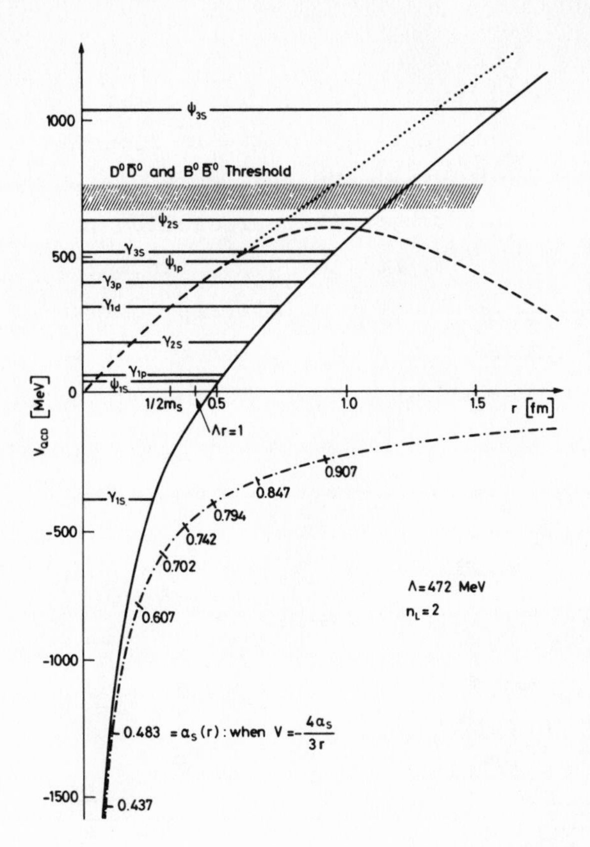

Fig. 1. Potentials for heavy quarks. Solid line displays the typical potential used in actual calculations. Dashed and dash-dotted lines are the two different QCD contributions, dotted lines are the phenomenological alternation. Along the short-range attractive dash-dotted line effective values of strong coupling constant are shown. Location of different charmonium and upsilon levels is shown within the quark potential.

tribution to the potential (5), the linear extrapolation of the first part (....) and the whole potential (——). The location of the bound states of charmonium and upsilonium (bottonium) as found by solving the Schrödinger potential is also indicated there.

The real test of the QCD comes with the upsilon states, since the $\psi(1s)$, $\psi'(2s)$ and the centre of gravity of the $P_c(1p)$ states is used to fit the parameters $\Lambda = 440$ MeV, $r_o = 0.378$ fm of the potential and the mass $m_c = 1525$ MeV, while other states are either unknown or above the $D^o\bar{D}^o$ threshold. Nonetheless we record here relatively good agreement with the $\psi''(3s)$ and the 3772 state as shown in Fig. 2. Looking at the position of the upsilon states we are amazed by the precision of the theoretical prediction for the 2s, 3s states. This suggests that the QCD potential is really a good point of departure for these relatively smaller states. We recall here that the Coulomb + linear potentials fail by c. 200 MeV in the description of the 2s, 3s upsilon states.

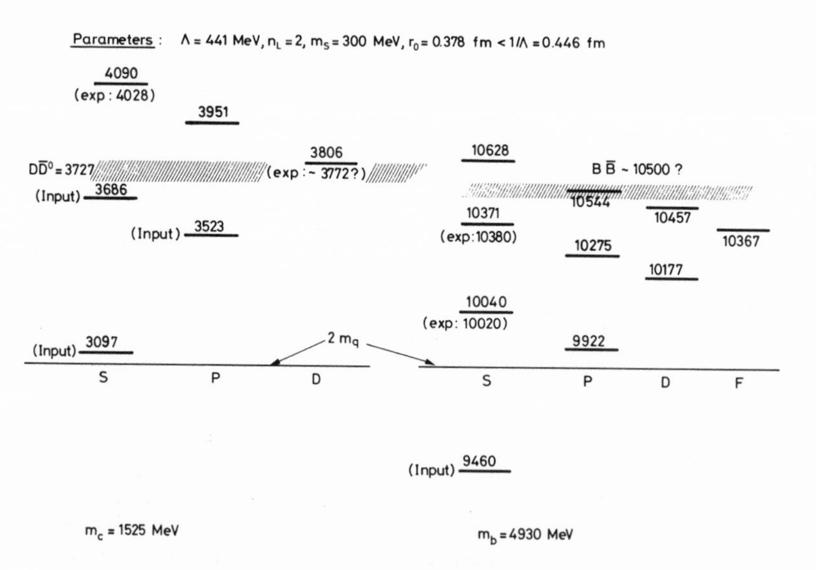

Fig. 2. Quantitative comparison between theoretical and experimental level scheme in charmonium and upsilon.

A quite different problem is the spin orbit and spin-spin splitting. In particular, the splitting of the 1p state in charmonium is experimentally well established. This splitting is governed by the first and second derivatives of the quark-quark potential and is therefore much more sensitive to its detailed form, in particular at ~0.4 fm for charmonium. Therefore we do not attempt a detailed fit ; however, we consider the following results as being adequate considering the crudeness of the potential in this intermediate domain.

In principle, the quark-antiquark potential in the non-relativistic approximation suitable for the Schrödinger equation should read

$$V(r) = V_0(r) + V_1(r)\,\vec{L}\cdot\vec{S} + V_2(r)\,S_{12} + V_3(r)\,\vec{\sigma}_1\cdot\vec{\sigma}_2 \tag{12}$$

where the total angular momentum $\vec{J}$ is the sum of the orbital momentum $\vec{L}$ and the spin $\vec{S} = \frac{1}{2}(\vec{\sigma}_1 + \vec{\sigma}_2)$. S_{12}, the tensor term, is $S_{12} = 3\vec{\sigma}_1\cdot\hat{r}\ \vec{\sigma}_2\cdot\hat{r} - \vec{\sigma}_1\cdot\vec{\sigma}_2$. V_1, V_2 and V_3 follow from the Breit reduction of Bethe-Salpeter equation in the non-relativistic approximation. This reduction is not quite consistent with a (v/c) expansion. Thus the contributions of these potentials to the eigenstates should be evaluated perturbatively. The potentials are [if no 1/r component is present in the potential $V_0(r)$] :

$$V_1 = \frac{1}{2m_c^2}\left(\frac{3}{r}\frac{dV_0}{dr} - \frac{1}{r}\frac{dS_0}{dr}\right) \tag{13a}$$

$$V_2 = \frac{1}{12m_c^2}\left(\frac{1}{r}\frac{dV_0}{dr} - \frac{d^2V_0}{dr^2}\right) \tag{13b}$$

$$V_3 = \frac{1}{6m_c^2}\,\Delta V_0 = \frac{1}{6m_c^2}\left(\frac{2}{r}\frac{dV_0}{dr} + \frac{d^2V_0}{dr^2}\right) \tag{13c}$$

The "centre of gravity" of the 1p state is split into three components J = 0,1,2, with the energies

$$E_{1^3P_J} = E_{1^3P} + \begin{cases} \langle V_1 \rangle - \frac{2}{5}\langle V_2 \rangle , & J=2 \\ -\langle V_1 \rangle + 2\langle V_2 \rangle , & J=1 \\ -2\langle V_1 \rangle - 4\langle V_2 \rangle ; & J=0 \end{cases} \tag{14}$$

From the known energies 3554, 3508, 3413 for $J = 2,1,0$, resp., one deduces easily $\langle V_1 \rangle = 35$ MeV, $\langle V_2 \rangle = 10$ MeV and the already used value $E_{1^3P} = 3523$ MeV.

Computing the matrix elements $\langle V_1 \rangle$ and $\langle V_2 \rangle$ with our potential is straightforward. In the 1p state we find a contribution of 41 MeV [from the last term in Eq. (5)], and 35 MeV [from the first term extrapolated linearly] to the matrix ele-element of V_1. The sum is 76 MeV ; if we assume the linear potential to be a scalar S_0, Eq. (13a), then we find 29 MeV. For V_2 we find 6 and 2.4 MeV, respectively, leading to 8.4 or 6 MeV total, the latter number being valid if the linear potential is scalar. Thus we find qualitative agreement between theory and experiment, in particular in view of our earlier remarks concerning our ignorance of the derivatives of the potential in the transition region, 0.4-0.8 fm. While the scalar potential hypothesis seems to be favoured by the experiment, it is not quite obviously correct.

The ratio of e^+e^- decay widths of ψ and ψ' is well fit by our potential as well as by all potentials that succeed in obtaining the proper energy of the 1p state [3]. We have the usual problem with the η_c if it is assumed to be the $S = 0$ companion of charmonium [see Ref. 1) for further discussion].

In conclusion we would like to emphasize that our results clearly indicate that the one (polarized) gluon exchange potential is well capable of describing the properties of non-relativistic bound states of heavy quarks. It seems that the detailed agreement between the computed spectrum and the experimental data not only confirm the reliability of the perturbative approach to QCD and asymptotic freedom but also establishes a strict quantitative relation between the parameters of the theory and experiment. The dispersive approach outlined seems to allow the study of the analytic properties of the running coupling constant when precise data on the upsilon states, and eventually bottonium, begin to be available. It appears that quite a new domain of exotic atoms will be opened in such experiments.

After completion of this work we have learned from Dr. T. Applequist about a related work by J.L. Richardson [4].

REFERENCES

1) We rely here heavily on the general presentation of the subject given at this Conference by Thomas Applequist. See also : T. Applequist, R.M. Barnett and K. Lane, "Charm and Beyond", Ann.Rev.Nuc.Part.Sci. 28 (1978) 387.

2) J.G.H. de Groot et al., Phys.Letters 82B (1979) 292, 456. It is amusing to note here that without strange quarks we find the identical value $\Lambda = 470$ MeV (compare with our Fig. 1) as the best fit to the 'onium spectrum.

3) See, for example : M. Krammer and M. Krasemann, "Quarkonium", DESY Preprint 78/66 (1978).

4) J.L. Richardson, "The Heavy Quark Potential and the Υ, J/ψ Systems", SLAC PUB 2229 (December 1978). Phys.Letters 82B (1979) 272.

BARYONIUM - NUCLEAR ATOM OR COLOUR MOLECULE?

CHAN Hong-Mo

CERN, Geneva, Switzerland

1. INTRODUCTION

An atom is a bound state of a nucleus with electrons held together by electromagnetic forces. When some of the electrons are replaced by other particles such as μ^- or $\bar{p}$, one gets an exotic atom which is the subject of this school. If one now further generalizes the concept by replacing the particles by nucleons or quarks, and the forces as well by nuclear or quark confinement forces, one obtains nuclei or hadrons, respectively. Although these latter systems have very different properties from atoms because of the very different dynamics, there are some similarities in both the experimental and the theoretical techniques involved in their study which are well worth a careful comparison. It is the foresight of the organizers in inviting such a comparison that gives me the pleasure of speaking to you now.

The subject of my lectures is "baryonium". As now used in the literature the name "baryonium"*) refers to a group of empirical

*) The name unfortunately also carries with it some theoretical prejudices which are not entirely appropriate and which we shall do our best in these lectures to ignore.

resonances with baryon number zero which have the unusual property of preferring to decay into baryon-antibaryon pairs instead of the kinematically more favourable mesonic final states. For a summary of the present experimental situation, I refer you to the excellent reviews by Montanet[1). Here I shall content myself just with quoting some examples as illustrations.

The best-established of these states were found in formation experiments in $p\bar{p}$ collisions. From accurate measurements of the total and elastic cross-sections, structures are seen at several energies which are usually referred to as the S, T and U bumps. Their properties are summarized in Table 1. One sees that if one interprets these structures as resonances they have large branching ratios into the elastic channel $p\bar{p}$. This is very surprising in view of the large number of competing mesonic channels which are more favourable kinematically. The fact that they prefer instead to decay into $p\bar{p}$ presumably means that their couplings to mesons

Table 1

	S	T	U
Mass (MeV)	1936 ± 1	2185 ± 5	2355 ± 5
Width (MeV)	8 - 4	130 ± 30	180 ± 20
σ_T(mb)	10.6 ± 2.4	5.0	3.1
σ_{el}(mb)	7.0 ± 1.4	2.9	2.8
$\sigma(p\bar{p}\rightarrow n\bar{n})$(mb)	0.3 ± 0.3	1.2	0.4
$\sigma(p\bar{p}\rightarrow \pi^+\pi^-)$(mb)		$\lesssim$ 0.24	$\lesssim$ 0.12

are for some reason strongly suppressed. The effect is particularly striking for the S bump, which, being close to the $N\bar{N}$ threshold, should have little chance for $p\bar{p}$ decay and yet was found in this channel with a branching ratio of at least 50 per cent. Such limits as are known for the cross-sections into specific meson channels are indeed smaller than one would have expected (see Table 1) which, if not due to selection rules forbidding decays into these channels, give a measure for these suppressed mesonic couplings. On the other hand, the total widths of the T and U -bumps well above the $N\bar{N}$ threshold are big, of the order 100-200 MeV, which, when combined with the large branching ratios into $N\bar{N}$, implies that they have normal hadronic couplings to $N\bar{N}$ channels of strengths similar, for example, to $f \to \pi\pi$.

In order to establish that these structures are indeed resonances, one would need to perform some amplitude analysis on the data which is at present impossible for $p\bar{p}$ elastic scattering because of the large number of partial waves involved. It can be done however for $p\bar{p} \to \pi^+\pi^-$ whose quantum numbers are more selective and for which good data are available in spite of the small cross-sections. The result is listed in Table 2. The masses of the first two states are in rough agreement with the T and U bumps seen in σ_T and σ_{el}, from which we may infer that these structures do indeed contain resonant components in the partial waves $1^+(3^-)$ and $0^+(4^+)$, respectively.

Table 2

$I^G(J^P)$	Mass (GeV)	Width (MeV)
$1^+(3^-)$	2.15	200
$0^+(4^+)$	2.31	210
$1^+(5^-)$	2.48	280

It is clear however that these S, T, U structures are not pure resonances in a single partial wave. If they have pure isospin, whether 0 or 1, then their branching ratios into $p\bar{p}$ and $n\bar{n}$ must be the same, whereas they are either not seen at all (S) in the charge exchange reaction ($p\bar{p} \to n\bar{n}$), or, if seen, much smaller (T,U) than in elastic scattering (Table 1) implying that they each contain both isospins 0 and 1. Indeed, from a tentative analysis of some data on the reaction $p\bar{p} \to K^+K^-$, Carter was able to show that in addition to the states in Table 2, there are probable states also in the partial waves $0^-(3^-)$ and $0^-(5^-)$ at about the same masses as the $1^+(3^-)$ and $1+(5^-)$, respectively. Moreover, there are indications from the reaction $p\bar{p} \to \pi^0\eta$ that the state $0^+(4^+)$ listed in Table 2 may also have an isospin partner with $I^G(J^P) = 1^-(4^+)$ at a similar mass. We have thus here an apparent isospin degeneracy which is reminiscent of the familiar degeneracy between ρ and ω, f and A_2, etc., in ordinary meson spectroscopy.

Apart from the S, T, U and V structures just discussed there are suggestions of further states both above and below the $p\bar{p}$ threshold obtained mainly from $\bar{p}d \to pX$ and $\bar{p}p \to \gamma X$ experiments. The impression that one gets is a great richness of new states with a particular affinity to the $B\bar{B}$ channel.

That there should exist mesonic states preferring to decay into $B\bar{B}$ is already of considerable interest but this is not all. An examination of the $p\bar{p}$ spectrum from production experiments reveals further surprises. The most popular of these spectra is shown in Fig. 1 which was obtained from backward production in π^-p collisions[2]. Apart from a small narrow peak at 1930 MeV which presumably corresponds to the S bump already discussed, there are two further spikes with the parameters listed in Table 3. They have appreciable branching ratios into $p\bar{p}$ and are not seen in mesonic channels. In particular, the upper limits for their decays into $\pi^+\pi^-$ and K^+K^- as shown in Table 3 are small, which if

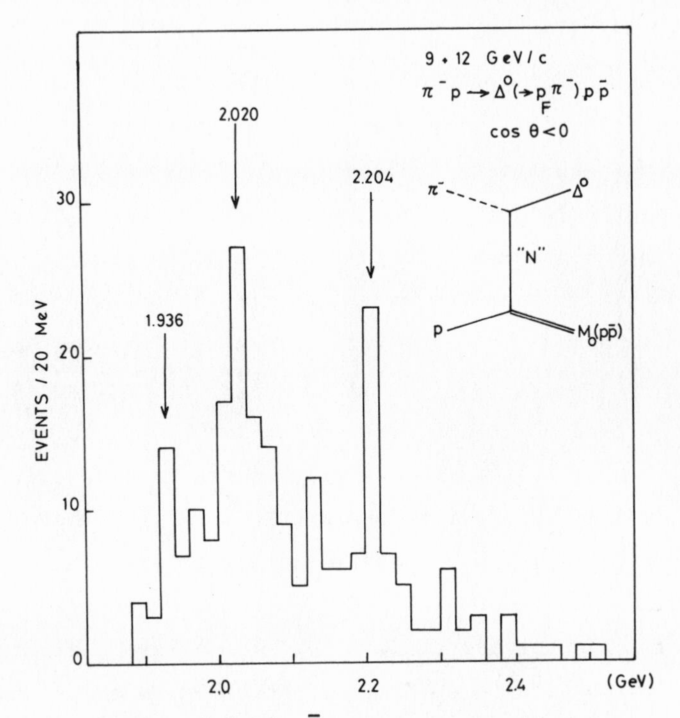

Fig. 1 The empirical $p\bar{p}$ spectrum obtained from a $\pi^- p$ backward production experiment

Table 3

Mass (MeV)	Width (MeV)	$\frac{M^0 \to p\bar{p}}{M^0 \to \text{total}}$	$\frac{M^0 \to \pi^+\pi^-}{M^0 \to p\bar{p}}$	$\frac{M^0 \to K^+K^-}{M^0 \to p\bar{p}}$
2020 ± 3	24 ± 12	> 14%	< 15%	< 11%
2204 ± 5	16^{+20}_{-16}	> 16%	< 17%	< 16%

not due to their quantum numbers being just such as to forbid these decays, imply that they have suppressed coupling to mesons as compared to $B\bar{B}$. These characteristics are sufficient to qualify them as candidates for baryonium.

However, they have some further peculiar properties which distinguish them from the baryonium states considered above. First, they have small total widths even though they are well above the $N\bar{N}$ threshold. For example, the peak at 2204 has a width of only $\lesssim$ 20 MeV as compared with 130 MeV for the T-state of similar mass in Table 1. This means that they must have suppressed couplings also to $B\bar{B}$ channels as compared with normal hadron couplings such as $f \to \pi\pi$. Second, the production cross-sections of these two states are comparable to that for the S(1936), as can be seen from the experimental numbers quoted in Table 4. Yet, these states have not been seen in formation experiments from $p\bar{p}$ collisions, the upper limits on their formation cross-sections being < 0.4 mb in both σ_{el} and σ_T. When compared with the formation cross-sections for S, T and U quoted in Table 1, one has a suppression factor of at least an order of magnitude. Now, in formation experiment, a resonance is coupled to an $N\bar{N}$ pair where both N and $\bar{N}$ are on mass-shell, whereas in the backward production experiment in Fig. 1, the

Table 4

	$\sigma : \pi^- p \to \Delta^0 M^0$ $\Delta^0 \to p\pi^-$; $M^0 \to p\bar{p}$	$\sigma : \pi^- p \to N^*(1520)M^0$ $N^*(1520) \to p\pi^-$, $M^0 \to p\bar{p}$
S(1936)	9 ± 5 nb	7 ± 5 nb
(2020)	10 ± 4 nb	26 ± 8 nb
(2204)	21 ± 5 nb	

resonance M^0 is coupled to an off-shell "$\bar{N}$" which is exchanged. The phenomenon noted above may thus be interpreted as saying that the couplings of these states to on-mass-shell nucleons are indeed suppressed, as expected already from their narrow decay widths, but that their couplings to off-mass-shell nucleons are comparable to that of the S. These properties are not only intriguing in themselves but are also quite dramatically different from those of our earlier examples in Table 1. It would thus appear that we are dealing here with yet another new type of objects.

Further narrow states are reported in other production experiments with masses up to 3 GeV with widths comparable to the experimental resolution of order 20 MeV, and decaying into $p\bar{p}$ accompanied in some cases by pions. One of them was even seen to cascade by pion emission into the states (2020) and (2204) just discussed. If taken seriously, they seem to form another family of baryoniums quite as numerous and at least as rich in interest as the first found in formation experiments.

It should be emphasized that few, or perhaps even none, of the baryonium states reported in experiment can yet be regarded as firmly established. Indeed, every one of them is still being hotly disputed among experimenters, not only by opponents but sometimes even by the discoverers themselves. However, assuming even that only some of them survive, they would still represent an exciting new branch of spectroscopy whose unusual characteristics are likely to provide us with new insight into hadron dynamics.

In order to appreciate more fully the challenge presented, let us summarize what we have learned. The two outstanding features of baryoniums are:

a) They have appreciable branching ratios into $B\bar{B}$ and suppressed couplings to mesonic final states. Yet the $B\bar{B}$ and purely mesonic channels are not, as far as we know, distinguished by any selection rules.

b) There appears to be two families of baryoniums. One family has large couplings to $B\bar{B}$ with widths above $N\bar{N}$ threshold of order 100-200 MeV. The other has widths of order only 10 MeV even high above $N\bar{N}$ threshold indicating suppressed couplings to $N\bar{N}$ pairs as well as to mesons, but yet these resonances can be readily produced.

Apart from these outstanding characteristics there are two further hints which may help in unravelling the mystery:

c) Those few baryonium states with identified quantum numbers all have fairly high spins ($J \geq 3$).

d) Some of them are degenerate in isospin 0 and 1.

Obviously, the first question to answer is: "What are these baryonium states?" Presumably they are not ordinary $Q\bar{Q}$ hadrons since these are known to decay into mesons copiously unless forbidden to do so by kinematics or selection rules. They will have to be something else. There are two suggestions:

A) They are bound or resonant states of a baryon and an antibaryon held together by ordinary nuclear forces ("nuclear atom") in much the same way as a deuteron is a nuclear bound state of a proton and a neutron.

B) They are hadrons with constituents $QQ\bar{Q}\bar{Q}$ held together by confinement forces which are supposedly responsible for the formation of all hadronic particles from their quark constituents, e.g., $Q\bar{Q}$ (mesons) and QQQ (baryons).

Here, we have to emphasize one point which is often not made clear in the literature. The assumptions (A) and (B) do not by themselves imply the existence of baryoniums. For example, the $B\bar{B}$ nucleus in (A) may annihilate readily into mesons thus giving it a large mesonic width in contrast to what is experimentally observed (see (a)). Indeed, a low energy $N\bar{N}$ system is known empirically to have large annihilation cross-sections. Similarly, the $(QQ\bar{Q}\bar{Q})$ system in (B) may also simply recombine into $(Q\bar{Q})(Q\bar{Q})$ dissociating thus into two $(Q\bar{Q})$ mesons. Since no quark pairs need be created, in contrast to decays such as $f \to \pi\pi$, i.e., $(Q\bar{Q}) \to (Q\bar{Q}) + (Q\bar{Q})$, one may even expect mesonic widths much in excess of the normal hadronic widths of order 100 MeV.

Therefore, in order to make either (A) or (B) into baryoniums, they have to be supplemented by further conditions. Again, two solutions have been suggested:

α) There is a new selection rule hitherto unrecognized which forbids those resonances decaying into mesons.

β) Baryoniums have "high" spins; mesonic decays are suppressed kinematically by the angular momentum barrier effect.

Now the solution (α) cannot apply to the $B\bar{B}$ nuclear model (A) simply because nuclear forces at long and medium ranges are quite well known and there is no room for the assumption of a new selection rule. **It can, however, be imposed on the $(QQ\bar{Q}\bar{Q})$ hadron model** (B) if one so wishes, since "confinement forces" are in any case so badly understood that assumptions made here cannot easily be disproved. In essence, what one would say here is that the "diquark" QQ in the $(QQ\bar{Q}\bar{Q})$ hadron is not simply just two quarks, but is a two-quark system distinguished by a new characteristic, e.g., a "junction". This characteristic is conserved, thus behaving like a new quantum number and giving rise to a new selection rule. With such a device one can clearly suppress the mesonic decays of baryoniums -- in other words, the effect is not explained but simply assumed.

Now if there exist baryonium states with low spins very stable against mesonic decays which cannot be explained by the angular momentum barrier alone, we may be forced to introduce a new selection rule as in (α). Notice that what actually matters is not the total spin J but the orbital angular momentum L between B and $\bar{B}$ in (A) and between (QQ) and ($\bar{Q}\bar{Q}$) in (B). For example, for L = 2, one can still build a J = 0 state from a spin 1 diquark and a spin 1 anti-diquark, which can already be quite stable due to the angular momentum barrier. In any case, as noted in (c) above, all those baryoniums with identified quantum numbers have $J \geq 3$. It seems therefore uneconomical at this stage to introduce a new selection rule which has also little theoretical justification. For this reason, I shall not discuss the solution (α) any further and concentrate on the alternative (β).

2. NUCLEAR FORCES VERSUS QUARK CONFINEMENT FORCES

In order to distinguish $B\bar{B}$ nuclei from $QQ\bar{Q}\bar{Q}$ hadrons as candidates for baryoniums, let us first examine the forces responsible for holding them together in each case.

A) Nuclear forces between nucleons are quite well understood for distances $\gtrsim 0.8$ fermi[3]. They can be described in terms of a potential due to the exchange of particles, e.g., pions, as illustrated in Fig. 2. The longest range part due to single pion exchange is familiar. The 2π exchange term, effective at medium ranges is sometimes approximated by the exchange of a fictitious σ resonance, but can be more accurately calculated using dispersion relations which has been done, for example, in great detail by the "Paris Group". The potential due to 3π exchange is much harder to calculate and is usually just approximated by ω exchange. Taking

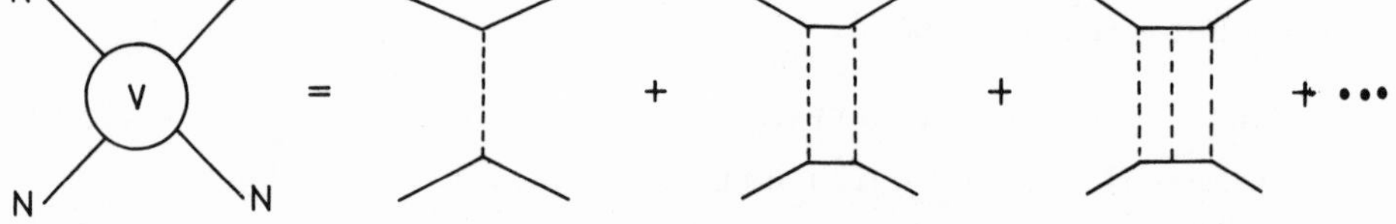

N N V N N = + + + •••

Fig. 2 The nucleon-nucleon potential due to particle-exchange

these exchanges into account, one believes that a fair description is obtained for the interaction between nucleons for $r \gtrsim 0.8$ fermi, but it is unlikely that the procedure can be extended further to smaller distances.

The potential between a nucleon and an antinucleon can be obtained from that between two nucleons by the "G-parity" rule, as illustrated in Fig. 3. Notice that since ω has G = −, the potential due to ω exchange though repulsive in NN becomes attractive in $N\bar{N}$. Together with the 2π exchange term which remains attractive here, one has then a very strong attraction at intermediate ranges. As example, we show in Fig. 4 the central potential calculated in this way by the Paris group.

The interaction in $N\bar{N}$, however, is complicated by the presence of annihilation channels as illustrated in Fig. 5. The range of annihilation forces is expected naïvely from the mass of the two nucleons exchanged to be $\sim 1/2m_N \sim 0.1$ fermi, i.e., well below the region of validity of the potential in Fig. 3, but they are likely to be very strong. In any case, because of relativistic effects at short distances, one cannot hope to describe them by a potential. For this reason, they remain largely unknown, and, as we shall see, are the main source of uncertainty in the nuclear model for baryonium.

We note in particular the following points relevant for baryonium spectroscopy.

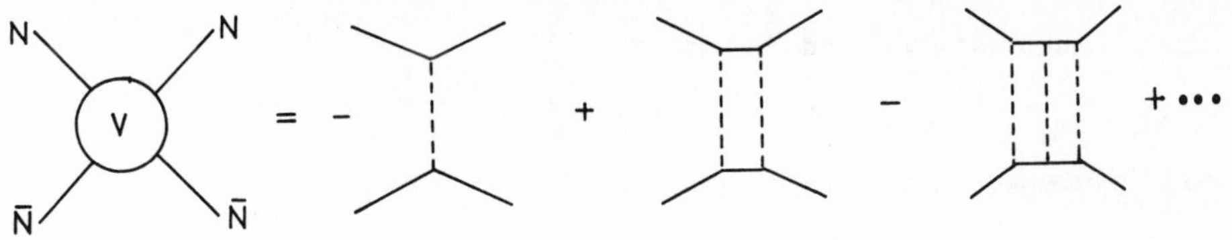

Fig. 3 The nucleon-antinucleon potential due to particle exchange.

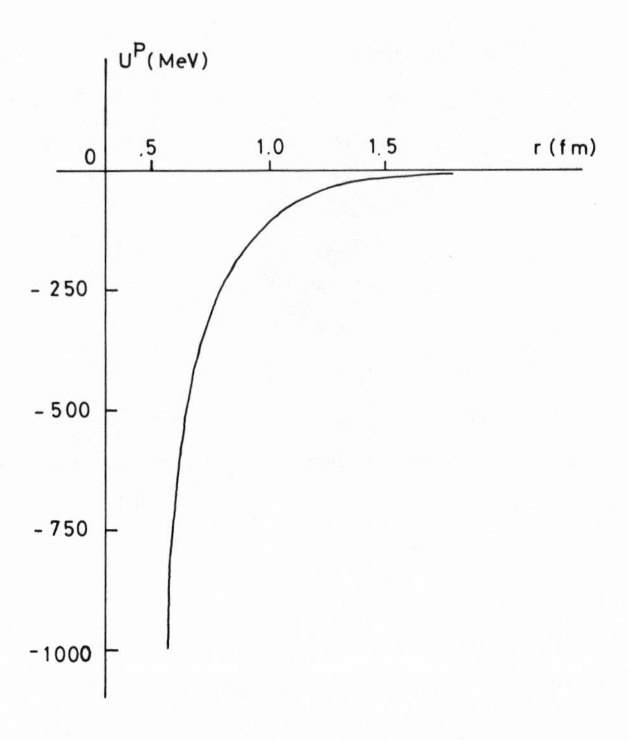

Fig. 4 The central component of the "Paris" $N\bar{N}$ potential.

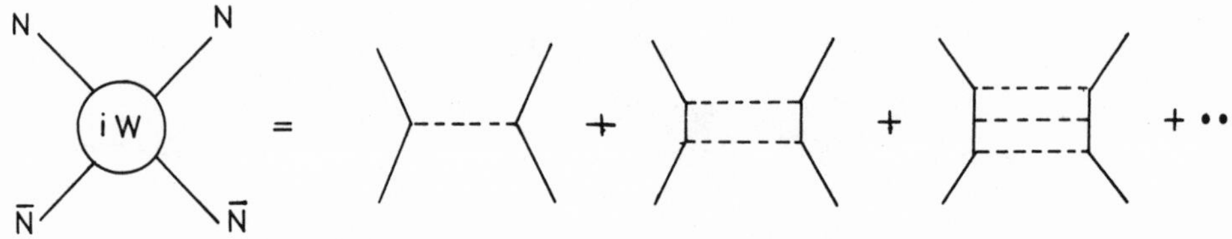

Fig. 5 Annihilation effects in nucleon-antinucleon interactions.

A(i) The strong attraction at intermediate ranges can give rise to many bound states and resonances, <u>but</u>

A(ii) Some (or all?) may be very unstable because of strong annihilations at short distances.

A(iii) At large distances, the potential vanishes exponentially, $V \sim \exp[-m_\pi r]$.

A(iv) Since the forces arise from exchange of particles carrying isospin, they are strongly isospin dependent.

A(v) They are also strongly spin dependent due especially to the tensor force.

B) The idea of confinement forces comes of course from the seemingly absurd assumption that quarks exist but are not seen. The only reason why one should entertain such a conjecture at all seriously is that there is now a quite attractive theory for strong interaction which makes confinement somewhat plausible. This theory, called quantum chromodynamics (QCD) is modelled on QED and based on a new quantity called "colour" which takes the place of the electric charge.

The concept of colour was first proposed to resolve an inconsistency in the quark model of hadron spectroscopy. Take, for example, Δ^{++} with $J_z = 3/2$ which in the quark model is a state consisting of uuu all in an s-wave and all with spin up. The total wave function is then symmetric under the interchange of any two quarks, whereas quarks being spin ½ fermions, ought to have antisymmetric wave functions. One can bypass this difficulty by assuming that quarks have an additional degree of freedom, namely colour, which can take three values. One further assumes that quark interactions are invariant under the colour group SU(3) in which quarks belong to the fundamental representation 3, and hadrons to the identity 1. The wave function of the $\Delta^{++}(J_z = 3/2)$ considered above has now to be multiplied by a colour wave function: $3 \times 3 \times 3 = 1$ which is antisymmetric, making thus the total wave function also antisymmetric.

At first this assumption looked unattractive and artificial, but gradually it was realized that one can now construct a theory of quark interactions in close analogy to QED. In electrodynamics, electrons interact by exchanging photons; analogously, quark interactions in chromodynamics are mediated by vector particles called gluons which belong to the adjoint representation 8 and carry colour charges. Further, QED is invariant under local phase transformations exp i $\Lambda(x)$; similarly, QCD has local gauge invariance under transformations belonging to the group SU(3).

The resultant theory has the following attractive features:

i) It is renormalizable.
ii) Due to renormalization effects, the effective coupling α_s decreases with q^2 (asymptotic freedom) so that at high q^2 (i.e., short distances) quarks are effectively free inside hadrons, as suggested by the parton model of deep inelastic scattering.
iii) Conversely, for low q^2 (i.e., large distances) α_s becomes large (infra-red slavery) so that there is a possibility that quarks may indeed be confined.

Unfortunately, the theory is complicated and one has not yet learned to calculate with it many observable quantities so that at present it has few direct experimental tests. This is particularly true for the quark confinement forces which interest us. Nevertheless, a number of their properties have been conjectured based partly on theoretical arguments and partly on phenomenological observations. We list some of their properties below which are relevant for baryonium spectroscopy.

B(i) Confinement forces are by assumption attractive only in colour singlet states.

B(ii) At large distances, the confining potential increases indefinitely (infra-red slavery), e.g., $V(r) \sim a|r|$, which means that it can lead to an infinite number of bound states and resonances with unlimited angular momentum. This has some support from the empirical hadron spectrum which was indeed at least partly responsible for such a conjecture.

B(iii) Since gluons carry no flavour, confinement forces due to their exchange are independent of isospin so that for example, the meson spectrum is expected to be isospin degenerate unless modified by other effects such as quark-pair annihilations which are isospin dependent. This also has some support from experiment for resonances with high angular momentum for which the short-ranged annihilation forces are presumably negligible.

B(iv) Confinement forces are probably spin-independent at large distances. Phenomenologically this may be demonstrated as follows. In the quark model, π and ρ are both s-wave $q\bar{q}$ states ($q = u,d$), differing only by the total quark spin S: for π, $S = 0$ (singlet) for ρ, $S = 1$ (triplet). The mass difference $\Delta m(0) = m_\rho - m_\pi$ measures therefore the strength of the spin-dependent interactions between q and $\bar{q}$ in the s-wave, i.e., $\ell = 0$. Similarly, B and A_2 are p-wave $q\bar{q}$ states differing only by the total quark spin S. $\Delta m(1) = m_{A_2} - m_B$ measures the strength of the spin-dependent interaction between q and $\bar{q}$ for $\ell = 1$. Hence, plotting the mass difference Δm between the triplet and singlet states along Regge trajectories gives the variation of spin-dependent forces as functions of ℓ, or, equivalently, of the distance between q and $\bar{q}$. Figure 6 shows such a plot for the $q\bar{q}$ ($q = u,d$) and $q\bar{s}$ mesons. In both cases Δm is seen to decrease rapidly with ℓ.

These properties are very different from those of nuclear forces listed before and the differences will be reflected in the bound state and resonance spectrum. For instance:

i) Nuclear forces are short ranged and cannot bind resonances with high orbital angular momentum, whereas confinement forces are supposedly under no such restrictions. Hence if there exist high mass baryoniums with very high spin, then they are most likely $QQ\bar{Q}\bar{Q}$ hadrons rather than $N\bar{N}$ nuclei.

ii) Nuclear forces are strongly isospin dependent so that any isospin degeneracy in the spectrum of their bound and resonant states is purely accidental, whereas for confinement forces such a degeneracy is, in general, to be expected. The empirical observation (d) in §1 is therefore a point in favour of the $QQ\bar{Q}\bar{Q}$ hadron model.

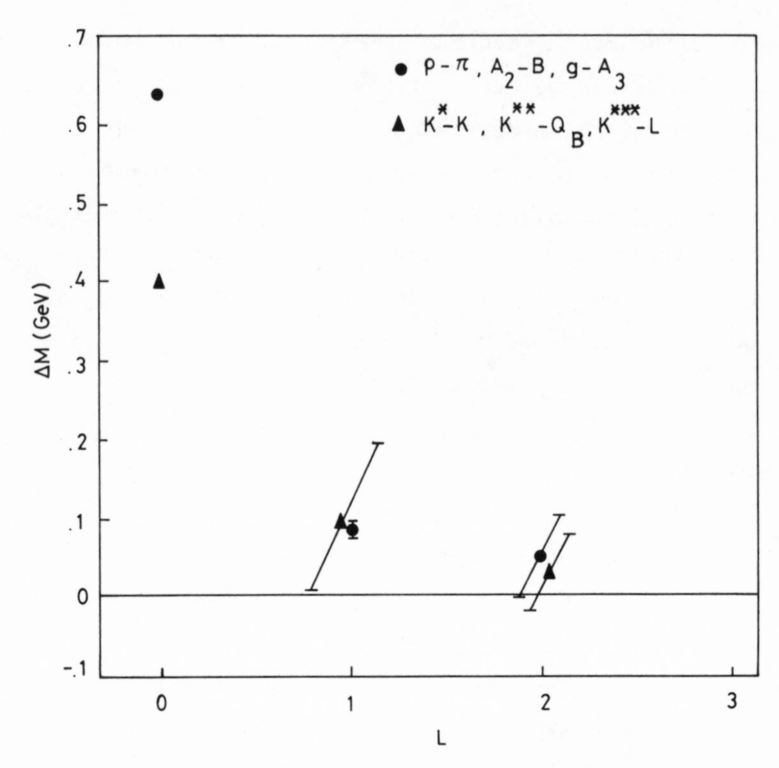

Fig. 6 The variation of spin dependent $Q\bar{Q}$ interactions as a function of the orbital angular momentum ℓ.

In order to distinguish further the two models, we shall have to calculate the spectrum of baryoniums in each case and to compare in detail the result with experiment.

3. BARYONIUMS AS NUCLEAR ATOMS

As the $N\bar{N}$ potential is well understood only for $r \gtrsim 0.8$ fermi some assumption has to be made for short distances before one can proceed to a calculation of the resonance spectrum. On this point there are some differences between various authors in this field, but hopefully such differences are not of vital importance for the states of interest with appreciable orbital angular momentum ℓ. For example, the Paris group assumed simply[3]:

$$U = \begin{cases} U^P(r) & r \geq r_c \\ U^P(r_c) & r < r_c \end{cases}$$

where $U^P(r)$ is the "Paris potential" the central part of which was shown in Fig. 4, and r_c is treated as a free parameter expected to be ∿ 0.8 fermi. With such an assumption, the calculation of the spectrum is then explicit.

The result for $r_c = 0.7$ fermi is quoted in Table 5 as an example. Because of the strong attraction at intermediate ranges there are many states both below and above the $N\bar{N}$ threshold. One notes however the following:

i) The spectrum is quite sensitive to r_c although the ordering of the states is not.

ii) The masses and even the existence of the tightly bound states are unreliable because of the strong annihilation effects at short distances.

iii) States far above the $N\bar{N}$ threshold are probably too unstable to be regarded as resonances.

Experimentally, there are indeed a large number of candidates for baryoniums reported near the $N\bar{N}$ threshold. Unfortunately, none of them has yet had its quantum numbers identified so that no detailed comparison is at present possible. It is obvious however that all states obtained in this way are strongly coupled to $N\bar{N}$ and cannot be associated with the narrow states high above the $N\bar{N}$ threshold found in production experiments as discussed in (b)§1. There are some theoretical suggestions that such states may still be nuclear bound states not of $N\bar{N}$ but of nucleon resonances such as $\Delta\bar{\Delta}$. However, since nucleon resonances themselves decay thus: $\Delta \to N\pi$, with widths ∿ 100 MeV, narrow states obtained in this way cannot be a general occurrence.

Table 5

$^{2I+1,2S+1}L_J$	Binding energy (MeV)	Mass (MeV)	$^{2I+1,2S+1}L_J$	Binding energy (MeV)	Mass (MeV)
$^{31}S_0$	-262	1616	$^{13}P_0$ (n = 0) (n = 1)	< -800 -102	< 1078 1776
$^{31}P_1$	-55	1823	$^{13}S_1$ - $^{13}D_1$ (n = 0)	< -800	< 1078
$^{33}P_1$	-116	1762	$^{13}S_1$ - $^{13}D_1$ (n = 1)	-39	1839
$^{33}S_1$ - $^{33}D_1$	-202	1676	$^{13}P_2$ - $^{13}F_2$ (n = 0)	-788	1090
$^{31}D_2$	165	2043	$^{13}D_3$ - $^{13}G_3$	-457	1421
$^{33}D_2$	85	1963	$^{13}F_4$ - $^{13}H_4$	-89	1789
$^{33}F_3$	> 200	> 2078	$^{13}G_5$ - $^{13}I_5$	307	2185

Cf., Experimental states: 1394, 1648, 1680, 1794, 1897, 1936, 1940, 2185 ...

4. BARYONIUMS AS COLOUR MOLECULES[4),5)]

We shall denote the $QQ\bar{Q}\bar{Q}$ hadron states of interest by

$$(QQ)^{X}_{2S+1} \overset{L}{\longrightarrow} (\bar{Q}\bar{Q})^{\bar{X}}_{2\bar{S}+1} \tag{1}$$

where (QQ) denotes a QQ system in an s-wave state (diquark) and X = its total colour representation and S = its total spin. Clearly X can be $3 \times 3 = \bar{3}$ or 6 and S = 0 or 1. Also since the hadron (1) is a colour singlet, $\bar{X}$ must be the conjugate representation of X.

In general, states with different diquark colours and spins can mix since there is a priori no reason why the total Hamiltonian should be diagonal in the diquark colour or spin. One believes however that such mixing becomes negligible when L is large. The arguments go as follows: Since quarks are fermions, the diquark wave function must be totally antisymmetric. Hence for q = u,d one must have the following isospin assignments for our diquark states:

$$\begin{array}{cccc} (qq)^{\bar{3}}_{1} & (qq)^{\bar{3}}_{3} & (qq)^{6}_{1} & (qq)^{6}_{3} \\ I = 0 & I = 1 & I = 1 & I = 0 \end{array} \tag{2}$$

Now the gluons which mediate interactions between the diquark and antidiquark carry no flavour and can therefore mix only those states with the same isospin, i.e.,

$$(qq)^{\bar{3}}_{1} \longleftrightarrow (qq)^{6}_{3} \quad ; \quad (qq)^{\bar{3}}_{3} \longleftrightarrow (qq)^{6}_{1} \tag{3}$$

In other words, to change the colour of a diquark one must also change its spin. However, by (B)(iv) in §2, spin dependent forces decrease rapidly with the separation L so that for large L the mixing between states with different X and S becomes negligible. We may then regard states with different X and S as essentially distinct.

At this stage it is convenient to draw an analogy between our hadron states (1) with ionic molecules such as $Na^+ - Cl^-$, where two ions of opposite charges are neutralized by an ionic bond linking them. Here we have two quark clusters with opposite colour charges ("chromions") neutralized by a tube of colour flux lines ("chromionic bond") and linked together to form a ("chromionic") colour molecule. The analogy is clearly only formal since the dynamics is very different, but it teaches us the wisdom of the chemists in treating ions and bonds as separate entities thereby reducing labour and simplifying concepts when considering the properties of molecules. We shall do the same for our colour equivalents.

Consider then first the diquark chromions. Since they cannot exist free, they cannot be assigned a mass, but one can still talk about mass splittings between, for example, their various spin states. Since spin-dependent effects are supposed to be short-ranged, B(iv), one might argue by asymptotic freedom that perturbation theory applies so that to leading order in α_s the "hyperfine splitting" between diquarks is given just by the one-gluon exchange diagram of Fig. 7. This assumption has some support in the spectroscopy of $Q\bar{Q}$ mesons and QQQ baryons. Now the diagram

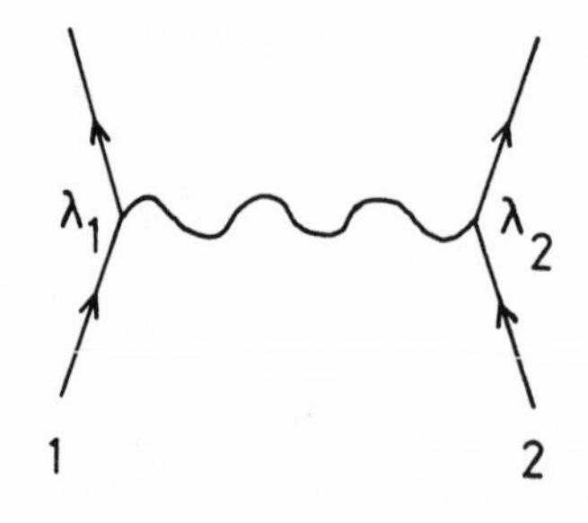

Fig. 7 The one-gluon exchange term in quark-quark interactions

is very similar to one-photon-exchange between electrons in QED except that the electric charge is now replaced by the colour charge λ_i which is a Gell-Mann 3 × 3 matrix of SU(3). By means of a Breit reduction as in QED, one obtains from Fig. 7 a spin-spin, a spin-orbit, and a tensor term, of which only the spin-spin term is relevant for our diquark in s-wave. This can be written as

$$H_1 = -C \sum_{a=1}^{8} \sum_{r=1}^{3} (\lambda_1^a \sigma_1^r)(\lambda_2^a \sigma_2^r) \tag{4}$$

where σ_i are Pauli matrices for the quark spins. The coefficient C is essentially $\alpha_s|\psi(0)|^2$ where ψ is the relative diquark wave function. Since the details of confinement forces are unknown, C cannot be calculated in a model-independent manner, but it can be estimated phenomenologically from QQQ baryon spectroscopy, for example: Δ-N = 16 C $\sim$ 300 MeV gives C $\sim$ 20 MeV. The hyperfine splitting between diquark ions can then be obtained by diagonalizing H_1, which is easily done and the result is listed in Table 6.

Table 6

Diquark	$\langle H_1 \rangle$
$(Q_1Q_2)^{\bar{3}}_1$	-8 C
$(Q_1Q_2)^{\bar{3}}_3$	$\frac{8}{3}$ C
$(Q_1Q_2)^6_1$	4 C
$(Q_1Q_2)^6_3$	$-\frac{4}{3}$ C

Next, the chromionic bond also carries a mass by virtue of the colour field energy confined inside the flux tube. Unfortunately this field energy cannot be calculated without an explicit model for confinement, which at present means essentially the MIT bag model. The result is best stated in terms of the angular momentum L carried by the bond, which is asymptotically linear in the bond mass squared thus

$$L = L_0 + \alpha' M_B^2 \tag{5}$$

where both L and α' depend on the colour X. The Regge slope α'_X is found to be inversely proportional to the square root of the quadratic Casimir operator $\mathcal{C}_X$, thus

$$\alpha'_X \sim 1/\sqrt{\mathcal{C}_X} \tag{6}$$

which, when combined with the information $\alpha'_{\bar{3}} \sim 0.9\ \mathrm{GeV}^{-2}$ obtained phenomenologically from, for example, the ρ trajectory, gives α'_X for other colours X also, e.g., $\alpha'_6 \sim 0.6\ \mathrm{GeV}^{-2}$. The intercept L_0 however is largely unknown and has to be determined phenomenologically from the baryonium spectrum.

Combining (5) with Table 6, one then has the following mass formula for the colour molecules (1):

$$M = \langle H_1 \rangle_{QQ} + \langle H_1 \rangle_{\bar{Q}\bar{Q}} + \sqrt{(L - L_0^X)/\alpha'_X} \tag{7}$$

depending on one parameter L_0^X for each family with colour X. One sees then that in spite of our ignorance of confinement forces compared with nuclear forces, one has managed by various means to construct a spectrum for $QQ\bar{Q}\bar{Q}$ hadrons quite as detailed as for the $N\bar{N}$ nuclei of Table 5.

Before we examine the predicted spectrum and compare it with experiment, let us first consider the decay properties of our colour molecules. By construction, all the colour molecules (1) are inhibited in decaying into mesonic final states by virtue of the angular momentum barrier. Their decays into $B\bar{B}$ pairs, however, depends on the colour X of the diquark. For X = 3, the molecule can decay by rupturing the colour bond linking the two chromions. Since colour is confined, a $Q\bar{Q}$ pair has to be created so as to neutralize the colour charge exposed. The result is a $B\bar{B}$ pair, thus:

$$(QQ)^{\bar{3}} -\!\!\not\!- (\bar{Q}\bar{Q})^{3} \longrightarrow ((QQ)^{\bar{3}}Q^{3})^{1} + ((\bar{Q}\bar{Q})^{3}\bar{Q}^{\bar{3}})^{1} = B+\bar{B} \tag{8}$$

For X = 6, however, the creation of one $Q\bar{Q}$ pair is insufficient to neutralize the colour 6 exposed on rupture of the bond. Indeed, $6 \times 3 = 8 + 10$ does not contain a colour singlet. At least two $Q\bar{Q}$ pairs have to be created in order to neutralize the colour, thus:

$$(QQ)^{6} -\!\!\not\!- (\bar{Q}\bar{Q})^{\bar{6}} \longrightarrow ((QQ)^{6}(\bar{Q}\bar{Q})^{\bar{6}})^{1} + ((\bar{Q}\bar{Q})^{\bar{6}}(QQ)^{6})^{1} \tag{9}$$

and the result is not $B\bar{B}$ but two baryoniums. Therefore, for decaying into $B\bar{B}$, a molecule with X = 6 has first to mix with X = 3 and then decay. Since the mixing effect is believed to be strongly suppressed for large L, the coupling to $B\bar{B}$ of X = 6 molecules has to be suppressed accordingly.

We therefore distinguish two families as candidates for baryoniums. By virtue of the large coupling to $B\bar{B}$ colour molecules with X = 3 acquire large widths above the $B\bar{B}$ threshold. Their decay into $B\bar{B}$ is similar in mechanism to the decay of, e.g., $f \to \pi\pi$ and is expected to have a similar width of order 100 MeV. For the same reason they are also readily formed in the direct channel in $N\bar{N}$

collisions. On the other hand, molecules with X = 6 couple reluctantly to $B\bar{B}$ as well as to mesons and therefore remain narrow even above the $B\bar{B}$ threshold. In addition, they cannot easily be formed in $N\bar{N}$ collisions but have to be produced along with other particles. This is strikingly reminiscent of the experimental observations in § 1.

In Figs. 8 and 9 we quote the spectrum of molecules with X = 3 and 6 as calculated in Ref. 4) by normalizing L_0 for each X to an empirical baryonium state, and compare them with the experimental structures observed. Further, in Table 7, the predicted quantum numbers of the dominant X = 3 states are compared with whatever is known experimentally. There is a rather close correspondence between the experimental and predicted spectra, especially for the X = 3 sector.

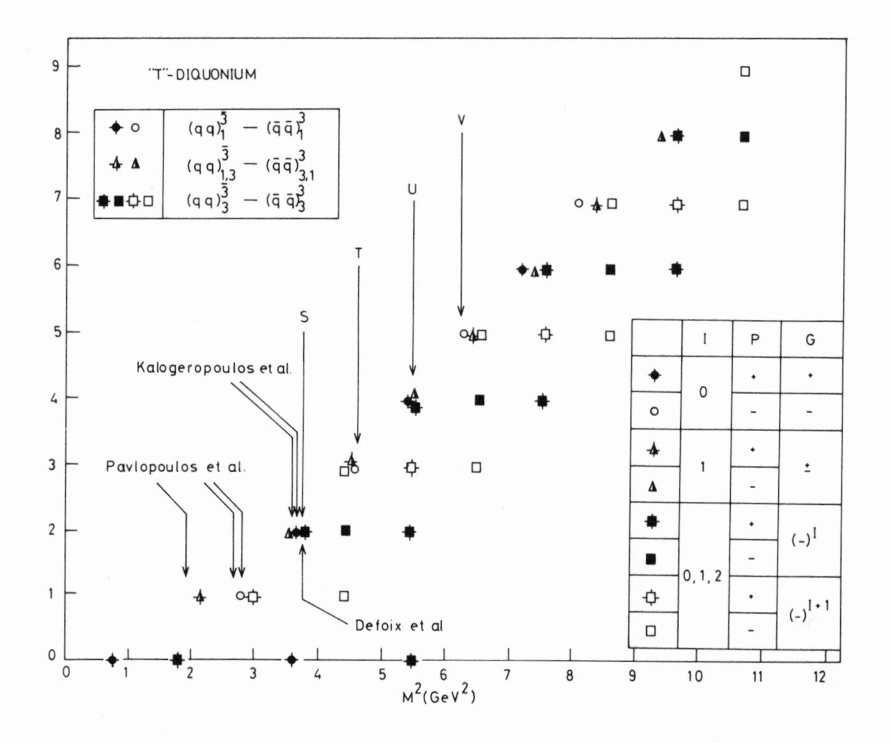

Fig. 8 The predicted spectrum of baryoniums in the colour molecular model with X = 3 compared with experimental structures in formation experiments.

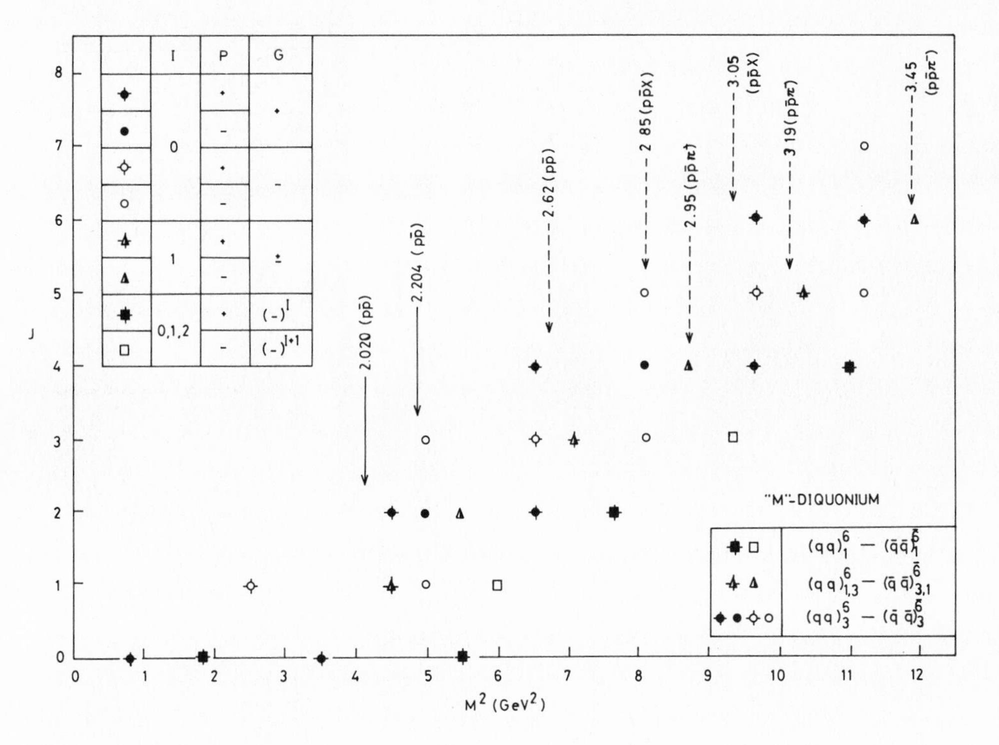

Fig. 9 The predicted spectrum of baryoniums in the colour molecular model with X = 6 compared with experimental structures observed in production experiments.

7. NUCLEAR ATOMS OR COLOUR MOLECULES?

It should be emphasized once more that the experimental evidence for baryonium states is still far from established especially for those belonging to the narrow family found in production experiments. It may well happen that under scrutiny, most, if not all, will disappear. Under such circumstances it would be unwise at present to conclude definitively in favour of either alternative. Superficially, however, at least to my perhaps biased mind, the existing experimental picture seems to favour slightly the colour molecules.

However, the two pictures need not be mutually exclusive and $N\bar{N}$ nuclear atoms could co-exist with $QQ\bar{Q}\bar{Q}$ colour molecules. We have

Table 7

Energy Region		S			T			U			V		
	Traj.	$(qq)^{\bar{3}}_1-(\bar{q}\bar{q})^3_1$	$(qq)^{\bar{3}}_3-(\bar{q}\bar{q})^3_3$		$(qq)^{\bar{3}}_1-(\bar{q}\bar{q})^3_1$	$(qq)^{\bar{3}}_3-(\bar{q}\bar{q})^3_3$		$(qq)^{\bar{3}}_1-(\bar{q}\bar{q})^3_1$	$(qq)^{\bar{3}}_3-(qq)^3_3$		$(qq)^{\bar{3}}_1-(qq)^3_1$	$(qq)^{\bar{3}}_3-(qq)^3_3$	
	L	2	0		3	1		4	2		5	3	
	$I^G(J^P)$	$0^+(2^+)$	$0^+(2^+)$	$1^-(2^+)$	$0^-(3^-)$	$0^-(3^-)$	$1^+(3^-)$	$0^+(4^+)$	$0^+(4^+)$	$1^-(4^+)$	$0^-(5^-)$	$0^-(5^-)$	$1^+(5^-)$
	Mass	1.92	1.94	1.94	2.13	2.11	2.11	2.33	2.34	2.34	2.51	2.56	2.56
σ_{el}	I,G,J,P	$J^{PC} = 2^{++}$	$J^{PC} = 2^{++}$ or 4^{++}		I = 0,1			I = 0,1			I = 0,1		
$\sigma_{ch.ex.}$	Mass	1.936	1.9 - 2.0		2.155			2.345			2.5		
$p\bar{p} \to \pi^+\pi^-$	$I^G(J^P)$	——	$[0^+(2^+)]$				$1^+(3^-)$	$0^+(4^+)$					$1^+(5^-)$
	Mass	——	[2.0]				2.150	2.310					2.5
$p\bar{p} \to \pi^0\pi^0$	$I^G(J^P)$	——	——					$0^+(4^+)$					
	Mass	——	——					2.35					
$p\bar{p} \to K^+K^-$	$I^G(J^P)$	——	——	——	$0^-(3^-)$		$1^+(3^-)$	$?(4^+)$			$0^-(5^-)$		$1^+(5^-)$
	Mass	——	——	——	2.15		2.15	2.34			2.35		2.5
$p\bar{p} \to \pi^0\eta$	$I^G(J^P)$			——						$1^-(4^+)$			
	Mass			——						2.32			
$p\bar{p} \to \rho^0\pi^0$	$I^G(J^P)$			$1^-(2^+)$						——			
	Mass			1.940						——			
SUM TOTAL	$I^G(J^P)$	$0^+(2^+)$	$0^+(2^+)$	$1^-(2^+)$	$0^-(3^-)$		$1^+(3^-)$	$0^+(4^+)$		$1^-(4^+)$	$0^-(5^-)$		$1^+(5^-)$
	Mass	1.936	∼ 2.0	1.940	2.15		2.15	2.34		2.32	2.5		2.5

deliberately taken a controversial approach here only for the sake of economy. Also, the view has sometimes been expressed that the two pictures may be just different descriptions of the same thing. I do not understand this viewpoint. It is of course true that since nucleons are made of quarks, an $N\bar{N}$ nucleus can also be described by a multiquark wave function. Conversely, so long as quarks are confined, a multiquark state can also be expressed as a composite of the hadrons from which it is formed and into which it decays. In other words, there has to be an equivalence transformation relating the two state vector spaces in terms of quarks on the one hand and of hadrons on the other. However, the baryonium states we are considering are not just any states in the state vector space. They have well-defined structures which are very different for the $N\bar{N}$ nuclei and the $QQ\bar{Q}\bar{Q}$ hadrons, so that their overlap in either picture must be very small. It just so happens that the $N\bar{N}$ nucleus is easy to describe in the hadron representation and the $QQ\bar{Q}\bar{Q}$ hadrons in the other. But in either representation they can be regarded as different states for all practical purposes. After all, a helium atom contains the same number of nucleons and electrons as a deuterium molecule, and both can be described in terms of a four-nucleon four-electron wave function, but no one has yet claimed that they are just different descriptions of the same thing.

In any case, both the models are sufficiently explicit that when the data clarify, one should be able to determine which, if any, of the pictures is correct.

I am grateful to the Directors of the School, especially Dr G. Fiorentini, for their kind hospitality.

REFERENCES

1) L. Montanet, Vth Int. Conf. on Experimental Meson Spectroscopy, Boston (1977), CERN preprint EP/PHYS 77-22; Proc. XIII Rencontre de Moriond "Phenomenology of Quantum Chromodynamics", ed. J. Tran Thanh Van, Editions Frontières, France (1978).

2) P. Benkhieri et al., Phys. Letters 68B (1977) 483; 81B (1979) 380;
J. Six, Proc. XIII Rencontre de Moriond, ibid (1978).

3) R. Vinh Mau, Proc. XIII Rencontre de Moriond, ibid (1978).

4) Chan Hong-Mo and H. Högaasen, Nuclear Phys. B136 (1978) 401.

5) Chan Hong-Mo, CERN preprint TH.2540 (1978) and Rutherford Laboratory preprint RL-78-089 (1978) to appear in Proc. IV European Antiproton Symposium, Barr (Strasbourg) (1978).

Further details can be found in the literature referred to in these papers.

Part III
The Chemical Physics of Mesic Atoms and Molecules

ATOMIC CAPTURE OF NEGATIVE MESONS IN HYDROGEN

M. Leon

University of California
Los Alamos Scientific Laboratory
Los Alamos, New Mexico 87545

INTRODUCTION

I will begin by giving a brief description of the present state of theoretical understanding of atomic capture of negative mesons.[1] In many ways, our ignorance here parallels that on the μ^+ location in μ^+SR studies that we will be hearing about. Then I will describe a very simple model calculation of μ^- capture by the simplest of all atoms, atomic hydrogen.

Finally, I will say a few words about the possibility of generalizing this approach to more complicated atoms and even molecules.

Baker[2] addressed this problem many years ago with an elaborate Born approximation calculation of the capture rates, and his approach has since been used for both hydrogen and heavier atoms by a number of workers.[3] However, the capture involves slow mesons, which cause quite severe distortions of the electronic wave functions. Thus, the basic requirement for the Born approximation is not met, and there is no reason to believe its predictions.

A different method has been applied more recently which treats the meson classically and the electrons as a Fermi gas.[4] Besides being clearly inapplicable to hydrogen, this approach offers little hope of understanding, in any basic way, the interesting observed chemical effects, which evidently have to do with the detailed response of the valence electrons to the invading meson.

Gershtein[5] has treated the hydrogen problem from a very different point of view. Using the fact, first noted by Fermi and Teller[6] and discussed at length by Wightman,[7] that the electron of the hydrogen atom becomes unbound when the meson-proton separation decreases to the critical distance $a_c = 0.64$ atomic units (1 a.u. = 0.53 Å), Gershtein calculated the distribution in n and ℓ for purely adiabatic ionization. That is, if the energy and angular momentum of the incident meson are such that the distance of closest approach of its (classical) orbit in the attractive Coulomb field of the proton is ≤ 0.64 a.u., the electron is expelled and therefore, the meson loses an energy equal to the ionization energy I. If the initial energy E is $< I$, capture results. Much earlier, Rosenberg[8] had investigated the effect of the finite meson velocity (non-adiabaticity) in causing ionization, but neglected the deflection (falling-in) of the meson orbits. Very recently, Baird[9] calculated the ionization probabilities including both the finite velocity and the deflection of the orbits. However, he did not calculate the n and ℓ distribution following capture. In the present paper, we use a simplified version of Baird's results to determine the (n,ℓ) distribution of the captured mesons.

THE MODEL

Our calculations are for atomic hydrogen; the proton and negative meson are both treated classically, while the electron must, of course, be treated quantum-mechanically. The first question to be addressed is the potential seen by the approaching meson. The (elastic channel) potential is simply (atomic units are used throughout)

$$V(r) = 0.5 - E_B(r) - \frac{1}{r} ; \tag{1}$$

r is the meson-proton separation and $E_B(r)$ the electronic binding energy in the proton-meson dipole field. Baird[9] has developed an efficient method of computing $E_B(r)$ and gives a table of E_B values for a discrete set of r values. We use these values and cubic spline interpolation to give $E_B(r)$.

The next required ingredient is the distribution of energy losses for each (elastic) orbit (following Baird, we neglect the possibility of discrete excitation). Baird has found that the differential cross section $\frac{d\sigma}{d\varepsilon}(E,\varepsilon)$ (ε is the energy loss) falls off rapidly as the electronic wave number κ (and therefore ε) increases; e.g., $\frac{d\sigma}{d\varepsilon}$ is down by a factor of two as κ goes from 0.1 to 0.2, for the relevant values of meson velocity ($E \sim I$). Therefore, most of $\frac{d\sigma}{d\varepsilon}$ is concentrated at very low electron energy, and the approximation

$$\frac{d\sigma}{d\varepsilon} \propto \delta(\varepsilon - I) \tag{2}$$

is a very reasonable one.

We also need the ionization probability P(E,b) (related to $\int \frac{d\sigma}{d\varepsilon}\, d\varepsilon$) as a function of E and impact parameter b. Baird has found that, as one expects, the ionization probability is mainly a function of the distance of closest approach r_o, with little remaining E-dependence. (Of course, r_o itself depends very strongly on E as well as b). A good approximation to the dependence on r_o is given by

$$P(r_o) = g(E) \begin{cases} e^{-[(r_o - R_o)/\sigma]^2} & r > R_o \\ 1 & r \leqslant R_o \end{cases} \tag{3}$$

with $R_o = 1.00$ and $\sigma = 0.57$; the actual form of g(E) is irrelevant. Note that this is drastically different from purely adiabatic ionization (as assumed by Gershtein)[6] which would imply

$$P(r_o) = \theta(R_o - r_o) \qquad \text{(adiabatic)} \tag{4}$$

with $R_o = 0.64$. Thus the finite velocity (due in large part to the "falling-in" of the meson) plays a dominant role in the ionization; this is as expected, since, e.g., the $E_B(r)$ for r = 1.0 is only 10^{-3}[9]. For P(E,b) we therefore take

$$P(E,b) \approx P[r_o(E,b)] \tag{5}$$

As long as the decay probability of the meson is negligible compared with the probability of ionizing an H atom, a weaker assumption than Eq. (2), that the shape in ε of $\frac{d\sigma}{d\varepsilon}$ is independent of E, is sufficient to ensure that the arrival probability is uniform in energy - i.e., a "white" spectrum.[10]

A final point to be mentioned is the possible angular momentum loss by the capturing meson. Since the electron has the largest probability of being ejected into an asymptotic p-state, we expect a maximum angular momentum loss $\Delta\ell_{max}$ of about one unit. Now for a given initial meson energy E there is a maximum angular momentum ℓ_{max} allowed for a bound state of the final energy E-I; we assume that for this angular momentum the angular momentum loss is $\Delta\ell_{max}$ and is proportionally smaller for smaller angular momentum:

$$\Delta\ell = \ell_{final} - \ell_{initial} = - \frac{\ell_{initial}}{\ell_{max} + \Delta\ell_{max}} \cdot \Delta\ell_{max} \; . \qquad (6)$$

In fact, since $\Delta\ell_{max} \ll \ell_{initial}$, this angular momentum loss changes the distribution very slightly; we discuss the effect of arbitarily assuming $\Delta\ell_{max} = 1$ below.

These ingredients enable us to deduce the (n,ℓ) distribution of captured mesons. The n-distribution is trivial, since the uniform (in E) distribution of incident mesons plus the assumption of energy loss $\varepsilon = I$ immediately implies a uniform distribution in binding energy, from zero to I. To find the distributions in ℓ for a given binding energy, we must, of course, determine the capture probability as a function of ℓ. This is straight forward since the distance of closest approach r_o (classical turning point) is easily deduced from Eq. 1, and then determines the probability from Eq. 3; the capture cross section $\sigma_{capt}(E)$ is determined at the same time. In this way, distribution function $\mathfrak{f}(B,\ell)$ in the binding energy-angular momentum plane is computed for a suitably fine mesh of points. The probabilities for given quantum numbers, $P_{n\ell}$, are then formed from integrals of the continuous distribution function,

$$P_{n\ell} = \int_{\ell}^{\ell+1} d\ell \int_{B_{n+}}^{B_{n-}} dB \; \mathfrak{f}(B,\ell) \qquad (7)$$

using a convenient spline-interpolation-integration technique. Here

$$B_{n\pm} \equiv \frac{M}{2(n\pm\frac{1}{2})^2} \qquad (8)$$

and M is the meson-proton reduced mass.

RESULTS

Results have been calculated for negative muons, taking I = 0.56 instead of 0.5 since H_2 molecules rather than H atoms are actually being ionized. A two-dimensional plot of the distribution is given in Fig. 1. The n-axis projection, of course, has just the n^{-3} behavior given by the density of states and the uniformity in energy. The distribution in recoil energy of the mesonic hydrogen atom is just given by the distribution of initial meson energy (and is hence uniform), since the electron carries off negligible momentum. Of course, these distributions are drastically different from those that would result from the Born-approximation calculations mentioned above[2,3] and from what is conventionally assumed.[11] In particular, the ℓ-distributions for a given n are remarkably close to statistical (i.e., $\propto (2\ell + 1)$), at least up to

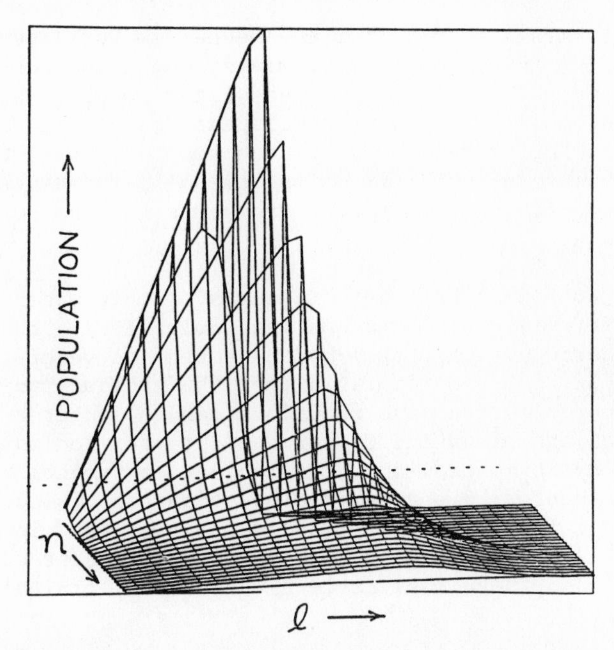

Figure 1. A plot of the population distribution as a function of n and ℓ.

$n \approx 25$; above this n-value, the ℓ distributions tend to cut off significantly below $\ell_{max} \equiv n - 1$.

DISCUSSION AND CONCLUSIONS

The experimentalist must, of course, deal with capture by H_2 molecules rather than H atoms. The uniformity in binding energy will hold just as well for H_2 as H. The ℓ-dependence of the capture _is_ subject to change and the ionization probability will now depend on the orientation of the meson orbit in relation to the p-p axis: $P \equiv P(\vec{r}_o)$. For H_2 there _is_ no critical radius, since the electrons can retreat to orbits around the second proton when the meson approaches the first (the H^- ion is bound by 0.75 eV). However, we still expect the finite velocity effects to be dominant and therefore significant ionization probability for $r_o \gtrsim 1$. Hence, we may reasonably conjecture that the distribution of captured mesons $\mathfrak{f}(B,\ell)$ is not drastically different from H_2. A calculation to verify this, since it would involve three-center wave functions, would be quite difficult.

We wish to point out explicitly that the energies from which capture takes place in our model (and Gershtein's[6]), up to 15 eV, are drastically smaller than the several hundred eV values of Haff and Tombrello, and Korenman and Rogovaya.[3] Our result, capture energy ~ I, comes directly from the mechanism for capture (electron ejection) being the same as the dominant mechanism for the slowing down of the meson. The higher values cited come, we believe, from combining inconsistent slowing and capture models; in effect, the slowing cross sections assumed are too small compared to the capture cross sections. Our predictions for the n- and ℓ-distribution of captures are also quite different from those of the Born approximation calculations.[3,4]

We can hope that the basic features of this model are applicable even for more complex atoms (and molecules): the meson is speeded up during its approach by the potential generated by the polarized atom. At some distance from the nucleus, adiabatic distortion of the electron wave functions can no longer keep up with the motion of the meson and the probability for ionization becomes large. The approximation that the ionization probability depends only on the (vector) distance of closest approach, independent of the asymptotic meson energy (at least for E ~ I and lower), is probably still valid. The problem is then to construct reasonable $P(\bar{r}_0)$ functions; these must reflect the valence orbitals.

REFERENCES

1. For a review of this subject, see L. I. Ponomarev, in Muon Physics, eds. W. V. Hughes and C. S. Wu (Academic Press, 1975) Vol. 3.
2. Baker, G. A., Phys. Rev. 117, 1130 (1960).
3. Haff, P. K. and Tombrello, T. A., Ann. Phys. 86, 178 (1974);
   Korenman, G. Ya. and Rogovaya, S. I., Sov. J. Nucl. Phys. 22, 389 (1976);
   Cherepkov, N. A., Proc. Inst. Symp. on Meson Chem., Dubna, 1977;
   Martin, A. D., Nuovo Cim. 27, 1359 (1963);
   Au-Yang, M. Y. and Cohen, M. L., Phys. Rev. 174, 468 (1963).
4. Leon, M. and Seki, R., Nucl. Phys. A282, 445 (1977).
   Leon, M. and Miller, J. H., ibid., 461 (1977).
   Haff, P. K. et al., Phys. Rev. A10, 1430 (1974).
   Vogel, P. et al., Nucl. Phys. A254, 445 (1975).
5. Gershtein, S. S., J.E.T.P. 12, 815 (1961).
6. Fermi, E. and Teller, E., Phys. Rev. 72, 399 (1947).
7. Wightman, A. S., Thesis, Princeton University, 1949, unpublished
8. Rosenberg, R. L., Philos. Mag. 40, 759 (1949).
9. Baird, T. J., LASL preprint LA-6619-T, unpublished.
10. Leon, M., Phys. Rev. A17, 2112 (1978).
11. Fiorentini, G. and Pitzurra, O., Nuovo Cim. 43A, 396 (1978).

NEW DEVELOPMENTS IN THE STUDY OF MESIC CHEMISTRY

Hubert Schneuwly

Institut de Physique
Université de Fribourg
CH-1700 Fribourg, Switzerland

1. INTRODUCTION

Experiments and calculations seem today to establish that the atomic capture of exotic particles proceeds predominantly through Auger electron ejection and that, in comparison, the radiative atomic capture is negligible. This fact gives a priori the possibility of an influence of the chemical structure of matter in the Coulomb capture mechanism of exotic particles.

A good knowledge of the atomic capture mechanism could open practical applications of the so-called mesic chemistry. Such an interest exists[1] and at most of the meson factories proposals have been submitted for the study of, e.g., material analysis[2], alloys, ionicities of chemical bonds, in vivo diagnostics in tissues, hydrogen bonds, etc., with negative muons or pions. No doubt, one has first to determine the parameters to which the atomic capture is sensitive.

2. THE BASIC PROBLEMS

One of the important problems in the capture of exotic particles (μ, π, K, Σ, $\bar{p}$) is the first bound state. One does not even know if the first bound state is an atomic or a molecular state. Today, one generally assumes that the exotic particle is captured in a very high atomic state with main quantum number of the order of $n = 30$. A calculation shows[3] that under certain assumptions, if one has some distribution over angular momentum states ℓ in such a high capture state, this distribution remains practically unaltered in a lower state, e.g. in n = 14, when the particle cascades down. This means that from measured X-ray intensities it is difficult to deduce the capture distribution over n-states.

In any case, to deduce from measured mesic X-ray intensities the capture distribution over (n,ℓ)-states, one needs a reliable cascade calculation. A recently published computer program[4] seems to fulfil the requirements even if the differences in the calculated intensities compared to the older code[5] in general use are not spectacular for heavy elements. One of the delicate problems in the cascade calculations are the Auger transition probabilities. One can calculate the electron shell refilling times for tightly bound shells[6] and the experiment agrees with the calculations[7]. However, in the early stage of the cascade the outer electron shells play a dominant role and one may expect to find differences in the cascade intensities for a given element in a conducting and an insulating medium. Maybe such effects are not measurable. In gases at low pressures, however, the refilling must be slow and therefore one expects drastic changes at low and high pressures even for the lower transitions in the Lyman series[8].

Another important problem is set by the hydrogen compounds and mixtures. Transfer alters the mesic X-ray intensities and the relative capture rates are affected too. The aspects of pion capture in hydrogen have been extensively treated two years ago in Erice[9]. The processes a muon undergoes in hydrogen isotope mixtures are fascinating[10] and will be treated in a special lecture[11].

The oldest problem is that of the capture rates[12]. It seems therefore that it should also be the simplest one. But even the experimental determination of relative capture rates is not a trivial procedure. Many calculations and models have been proposed to predict relative capture rates, but their successes have remained limited.

3. MESIC X-RAY INTENSITIES

During the deceleration process of a charged particle in matter, electrons are ejected from atoms in the path of this particle. The electrons are preferentially emitted in the direction of the momentum of the incoming particle. These electrons in their turn ionize atoms sitting in front of the incident exotic particle. If the meson is captured while still having a relatively high kinetic energy, the host atom is then one of the highly ionized track and probably itself ionized. With the assumption of the capture proceeding through electron ejection, this capture has to proceed through tightly bound electrons for two reasons: the particle to be captured has a relatively high kinetic energy (far from thermalization) and the outermost electrons have already been ejected. In this case, no effect due to the outermost electrons should be observed neither in the exotic X-ray intensities not in the capture rates. As a consequence, mesic chemistry should not exist.

3.1. Muonic X-ray intensities in sodium

In a recent measurement[13] we have investigated the muonic Lyman series intensities in sodium in $NaNO_2$, $NaNO_3$, Na_2SO_3, Na_2SO_4, Na_2SeO_3, Na_2SeO_4, Na_2S, Na_2Se, NaCl, NaBr and Na(met). Within the experimental uncertainties no difference has been found. This result would confirm the hypothesis that muons are captures at relatively high kinetic energies unless other reasons could be found to explain this behaviour.

In all these compounds, the sodium bond is strongly ionic and metallic sodium, which is a very good conductor, can be considered as sodium ions embedded in an electron gas. For this reason alone one does not expect to find differences in the muonic X-ray intensity pattern.

Whereas the metallic sodium is a good conductor, the sodium compounds are good insulators. One could therefore expect that the refilling times of the outermost electron shells to be different, leading to differences in the Auger transition probabilities for the muon and consequently to changes in the muonic X-ray intensity pattern. No difference is observed, i.e. the electron refilling in insulators must be almost as fast as in the conductor compared to muonic transition probabilities.

One has imagined that the only presence of atoms with different charge numbers Z in a target could affect the energy spectrum of the muons just before capture. As a consequence, one would expect the initial (n,ℓ)-distribution and consequently also the X-ray intensities to be different. But this has not been observed.

3.2. Allotropic effects

If changes in the muonic X-ray intensities were observed between different allotropes of an element or a compound, one would be free from some other hypotheses about the origin of such differences and could assert that they can only be due to the structure of the valence electron shells, i.e. to differences in the bondings, what we call "chemical effects". Unfortunately, no difference has been observed in the muonic X-ray intensities between diamond and graphite structures of, e.g., carbon[14] even if the difference in the interatomic distance is of the order of 10%.

In a recent measurement of the muonic X-ray intensities in three modifications of phosphorus[15], the Lyman series showed clear systematic differences. One might imagine a correlation of this behaviour to atomic radii or internuclear distances similar to that shown for kaonic X-rays[16]. The structure of white phosphorus, e.g., is supposed to be that of a solidification vapor, a stochastic conglomerate of P_4 tetrahedrons of 2.21 Å length of side[17]. The shortest interatomic distance is 2.24 Å in red phosphorus[18] and 2.17 Å in the black modification[19]. It is obvious that a correlation to atomic radii or interatomic distances is problematic.

4. HYDROGEN COMPOUNDS

From the point of view of chemical effects in meson capture, the hydrogen compounds are the most interesting ones. The capture in hydrogen can indeed only proceed through its electron which is always a bonding electron. But the measurement of the hydrogen mesic X-ray cascade in compounds is, because of the low energies of the transitions, very difficult and practically impossible today. Even the effect of a hydrogen bonding on another element is hard to measure, because a meson in an atomic orbit of hydrogen is easily transferred to an atom of higher charge number modifying there the cascade intensities and therefore also the apparent capture rate[9,20,21].

4.1. Hydrogen isotope effects

The chemical bond of hydrogen and deuterium to another element Z are practically identical. Thus one does not expect direct

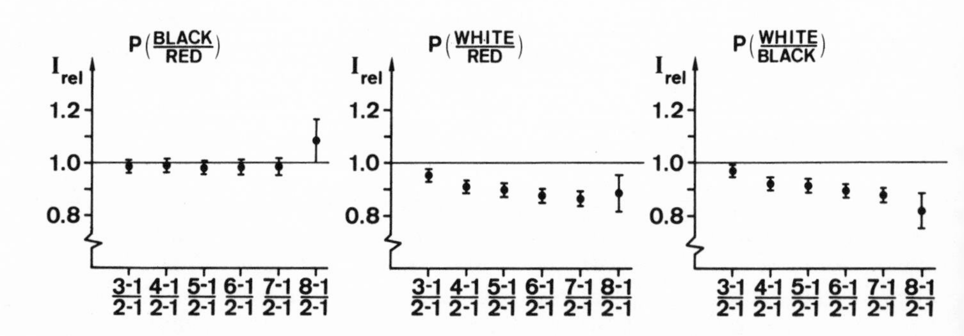

Fig. 1: Relative muonic X-ray intensity ratios of phosphorus in three allotropic modifications

chemical effects on the mesic X-rays of the element Z. If differences are observed in Z when bound in Z_nH_m or Z_nD_m, such differences should mainly be due to differences in the transfer mechanism[21]. However, according to a recent calculation[22] an isotopic effect even in the mesic atom formation in hydrogen isotope mixtures is expected.

Observations of hydrogen isotope effects in muonic X-ray intensities in water have been reported recently by Mausner et al.[23] and Bergmann et al.[24]. Unfortunately the experimental data which should be comparable do not match. However, this discrepancy has now been removed[24]. The isotopic effect[23] is explained using a model without muon transfer, which considers the effect of the hydrogen on the energy spectrum in the target and, hence, the angular momentum distribution of the captured muon[26].

In a recent experiment, Wiegand et al.[27] have investigated the hydrogen isotope effect on pionic Balmer intensities in oxygen in H_2O and D_2O and carbon (C_6H_6, C_6D_6). However, the statistical uncertainty is too large to be affirmative even if a tendency seems to exist.

4.2. Hydrogen effects

In hydrogen compounds, the initial (n,ℓ)-distribution of muons in an element Z of the compound is not only influenced by

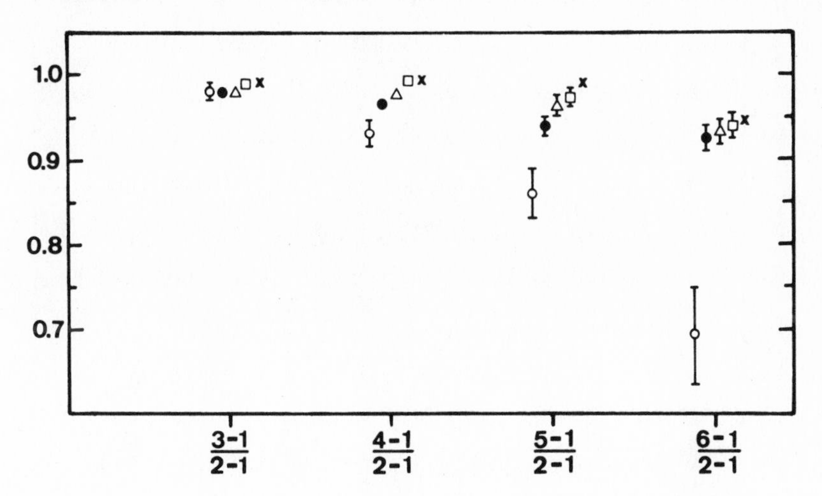

Fig. 2: Relative muonic X-ray intensity ratios of oxygen in water: o D_2O/H_2O (ref. 23), • D_2O/H_2O, Δ $\frac{1}{4}H_2O + \frac{3}{4}D_2O/H_2O$, □ $\frac{1}{2}H_2O + \frac{1}{2}D_2O/H_2O$, x $\frac{3}{4}H_2O + \frac{1}{4}D_2O/H_2O$ (ref. 24)

the chemical bond, but also by the transfer. One expects very light elements to be especially sensitive to such effects.

In a recent measurement, Cox et al.[28] investigated muonic X-ray intensities in low Z elements and their hydrides (Fig. 3). In beryllium, one observes the greatest chemical effect on muonic X-ray intensities ever seen. The (3-1)/(2-1) intensity ratio in BeH_2 is about 60% higher than in Be(met). Using the pion capture ratio H_2/Be measured in BeH_2[29], one can show that this factor cannot be due to transfer alone.

Measurements of muonic X-rays have also been performed in solutions, e.g. in sodium chlorine in a 4M solution and compared to crystalline sodium chlorine[30]. The intensity pattern of the negative aqueous chlorine ion exhibits enhanced intensity of the higher Lyman members in comparison to chlorine from crystalline sodium chlorine. To account for this intensity variation, one notes that the hydrogen atoms of the dipolar water molecules are oriented towards the chlorine atom. Transfer is therefore more probable to the chlorine than to the sodium atom. Actually one may imagine that the hydration of the chlorine ion produces a different valence electron distribution. With a bit of luck

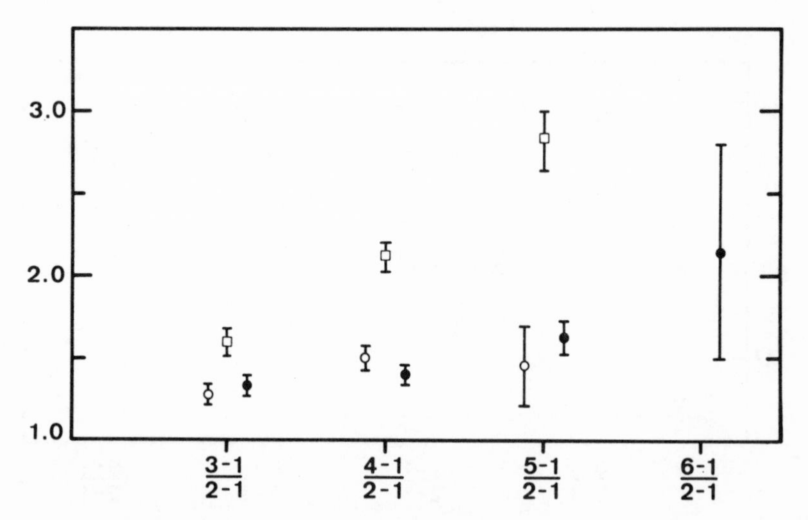

Fig. 3: Relative muonic X-ray intensity ratios of low Z elements in metallic form and their hydrides[28]: o LiH/Li, □ BeH_2/Be, ● $B_{10}H_{14}$/B

such an effect might already be observed in compounds containing water of crystallization, where the capture rate in hydrogen would be so low that the transfer should be negligible.

According to a recent calculation by Bracci and Fiorentini[31], the transfer might also be used to study chemical structures. The transfer cross section from the metastable 2s state depends strongly on the nature of the bond. They have found that

$$\sigma^{-} < \sigma^{cov} < \sigma^{is} < \sigma^{+}$$

where the σ are the transfer cross sections to the isolated element, the element with a pure covalent bond and the element as a positive or negative ion. The differences between two consecutive cross sections are of the order of 20%.

Another calculation shows that a low energy μp atom, e.g. formed in liquid hydrogen, cannot penetrate more than a few monolayers into a solid surface[32]. The muonic X-rays from atom Z subsequent to transfer must therefore be highly surface-correlated. The transfer reaction might be of interest from the point of view of surface analyses.

5. RELATIVE CAPTURE RATES OF MUONS

In 1966, one had the feeling that the atomic capture probabilities into metals in oxides show a correlation to their position in the Periodic Table[33]. Since then a lot of new capture ratios in oxides have been published[14,34] from which the periodicity becomes less clear. Coulomb capture ratios in fluorides[35] show perhaps periodic variations with the atomic number of the fluorinated element. If the muon capture mechanism is sensitive to chemical structures, the capture rates should depend on the chemical properties of the elements. As a consequence, capture rates should show something like a periodicity.

There is still a lack of systematic studies of the dependence of the relative capture rates upon the chemical structure. It is not clear to which degree the chemical bonding is nothing more than a perturbation to the main mechanism of Coulomb capture of muons. Presently most of the recently published calculations consider its influence only as a perturbation to the main process[3,36,39]. Others give more weight to the chemical bond[40]. In addition, although one supposes the chemical effects to become smaller with increasing charge number of the element, this has not yet been systematically confirmed.

For these purposes, we have investigated the relative atomic capture of muons in nitrogen, sulphur and selenium[41] in the compounds $NaNO_2$, $NaNO_3$, Na_2SO_3, Na_2SO_4, Na_2SeO_3 and Na_2SeO_4. The three elements have in the different compounds different formal valence states: N^{3+} and N^{5+}, S^{4+} and S^{6+}, Se^{4+} and Se^{6+}. The other elements of the compounds - sodium and oxygen - have in all cases the same formal valence states. In addition, the three investigated elements N, S and Se are in all compounds surrounded by the same elements so that their influence, e.g. the atomic size, the electron binding energies and other parameters which might influence the muon deceleration, is approximately the same for all compounds.

The experiment[41] has shown that an element in a higher positive valence state captures less muons than in a lower state. This is just the opposite of what one sometimes believed earlier and demonstrates that the capture rate is closely connected to the electron density in the vicinity of the atom. The chemical effect seen in the nitrogen compounds and which amounts to roughly 20% in the capture ratios is only a difference arising from a difference in the chemical structures. The role played by the chemical structure itself in the capture rates must therefore obviously be much more important and perhaps dominate the capture mechanism in light elements.

In view of these results, one might be interested to search for correlations of the capture rates to spatial valence electron distributions. The ionicity of a bond gives an indication for such a distribution expressed by a singly number. This is a pleasant property but the agreement between measured and calculated[42] ionicities is not always satisfactory. In addition, measurements are only possible for some particular molecules.

Ionic charges of elements in compounds are attractive quantities for such a correlation. Such charges are calculated theoretically or semi-empirically, e.g. from ESCA (Electron Scattering for Chemical Analyses). Depending on the definition of this charge q (q = 0 for the neutral atom) and the methods of calculation, this charge parameter has different values for a given element in a given compound. For sulphur in Na_2SO_4 one finds, e.g., q(eff) = 0.45 (ref. 43), q(eff) = 1.79 (ref. 44), q(P) = 1.12 (ref. 45) and q(HFS) = 0.746 (ref. 46). For the same element in Na_2SO_3 the same authors find: q(eff) = 0.30, q(eff) = 0.93, q(P) = 0.68 and q(HFS) = 0.586. The differences in the q-values are too large to try a direct correlation to the muon capture rates, but the relative differences of the charge parameters between the two compounds are on average the order of $\Delta q/q = 45\%$.

For nitrogen in the compounds $NaNO_2$ and $NaNO_3$, the mean relative difference in the charge parameters determined from q(EH) and q(CNDO) gives[47]: $\Delta q/q$ = 95%. For selenium in Na_2SeO_3 and Na_2SeO_4 one finds[46] $\Delta q/q$ = 13%.

Of course, one cannot pretend that a correlation between relative differences in the charge parameters and the muon capture ratios has been established (Fig. 4). Nevertheless, it is encouraging to see that these two quantities are not strangers for each other.

6. A MODEL FOR THE CAPTURE MECHANISM

One of the consequences of the experimental results is that the chemical structure cannot be treated as a perturbation in the capture mechanism and that the capture proceeds through bound electrons. A recently proposed model predicting atomic capture ratios in mixtures of atoms and in molecules takes these experimental facts into account[9,48]. Considering its simplicity, its success is encouraging (Fig. 5).

Hydrogen compounds provide a crucial test for all models and calculations. Unfortunately there exist no experimental data on muon capture ratios for hydrogen compounds. On the other hand, a considerable number of pion capture rates are available (cf. e.g. ref. 9). If one assumes that there is almost no difference in the atomic capture mechanism between muons and pions, one can simply use the same formula. For the relative capture in hydrogen in hydrocarbons this formula is:

$$\frac{W(H_n)}{W(C_mH_n)} = \frac{2(1 - \sigma)}{(4\frac{m}{n} + 1)(37 + 35\,\sigma)}$$

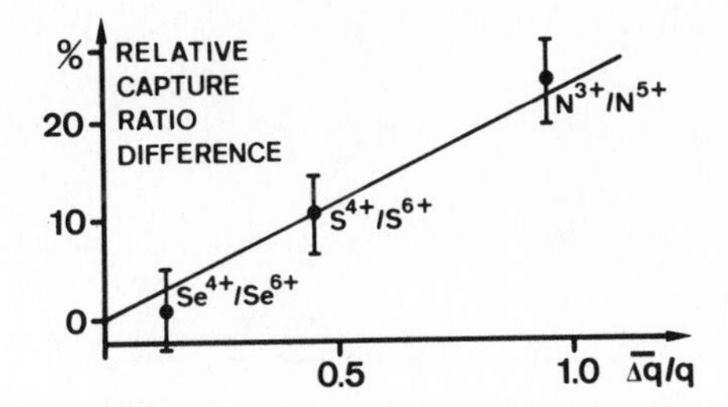

Fig. 4: Correlation between relative charge differences $\Delta q/q$ and relative capture probabilities in nitrogen in $NaNO_2/NaNO_3$, in sulphur in Na_2SO_3/Na_2SO_4 and Se in Na_2SeO_3/Na_2SeO_4

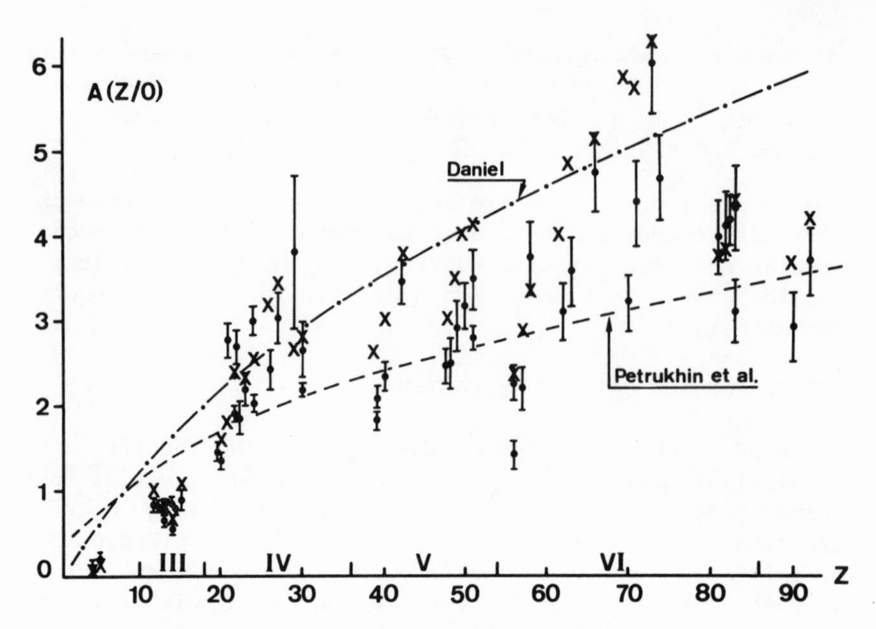

Fig. 5: Muonic per atom capture ratios A(Z/O) as a function of Z in oxides Z_mO_n. The dotted-dashed line corresponds to Daniel's formula[37], the dotted line to that of Petrukhin and Suvorov[38] and the crosses to the model calculations[9,48] in its simpler version

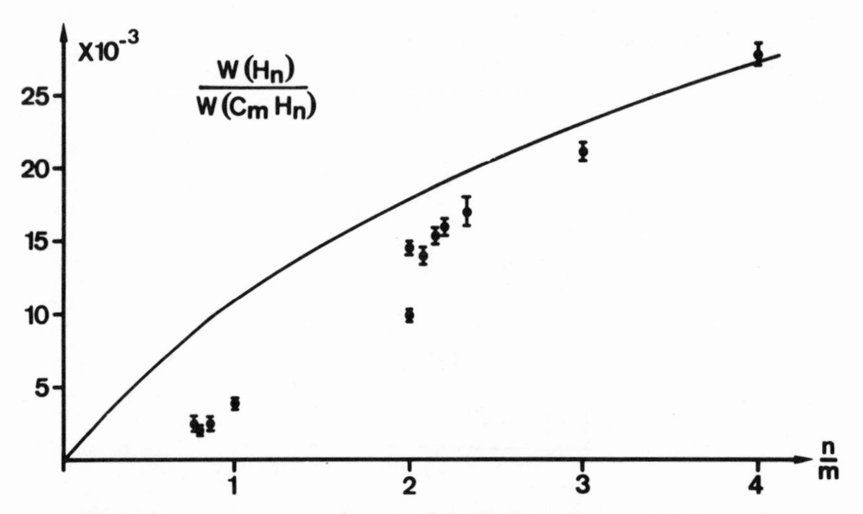

Fig. 6: Relative pion capture probabilities in hydrogen in hydrocarbons. The solid line represents the predictions of the model[9,48] assuming pure covalent C-H bonds.

where σ is the ionicity of the C-H bonds. Fig. 6 shows a comparison of our predictions with measured pion capture probabilities. There is a good agreement for CH_4. When going to lower hydrogen concentrations, i.e. for decreasing n/m values, the discrepancy increases. One knows that the C-H bonds are not identical in all hydrocarbons, but the pion transfer is also expected to increase with increasing carbon concentration. However, it has not yet been shown whether the transfer alone explains the discrepancy.

7. CONCLUSIONS

Even if one does not yet have a theory able to predict atomic capture probabilities and initial distributions, a lot of problems have already be solved. Further systematic experimental studies are necessary to shed light on the capture mechanism of negative particles in atoms and compounds.

I would like to thank Paul Sharman for a critical reading of the manuscript.

REFERENCES

1. *Mesons in Matter*, Proc. Int. Symp. on Meson Chemistry and Mesomolecular Processes in Matter, Dubna (USSR), June 7-10, 1977, Dubna Report D1,2,14-10908 (1977)
2. J.J. Reidy, R.L. Hutson, H. Daniel and K. Springer, Anal. Chem. 50 (1978) 40
3. M. Leon and R. Seki, Nucl. Phys. A282 (1977) 445
   M. Leon and J.H. Miller, Nucl. Phys. A282 (1977) 461
4. V.R. Akylas and P. Vogel, Comp. Phys. Comm. 15 (1978) 291
5. J. Hüfner, Z. Physik 195 (1966) 365
6. P. Vogel. Phys. Rev. A7 (1973) 63
   P. Vogel, Phys. Lett. 58B (1975) 52
7. W.D. Fromm, Dz. Gansorig, T. Krogulski, H.G. Ortlepp, S.M. Polikanov, B.M. Sabirov, U. Schmidt, R. Arlt, R. Engfer and H. Schneuwly, Phys. Lett. 55B (1975) 377
   J.L. Vuilleumier, W. Dey, R. Engfer, H. Schneuwly, H.K. Walter and A. Zehnder, Z. Physik 278 (1976) 109
   P. Vogel, A. Zehnder, A.L. Carter, M.S. Dixit, E.P. Hincks, D. Kessler, J.S. Wadden, C.K. Hargrove, R.J. McKee, H. Mes and H.L. Anderon, Phys. Rev. A15 (1977) 76
8. J.D. Knight, LAMPF Newsletter 11 (March 1979) 83
   C.K. Hargrove, private communication
9. H. Schneuwly, in *Exotic Atoms*, ed. by G. Fiorentini and G. Torelli, INFN Press (1977), p. 255

10. S.S. Gershtein and L.I. Ponomarev, Phys. Lett. 72B (1978) 80
11. J. Rafelski, contribution to this volume
12. E. Fermi and E. Teller, Phys. Rev. 72 (1947) 399
13. K. Kaeser, B. Robert-Tissot, L.A. Schaller, L. Schellenberg and H. Schneuwly, submitted to Helv. Phys. Acta
14. J.D. Knight, C.J. Orth, M.E. Schillaci, R.A. Naumann, H. Daniel, K. Springer and H.B. Knowles, Phys. Rev. A13 (1976) 43
15. K. Kaeser, T. Dubler, B. Robert-Tissot, L.A. Schaller, L. Schellenberg and H. Schneuwly, Helv. Phys. Acta (in press)
16. C.E. Wiegand and G.L. Godfrey, Phys. Rev. A9 (1974) 2282
    G.T. Condo, Phys. Rev. Lett. 33 (1974) 126
    G.L. Godfrey and C.E. Wiegand, Phys. Lett. 56B (1975) 255
17. H. Krebs, Z. Anorg. Allg. Chemie 266 (1951) 175
18. H. Krebs and H.U. Gruber, Z. Naturforsch. 22a (1967) 96
19. M.L. Huggins, J. Am. Chem. Soc. 75 (1953) 4126
20. A. Bertin, contribution to this volume
21. A. Vitale, contribution to this volume
22. G.Ya. Korenman and S.I. Rogovaya, preprint no. 9(208), Moscow (1978)
23. L.F. Mausner, J.D. Knight, C.J. Orth, M.E. Schillaci and R.A. Naumann, Phys. Rev. Lett. 38 (1977) 953
24. R. Bergmann, H. Daniel, F.J. Hartmann, J.J. Reidy and W. Wilhelm, p. 156 of ref. 1
25. J.D. Knight, private communication
26. H. Daniel, private communication
27. C.E. Wiegand, G.K. Lum and G.L. Godfrey, Phys. Rev. A15 (1977) 1780
28. C.R. Cox, G.W. Dodson, M. Eckhause, R.D. Hart, J.R. Kane, A.M. Rushton, R.T. Siegel, R.E. Welsh, A.L. Carter, M.S. Dixit, E.P. Hincks, C.K. Hargrove, H. Mes, Can. J. Phys. (in press)
29. Z.V. Krumshtein, V.I. Petrukhin, L.I. Ponomarev and Yu.D. Prokoshkin, Sov. Phys. JETP 27 (1968) 906
30. R.A. Naumann, J.D. Knight, L.F. Mausner, C.J. Orth, M.E. Schillaci and G. Schmidt, p. 110 of ref. 1
31. L. Bracci and G. Fiorentini, preprint IFUO/TH12, Pisa (1978)
32. P.K. Haff, preprint Yale-3074-421
33. V.G. Zinov, A.D. Konin and A.I. Mukhin, Sov. J. Nucl. Phys. 2 (1966) 613
34. H. Daniel, W. Denk, F.J. Hartmann, J.J. Reidy and W. Wilhelm, Phys. Lett. 71B (1977) 60
35. W. Wilhelm, R. Bergmann, G. Fottner, F.J. Hartmann, J.J. Reidy and H. Daniel, Chem. Phys. Lett. 55 (1978) 478
36. P. Vogel, P.K. Haff, V. Akylas and A. Winther, Nucl. Phys. A254 (1975) 445
37. H. Daniel, Phys. Rev. Lett. 35 (1975) 1649

38. V.I. Petrukhin and V.M. Suvurov, Sov. Phys. JETP 43 (1976) 595
39. H. Daniel, Z. Physik (to be published)
40. S.S. Gershtein, V.I. Petrukhin, L.I. Ponomarev and Yu.D. Prokoshkin, Sov. Phys. Uspekhi 12 (1969) 1
M.Y. Au-Yang and M.L. Cohen, Phys. Rev. 174 (1969) 468
L.I. Ponomarev, Annu. Rev. Nucl. Sci. 23 (1973) 395
P. Vogel, A. Winther and V. Akylas, Phys. Lett. 70B (1977) 39
S.S. Gershtein and L.I. Ponomarev, p. 17 of ref. 1
41. H. Schneuwly, T. Dubler, K. Kaeser, B. Robert-Tissot, L.A. Schaller and L. Schellenberg, Phys. Lett. 66A (1978) 188
42. L. Pauling, *The Nature of the Chemical Bond and the Structure of Molecules and Crystals,* Cornell University Press (1960)
43. I. Lindgren, Proc. Int. Symp. Roentgenstrahlen und Chem. Bindung, Peipzig (1965)
44. R. Manne, J. Chem. Phys. 46 (1967) 4645
45. B.J. Lindberg, K. Hamrin, G. Johansson, V. Gelius, A. Fahlmann, C. Nordling and K. Siegbahn, Physica Scripta 1 (1970) 286
46. W.E. Swartz, K.J. Wynne and D.M. Hercules, Anal. Chem. 43 (1971) 1884
47. D.N. Hendrickson, J.M. Hollander and W.L. Jolly, Inorg. Chem. 8 (1969) 2642
48. H. Schneuwly, p. 86 of ref. 1
H. Schneuwly, V.I. Pokrovsky and L.I. Ponomarev, Nucl. Phys. A312 (1978) 419

ELASTIC SCATTERING OF MUONIC HYDROGEN ATOMS AGAINST PROTONS: STATUS OF EXPERIMENTS

A. Bertin

Istituto di Fisica dell' Università di Bologna, and

Istituto Nazionale di Fisica Nucleare, Bologna, Italy

INTRODUCTION

Let us first recall that, when a negative muon has been slowed down to sufficiently low energy in a hydrogen target, it undergoes the atomic capture process by a hydrogen atom, which leaves the muon in an excited state of the newly formed μp system. Due to the fact that the muon interaction with the nucleus is weak, this system is allowed to attain the 1S ground state through different processes (radiative transitions, Auger effects, Stark-mixing collisions), which altogether are referred to as the atomic cascade, and mostly take place within a very short time ($\simeq 10^{-10}$ sec) with respect to the free muon lifetime. When they have de-excited to their 1S level, the electrically neutral μp atoms have exceptionally small atomic dimensions ($a_\mu = 2.56\times10^{-11}$ cm), are endowed with a kinetic energy E_0 (with which they begin to diffuse throughout the surrounding medium) and are initially formed in a statistical mixture of the hyperfine structure states (triplet and singlet, separated by an energy gap $\Delta E = 0.183$ eV).[1]

The present talk concerns the status of experimental results on the elastic scattering cross section of muonic hydrogen against protons, following the reaction

$$\mu p + p \rightarrow \mu p + p \qquad (1)$$

which the μp systems in the 1s state undergo against the surrounding hydrogen atoms. One may underline three reasons of interest in studying this process: (i) to get an effective comparison between the results of experiment and the theoretical predictions[2] on the problem of three bodies interacting through the Coulomb's law; (ii) to get information on E_o, a parameter to which (at least to some extent the features of process (1) can be related: E_o is an important quantity to establish for the feasibility of precision experiments on the QED effects on the levels of muonic hydrogen;[1] and (iii) to establish the relative population of the triplet and singlet states of the μp atoms formed in given experimental conditions, as is required for the interpretation of the muon nuclear capture experiments in hydrogen targets at low density.[3]

To clarify point (iii) some more, one would add that process (1 is also responsible, through the incoherent scattering channel

$$\mu p(1) + p \rightarrow \mu p(0) + p \qquad (2)$$

(the indexes 1 and 0 labeling triplet and singlet states respectivel of the depopulation of the triplet states in favour of the lower-lying singlet ones; and that such transitions become irreversible, onc the μp systems have been slowed down (through the diffusion process) to kinetic energies sufficiently smaller than the hyperfine separation ΔE. Regardless of the calculation technique, moreover, the theo retical expectations[2] agree to indicate that the cross section σ(1,0) (which is effective when the μp atoms are in the statistical mixtu-re of triplet and singlet states) is much larger than the cros section σ(0), which holds when the μp atoms have irreversibly attai ned the singlet state; and that the cross section σ(1→0) for proces (2) is of the same order as σ(1,0) (see Table I). One may also not that the various theoretical approaches foresee that the different[4-9] spin-dependent cross sections involved in process (1) can all be expressed in terms of common parameters (such as e.g. the scattering lengths for the states of different symmetries with respect to the interchange of the two nuclei). Once these parameters have been determined, the value of all the competing cross sections can be calcu lated, as well as the relative population of the hyperfine structure levels and their evolution in time in a given experimental condition

μp ATOM SCATTERING EXPERIMENTS IN THE SIXTIES

The results of the first experiments performed on process (1) a also summarized in Table I. The measurements were carried out by dif ferent techniques, namely within a hydrogen diffusion chamber (Dubna where known admixtures of alcoholic impurities were present, and b a counter technique (Bologna-CERN),[5] using a target of ultrapure gaseous hydrogen. The most direct results obtained from these measure-

TABLE I.- Summary of early experimental (E) and updated theoretical (Th) results on the scattering cross sections for the process

$$\mu p + p \rightarrow \mu p + p$$

Authors	Year	Reference	Cross-section symbol $\sigma(0)$	$\sigma(1,0)$	$\sigma(1\rightarrow 0)$
Dubna	1965	4 (E)	167 ± 30	24 ± 5.0	16.8 ± 3.0
Bologna-CERN	1967	5 (E)	7.6 ± 0.7	4.1 ± 0.5 8.8 ± 0.7	4.0 ± 0.5 8.7 ± 0.7
Zel'dovich and Gershtein	1959	6 (Th)	1.2	7.9	7.8
Cohen et al.	1960	7 (Th)	8.2	4.0	3.9
Matveenko and Ponomarev	1970	8 (Th)	2.5	4.5	4.45
Ponomarev et al.	1978	2 (Th)	35	27	27
Units (cm^2)			10^{-21}	10^{-19}	10^{-19}

ments are those reported for $\sigma(0)$: it is seen that the two relevant experimental values are affected by a very significant discrepancy. Nevertheless, within the theoretical frame available at the date of these experiments, one evaluates that: (a) if one trusts the result of the Bologna-CERN group, the μp atoms formed in a gaseous target at a pressure of about 10 atm are all landing on the 1S singlet state within a very short time (≃ 100 ns) following their formation; and (b) given the higher value for $\sigma(1\rightarrow 0)$ obtained from the Dubna group, this time would be even smaller if the result of the diffusion chamber experiment were assumed as the good one.

On the other hand, a yet unattained ambitious goal of low-energy muon physics is to measure the muon nuclear capture rate in hydrogen, starting from the triplet state of the muon-proton system, where the induced pseudoscalar term contribution to the interaction coupling is dominant;[3] so that one is also interested in identifying density conditions sufficiently low in order that the depolarization from the triplet to the singlet state through process (2) may occur at a reduced rate. In this case, then, assuming either one of the quoted experimental results would yield significantly different predictions. New measurements are apparently needed, with the first purpose of clarifying the discrepancy between the existing results. This need is confirmed by a renewed theoretical analysis by Ponomarev et al.,[2] (see again Table I), which, besides foreseeing for $\sigma(0)$ a somewhat larger value than any preceeding theoretical estimate, provides $\sigma(1\rightarrow 0)$ values which would result in a faster depletion of the triplet states than previously calculated.

THE NEW EXPERIMENT

A new experiment on process (1) has recently been performed by a Bologna group at the muon channel of the CERN SC2 600 MeV Synchrocyclotron; the results are being published.[10-13] The measurements were carried out stopping negative muons in ultrapure gaseous hydrogen. On the lines of the previous counter experiment, the cross section value for reaction (1) was obtained by analysing the time distribution of the delays with which the hydrogen muonic atoms formed in the gas attained one of a series of regularly spaced foils, placed within the hydrogen container itself. The time at which one μp atom attained one of the foils (which were made of aluminium in the present case) was determined by looking at the K-series of the muonic X-radiation which was promptly released following the muon transfer reaction

$$\mu p + Y \rightarrow (\mu Y)^{*} + p \qquad (3)$$
$$\qquad\qquad \hookrightarrow Y(1S) + X$$

to an atom Y of aluminium or oxygen, belonging to the thin natural layer of Al_2O_3 present on the aluminium foils. With respect to the previous Bologna-CERN experiment, the following advantages were introduced, and proved to be significant: (i)the data recording (carried out event-per-event) allowed to single out the prompt events from the delayed ones, in such a way that it was possible to reconstruct the energy spectrum of the accepted X-rays; and to verify that the events collected were indeed due to the transfer of muons to the aluminium foils;[12] (ii)the muon decay electrons coming in sequence to the X-ray signals were also detected, in such a way that the reliability of the event selection was also checked through the observation of the lifetimes of the μAl and μO atoms;[13] and (iii) a direct measurement of the purity level of the hydrogen gas before and after each run was carried out by a gas chromatographic technique (the residual level of accepted impurities was always below the sensitivity of the gas chromatograph, namely 1-2 p.p.m.).

The analysis of the data performed so far concerns the measurements performed at 26 atm of hydrogen pressure, and was carried out

TABLE II.- Summary of direct experimental results on the scattering cross section σ of μp muonic atoms against protons.

Authors	Reference	Hydrogen pressure (atm)	σ [a] (units of 10^{-21} cm^2)
Dubna	4	up to 23	173 ± 19
Bologna-CERN	5	26.2	6.5 ± 1.0
Bologna (1978)	12	26	14 ± 2

[a] Effective value of the cross section, obtained by analysing the data disregarding possible transitions between the hyperfine structure states of the μp system.

under the simplifying assumption that process (1) is governed by a single effective cross section (that is disregarding possible transitions between the triplet and singlet states of the μp atom). The preliminary result obtained in this way is compared in Table II to the corresponding values obtained in the previous experiments. In connection to the cross section value reported in the last row of the Table, the analysis provided for the initial kinetic energy E_o[11]

$$E_o = (0.11 \pm 0.03) \text{ eV} , \qquad (4)$$

a value which, due to the simplified procedure followed, should be considered as a very approximate information (possibly a lower limit) on the true initial energy of the μp atom. In any case, the value (4) is larger than the thermal energy at room temperature (0.038 eV). One should note that the central value reported in the Table for the cross section σ turns out from the data analysis to be quite independent on the choice of the initial energy, within a range which goes from the thermal value up to about 1 eV. The results are presently being analysed in more detail, allowing for the different (coherent and incoherent) scattering processes which compete within the general reaction (1), with the purpose of getting further indications both on the different cross sections involved (see Table I) and on the true value of E_o. With the same purposes, the same group has taken data at lower hydrogen pressures: also these data are under analysis.

CONCLUSIONS

If one compares the result obtained by the Bologna group to the previous ones reported in Table II, the following concluding remarks can be expressed:

(i)although the recent result is not in complete agreement with the one obtained in the previous Bologna-CERN experiment, it is far off the value obtained by the Dubna group.

(ii)orientatively, (see Table I) the present result would correspond to larger cross sections for process (2) than those derived leaning on the data of the previous Bologna-CERN experiment. This would lead to expect that the depletion of the triplet states for μp atoms formed in gaseous hydrogen occur more rapidly than it was hoped;[5,9] which would also be in support of the recent theoretical predictions.

As a final point, one might mention that in deuterium targets where negative muons are stopped, a scattering process of the same type as reaction (1),

$$\mu d + d \rightarrow \mu d + d \qquad (5)$$

occurs between μd muonic deuterium atoms and deuterons. Opposite to the case of hydrogen, the values of the cross section for process (5 are quite similar when this takes place in the initial statistical mixture of states of hyperfine structure (quartet and doublet) and in the lower-lying doublet state. Furthermore, the quartet-to-double transition is not fully irreversible at room temperature, since the hyperfine structure separation energy (0.049 eV) is in this case qui te comparable to the thermal energy (0.038 eV) . Nevertheless, measurements on the cross section for process (5) have an independent reason of interest, which lies in the fact that the theory (which provides fairly spin-independent predictions in all theoretical approaches)[1,2]seems in this case better established.

Experimental results on the cross section for process (5) are also discrepant, although in a less showy way with respect to those on process (1). The Bologna group has also taken data on this scattering reaction. The analysis is in progress.

ACKNOWLEDGEMENTS

The author is indebted to I. Massa, M. Piccinini, G. Vannini an A. Vitale for fruitful discussions on the subjects of the present report.

REFERENCES

1. See e.g. A. Bertin, A. Vitale, and A. Placci, La Rivista del Nuovo Cimento 5:423 (1975); and references therein.

2. L.I. Ponomarev, L.N. Romov, and M.P.Faifman, JINR preprint P4-11446 (1978); and references therein.
3. See e.g. A. Vitale, A. Bertin, and G. Carboni, Phys. Rev. 11D: 2441 (1975); and references therein.
4. V.P. Dzhelepov, P.F. Ermolov, and V.V. Fil'chenkov, Zurn. Eksp. Teor. Fiz. 49:393 (1965)(English translation: Sov. Phys. JETP 22:275 (1966)).
5. A. Alberigi Quaranta, A. Bertin, G. Matone, F. Palmonari, A. Placci, P. Dalpiaz, G. Torelli, and E. Zavattini, Nuovo Cimento 47 B: 72 (1967).
6. Ya.B. Zel'dovich and S.S. Gershtein, Usp. Fiz. Nauk 71:581 (1960) (English translation: Sov. Phys. Usp. 3:593 (1961).
7. S. Cohen, D.L. Judd, and R.J. Riddell, Phys. Rev. 119:384 (1960); and Phys. Rev. 119:397 (1960).
8. A.V. Matveenko and L.I. Ponomarev, Zurn. Eksp. Teor. Fiz. 58:1640 (1970) (English translation: Sov. Phys. JETP 31:880 (1970). A.V. Matveenko and L.I. Ponomarev, Zurn. Eksp. Teor. Fiz. 59:1593 (1970)(English translation: Sov. Phys. JETP 32:871 (1971)).
9. See also G. Matone, Lett. Nuovo Cimento 2:151 (1971).
10. A. Bertin,I. Massa, M. Piccinini, G. Vannini, and A. Vitale, Lett. Nuovo Cimento 21:577 (1978).
11. A. Bertin, I. Massa, M. Piccinini, A. Vacchi, G. Vannini, and A. Vitale, Phys. Lett. B78:355 (1978).
12. A. Bertin, F. Ferrari, I. Massa, M. Piccinini, G. Vannini, and A. Vitale, Phys. Lett. A68:201 (1978).
13. A. Bertin, I. Massa, M. Piccinini, G. Vannini, and A. Vitale, Lett. al Nuovo Cimento 23:401 (1978).

MUON TRANSFER PROCESSES FROM FREE MUONIC HYDROGEN AND DEUTERIUM: RECENT EXPERIMENTAL RESULTS

A. Vitale

Istituto di Fisica dell' Università di Bologna, and

Istituto Nazionale di Fisica Nucleare, Bologna, Italy

INTRODUCTION

One of the most important reactions occurring when negative muons are slowed down in a hydrogen (deuterium) target, contaminated by an even small admixture of an extraneous element Y, is the process of muon transfer from the primary μp (μd) atoms to a Y nucleus, following the reactions

$$\mu p + Y \rightarrow (\mu Y)^{*} + p \qquad (1)$$
$$\quad \mid\!\rightarrow (\mu Y(1S)) + X(p)$$

$$\mu d + Y \rightarrow (\mu Y)^{*} + d \qquad (2)$$
$$\quad \mid\!\rightarrow (\mu Y(1S)) + X(d)$$

where X is the muonic X-radiation characteristic of the element Y, which is promptly released in the de-excitation process of the μY atom formed in this way. Processes (1) and (2) represent an alternative procedure to produce μY muonic atoms, with respect to the ordinary way, i.e. stopping negative muons directly within a Y target,

$$\mu^{-} + Y \rightarrow (\mu Y)^{*} + e^{-} \qquad (3)$$
$$\quad \mid\!\rightarrow (\mu Y(1S)) + X(dt).$$

Concerning reactions (1) and (2), the following questions can a priori be advanced:

i) which is the initial state of the μp (μd) atom from which the transfer reaction occurs?

ii) Which is the rate at which both reaction proceeds for a given Y element?

iii) If a μY atom is formed , is there any particular feature which keeps memory of the formation channel?

iv) If the reactions (1) and (2) occur to a nucleus Y which is bound in a compound molecule, which is the capture ratio of the transferred muons between the different nuclei which constitute the compound itself?

All of these questions have been at least partially answered in the past, both on the theoretical side[1] and through experiments.[2] For the purpose of being short, one may summarize the results of first-generation experiments in the following:

a) Starting from $\mu p(1S)$ and $\mu d(1S)$ ground states, reactions (1) and (2) occur at large rates, which increase as a function of the atomic number Z of the element Y. The general trend of these observations is in agreement with the theoretical expectations.[3]

b) The rate of process (2) is systematically smaller than the corresponding rate for process (1) for a given element Y, with Z larger than 10. This fact is also theoretically understood in terms of the dependence of the rates on the mass of the initial μp (μd) atom.

Less systematical experimental observations, instead, are presently available on points i), iii), iv). These questions, however, are indeed to be clarified in order to get a full understanding on the muon transfer mechanism. I shall try here to recall some recent results of experiments, which provide useful information in different directions in the field of low-energy muon physics.

MUON TRANSFER FROM EXCITED STATES OF MUONIC HYDROGEN AND DEUTERIUM

The knowledge of the initial atomic state from which the transfer reactions (1) and (2) proceed is an essential feature of these processes. However, since the atomic cascade process of an excited (μp) or (μd) atom to its 1S ground level is in general very fast ($\simeq 10^{-10}$ sec), and anomalously long-living states (such as e.g. the 2S metastable state) are formed in a few per cent of the cases only,[4] to identify any particular process occurring from excited levels of muonic hydrogen and deuterium represents in practice a quite difficult experimental objective.

Nevertheless, in a theoretical work where estimates of the transfer cross sections from the $\mu p(2S)$ state to heavy atoms are obtained in semi-classical approximation, Fiorentini and Torelli[5] have shown that in the case of helium (where the transfer process from the $\mu p(1S)$ system is strongly suppressed) the ratio of the relevant rates (λ) is given by

$$\lambda(\mu p(2S),He) \approx 10^5\ \lambda(\mu p(1S),He). \qquad (4)$$

This circumstance obviously pointed at the process of muon transfer to helium as to an optimum candidate to observe muon transfer from excited states of the primary atom.

It is worth while recalling that most of the measurements performed so far on the rates of reactions (1) and (2) have been carried out stopping negative muons within suitable admixtures of small concentrations of the Y element in hydrogen (deuterium),[1,2] and by measuring either the time distribution of the X(p), (X(d)) radiation, or the time distribution of the muon decay electrons. In these cases, when the data were taken with an initial time cut of about 1 microsec following the time of stopping muons, one could realistically assume that processes (1) and (2) were occurring from μp and μd atoms which had already attained their 1S ground states.

If an appreciable fraction of the stopped muons undergoes either reaction (1) or (2) from an excited atomic level, however, the total number of both type of delayed events is sensitive to the initial population and to the effective lifetime of any possible metastable level.[4] The first experimental evidence of this type of processes was obtained by Bertin et al.,[6] by analysing the time distribution of the delayed electrons observed after stopping negative muons in a suitable mixture of deuterium + xenon + helium, and comparing it to the corresponding distribution obtained with a deuterium + xenon mixture. These data were first analysed[7] (when the existence of the metastable muonic states[4] had not yet been called into attention) only with regard to their time behaviour, getting information on the rate of reaction (2) occurring to helium starting from $\mu d(1S)$ systems. The same date were re-analysed 10 years later[6] looking at the total number of detected muon decay electrons. This second analysis shew that transfer processes from (μd) levels (n being the principal quantum number) with $n \geq 2$ to helium nuclei occurs for about 20% of the muons stopped in deuterium. Furthermore, few per cent appear to be transferred from levels having $n > 2$. This implies that the transfer rate from the μd atom to helium nuclei starting from the n=2 level, or from $n > 2$ levels, or both, are quite high, namely of the order of the rates of muon transfer from $\mu d(1S)$ states to nuclei other than helium.

By exploiting a similar method, a Dubna group[8] has observed the transfer process from excited states of muonic hydrogen to helium nuclei, which would occur starting from $n \geq 2$ levels for about 17% of the cases, in fair agreement with the result quoted for the μd system.

MUON TRANSFER EFFECT ON THE LYMAN SERIES OF μY ATOMS FORMED FROM FRE μp AND μd SYSTEMS

The most direct characteristic which may indicate whether a μY a tom formed by a given procedure keeps memory of its formation channel is the muonic X-radiation released in its de-excitation process. Some relevant observations were carried out in the past, which have shown that the X(p) radiation released in process (1) has a markedly different structure with respect to X(dt) (see process (3)). In particular, as was first reported by Budyashov et al.,[9] for the case Y=Ar, the higher members of the Lyman series transitions turn out to be significantly enhanced within X(p) with respect to the case of X(dt). Some relevant data are reported in Table I. Quite generally,[10] these results indicate that the atomic cascade process which takes the (μY) atom to its 1S ground state is somewhat changed by the transfer mechanism.[9-14] In a detailed analysis, Pfeiffer et al.[11] have proved that such effect can be reproduced by assuming angular momentum distributions for the initial formation level of the μY atoms which are more weighted to the lower values. For a complete picture of the effect pointed out by these authors, however, two additional pieces of information were required. First, it has to be clarified whether the effect is strictly connected to the physical phase in which the Y capturing element is present at the time of muon transfer; more specifically, since the features of the de-excitation process of the μY atom may well depend on the density and state of the surrounding medium, a natural progress in the information would be to establish whether this type of effect is also present when negative muons are transferred to Y elements in the solid phase. From the experimental point of view, however, investigations in this direction are more difficult, because elements in the solid phase, which are not easily available in dilute concentration within hydrogen (deterium), absorb in practice most of the stopping muons. The only possibility of observing X(p) or X(d), in this case, is to separate these radiations (which are delayed in time due to the μp and μd atom diffusion processes)[15] by a suitable time cut from the dominant component X(dt).

A most effective technique on this standpoint is obviously to form μp (μd) atoms in gaseous targets where a series of regularly spaced thin foils made of the element Y are immersed, the distance between the foils being of the order of the mean free path of the μp (μd) atoms at the chosen density condition.[13] This type of apparatus has been used at the CERN SC2 600 MeV Synchrocyclotron by a Bologna group, who transferred negative muons from μp (μd) atoms to a series of regularly spaced aluminium foils, and observed the struc-

TABLE I.- Summary of the experimental results of the muon transfer effects on the Lyman series of muonic atoms of different elements Y.

Y	(K_α/K_t) (X(dt))	(K_α/K_t) (X(p))	(K_α/K_t) (X(d))	$(K_\beta+K_\nu)/K_\alpha$ (X(dt))	$(K_\beta+K_\nu)/K_\alpha$ (X(p))	$(K_\beta+K_\nu)/K_\alpha$ (X(d))
Ar[a]	0.85 ± 0.01	0.42 ± 0.01		0.170 ± 0.016	1.40 ± 0.07	
Ar[b]	0.90 ± 0.01	0.44 ± 0.03		0.113 ± 0.013	1.25 ± 0.16	
F[c]	0.68 ± 0.01	0.33 ± 0.01		0.47 ± 0.03	1.99 ± 0.10	
Al[d]	0.79 ± 0.02	0.45 ± 0.05	0.59 ± 0.13	0.26 ± 0.03	1.22 ± 0.19	0.68 ± 0.24

[a]See ref. 9. [b]See ref. 11. [c]See ref. 12. [d]See references 13 and 14.

ture of the X(dt), X(p), and X(d) radiations from the μAl atoms formed on the foils themselves. As it is seen from Table I, a first significant result obtained by these authors is that a strong difference is observed in the structure of the X(dt) and X(p) radiations from μAl. If one looks at the previous experimental results, moreover, qualitative agreement is verified between the muon transfer effects on the Lyman series of μY atoms formed both in the gaseous and in the solid phase. In particular, the relative increase of the higher members of the series with respect to the K_α contribution is very significant in all the cases examined.

The second point which one may explore on this line is the possibility that the effect may depend to some extent on the nature of the primary muonic atom from which the transfer reaction proceeds; more specifically, one might wonder wether the effect is also present - and with which features - in the case of reaction (2), which was never investigated in this direction. From the recent Bologna results listed in the Table, it is apparent that the muon transfer effect on the Lyman series of μAl atoms is significantly different according to the hydrogen isotope muonic atom from which the reaction proceeds. In any case, the higher members of the K-series within the X(d) radiation are also enhanced with respect to the case of direct muon stopping in aluminium.

CONCLUSIONS

As a conclusion of the discussion which has been carried out in the previous paragraphs, the following remarks can be made:

(a) on the experimental line which has been opened by both theoretical and experimental results on the muon transfer from excited states of muonic hydrogen and deuterium, more precise observations of the muon transfer processes (1) and (2) proceeding both from the 1S and from excited levels of the primary μp or μd muonic atoms would be welcome. Results in this direction might also be obtained in experiments which are presently in progress [16] on the mean lifetime and population of the 2S metastable state in muonic hydrogen and deuterium. These measurements are related to the feasibility of spectroscopy experiments aiming to check the QED effects (in particular vacuum polarization) on the levels of the μp and μd atoms

(b) quite generally, one may expect that the results on- the muon transfer-generated alterations of the Lyman series of μY atoms formed through reactions (1) and (2) might both be explained by an atomic cascade model where the initial angular momentum of the μY atoms formed is weighted to lower l-states than in the case of direct muon stopping. Theoretical understanding of this type of isotopical effect on the muon transfer process might provide further insight on this type of reactions.

(c) improved cascade calculations referring to the de-excitation process of μY atoms should indeed keep into account the formation channel of the muonic atoms concerned.

(d) the muon transfer effect on the X-muonic radiation from μY atoms has been investigated only for some low atomic number elements (i.e. up to Z=18). Further insight on this phenomena would most likely be provided by more experimental work in the region of high atomic number elements.

(e) as a final information one might recall that quite a few data are available on point (iv) of the Introduction. However, some recent results [18] of the Bologna group indicate that (in the case of alumina - the atomic capture ratio W(Al)/W(O) of muons is somewhat different when muons are transferred to aluminium oxide with respect to the case of direct muon stopping (see Table II).

TABLE II.- Observed per-atom capture ratios for different compounds referring to different muon capture channels (See ref.12, 17, 18).

Compound	Capture ratio	Direct formation	Form. by muon transfer from μp systems
SF_6	W(F)/W(S)	0.96 ± 0.09	1.04 ± 0.24
Al_2O_3	W(Al)/W(O)	0.85 ± 0.06	1.20 ± 0.11

ACKNOWLEDGEMENTS

The author wishes to thank A. Bertin, I. Massa, M. Piccinini, and G. Vannini for some useful discussions.

REFERENCES

1. G. Fiorentini and G. Torelli, Preprint, Pisa, April 1975.
2. See e.g. A. Bertin, M. Bruno, A. Vitale, A. Placci, and E. Zavattini, Phys. Rev. A 7: 462 (1973); and references therein.
3. See e.g. P.K. Haff, E. Rodrigo, and T.A. Tombrello, Ann; Phys. 104:363 (1977); and references therein.
4. A. Placci, E. Polacco, E. Zavattini, K. Ziock, G. Carboni, U. Gastaldi, G. Gorini, G. Neri, and G. Torelli, Nuovo Cimento 1A:445 (1971); A. Bertin, G. Carboni, G. Gorini, O. Pitzurra, E. Polacco, G. Torelli, A. Vitale, and E. Zavattini, Phys. Rev. Lett. 33: 253 (1974).
5. G. Fiorentini and G. Torelli, Il Nuovo Cimento 36A:317 (1976).
6. A. Bertin, A. Vitale, and E. Zavattini, Lett. al Nuovo Cimento 18:381 (1977).
7. A. Placci, E. Zavattini, A. Bertin, and A. Vitale, Nuovo Cimento 52A:1274 (1967).
8. V.M. Bystristki et al.,in Mesons in Matter, Proc. Int. Symp. Meson Chem. and Mesomol.Processes in Matter, edited by V.H. Pokrovskij, Dubna 10908,p.220 (1977).
9. Yu.G. Budyaschov, P.F. Ermolov, V.G. Zinov, A.D. Konin, and A.I. Mukhin, Yad. Fiz. 5:599 (1967) (English translation: Sov.J. Nucl. Phys. 5:426 (1967).
10. The nP $\rightarrow$1S transitions are labeled here by K_α(2P$\rightarrow$ 1S), K_β(3P$\rightarrow$ 1S), ... $K_t = K_\alpha + K_\beta + K_\nu$.
11. H.J. Pfeiffer, K. Springer, and H. Daniel, Nucl. Phys. A 254:433 (1975).
12. H. Daniel, H.J. Pfeiffer, and K. Springer,Phys. Lett. 44A:447 (1973).
13. A. Bertin, F. Ferrari, I. Massa, M. Piccinini, G.Vannini, and A. Vitale, Phys. Lett. 68A:201 (1978).
14. A. Bertin, F. Ferrari, I. Massa, M. Piccinini, G. Vannini, and A. Vitale, to be published.
15. A. Bertin, I. Massa, M. Piccinini, A. Vacchi, G. Vannini, and A. Vitale, Phys. Lett. 78B:355 (1978).
16. G. Torelli,private communication.
17. V.G. Zinov, et al., Yad. Fiz. 2:859 (1965).
18. A. Bertin, et al., Lett. Nuovo Cimento 23:401 (1978).

HYDROGENIC MESOMOLECULES AND MUON CATALYZED FUSION

Johann Rafelski

Theoretical Physics Division
CERN
1211 Genève 23

1. - Introduction

With respect to many atomic and molecular processes the lifetime of the free muon, $\tau_\mu = 2.20\times10^{-6}$ sec, is sufficiently long to consider it as a stable object. So the possibility arises that mesomolecules of heavy hydrogen isotopes are formed. Such a mesomolecule is, up to recoil corrections, a scaled down model of the usual one-electron hydrogen molecule, H_2^+, with the scaling factor being the ratio of muon to electron mass, 206.8. In particular, the hydrogenic isotopes are only about $1\text{Å}/206.8 = 500$ fm apart. Consequently, the probability for spontaneous fusion is appreciable especially for fusion reaction channels governed by strong interactions. This is in particular the case for the dd and dt mesomolecules.

In view of these circumstances we must examine in more detail the processes leading to formation of mesomolecules - if they were sufficiently fast, one could think of a chain of reactions in which muons play the rôle of a catalyser : as soon as a mesomolecule is formed, a fusion process occurs, subsequently the freed muon forms a new mesomolecule, etc.

This reaction chain is interrupted either by the decay of the muon, or by its capture by an element other than hydrogen - als by its sticking to the fusion product, He.

This fascinating mechanism to catalyze "cold" fusion deman the understanding of many fundamental aspects of atomic and mol cular physics and for that reason it has attracted much interes during the past decades. The first fusion catalyzed by muons has been discovered in 1956 in Berkeley [1] but theoretical work [2-6] goes back to the original suggestion by Frank in 1947. Interest persisted through the early sixties, when most of the relevant rates seemed to have been established qualitatively [5,6] indicating that this process could not be of practical importance in the immediate future.

During the past years we have seen again growing interest in the μ catalyzed fusion mainly due to the discovery of the temperature dependence of the reaction rate [7] for μ catalyze fusion in deuterium, its theoretical explanation in terms of weakly bound mesomolecular states [8-10] and the suggestion to use the cold fusion as the heating mechanism to start the fusio reaction in compressed pellets [11]. In these lectures, we shal discuss in particular these new developments that depend on new insights into mesoatomic and mesomolecular physics. To readers looking for a longer introductory paper to the subject of mesomolecular physics we recommend the excellent review by Gershtei and Ponomarev [12].

Upon a short introduction to the present status of the μ catalyzed fusion in the next section, where we shall also prese the relevant data, we turn to the discussion of special questio particularly important for the process of the catalyzed fusion in the subsequent sections. In Section 3 we will describe the general approach to the three-body Coulomb problem when one of the bodies (muon) is (not much) lighter than the others. Then the process of resonant mesomolecule formation is discussed in Section 4. This is the new mechanism predicted by Vesman[8] that has reopened the field of μ catalyzed fusion. Then we turn to the discussion of the possible fate of the muon in liquid hydrogen. In particular in Section 5 we consider the probability that the muon will stick to the fusion product, helium, which would remove it from the catalytic chain of reactions. In the following section, we consider, assuming our present theoretical knowledge of the different rates discussed in Section 2, the necessary conditions to catalyze 100-200 fusions - a number that seems feasible today. In the last section we present an outlook devoted in the part to the discussion of fundamental physical problems involved.

In order to distinguish the different isotopes of hydrogen we will refer to ^1H as p or protium, to ^2H as d or deuterium and to ^3H as t or tritium. The word "(liquid) hydrogen" implies a mixture of these three isotopes.

2. - Overview

When a muon enters a hydrogen target, a chain of reactions begins - here we list those that dominate the history of the muon in a liquid $(\rho = 4.25\times10^{23}/cm^3)$ hydrogen target.

1 μ is stopped,
2 μ is captured by the dominant isotopic fraction,
3 μ cascades to the mesohydrogenic 1s state,
4 μ can be transferred to the heavy isotopes,
5 μ mesomolecule formation,
6 de-excitation to the mesomolecular ground state,
7 muon decay or nuclear fusion,
8 muon sticking to He, or at, point 1 or 2.

When fusion occurs, muon can be shaken off and is then available to repeat the chain of reactions 1 to 7. Naturally muons can decay at any stage between 1 and 7 but it is usually the mesomolecular state that involves a p that lives longer than the muon lifetime. It can be assumed that about, or less than, 10^{-9} sec are required for fast muons to slow down and reach the 1s state in liquid hydrogen [12]. The properties of the bound mesohydrogenic 1s states are given in Table 1.

The last column gives the hyperfine splitting which plays an important rôle in the moderation of mesonic hydrogen. We further notice the strong isotope effects ΔE on the binding energy of the 1s state. In particular, it leads to relatively high transfer rates from light to heavier isotopes. In Table 2 we give the results of the calculations by Ponomarev et al. [13] for liquid hydrogen at room temperature for these transfer rates. Experimental figures for the $p\mu + d$ process agree well with theoretical calculations and among themselves up to a factor of two.

Table 1 : Main characteristics of mesonic atoms

M	M/m_e	$-E_{1s}$ (eV)	ΔE (eV)	ΔE^{hfs} (eV)
μ	206.769			
p	1836.152	2528.437	$\Delta E_{pd} = 134.705$	0.183
d	3670.481	2663.142	$\Delta E_{pt} = 182.745$	0.049
t	5496.918	2711.182	$\Delta E_{dt} = 48.040$	0.241

Table 2 : Transfer rates of muons between hydrogen isotopes

$p\mu + d \rightarrow p + d\mu$	$\Lambda_T \cdot \tau_\mu \sim 3.7\times10^4$
$p\mu + t \rightarrow p + t\mu$	$\Lambda_T \cdot \tau_\mu \sim 1.6\times10^4$
$d\mu + t \rightarrow d + t\mu$	$\Lambda_T \cdot \tau_\mu \sim 220$

From the point of view of other atoms, a mesic hydrogen acts in many respects as if it were a "neutron" ; it is neutral and its size of ~500 fm is quite small when compared with the usual atomic sizes. So it can easily penetrate the electron cloud, without feeling the repulsive Coulomb potential of the atomic nucleus. However, when a mesic hydrogen approaches another hydrogen isotope relatively closely, a mesomolecule may be formed accompanied by conversion (Auger emission) of an electron. Such a mesomolecule is, from the viewpoint of other electronic atoms a very heavy isotope of hydrogen, being relatively small. Thus it can form an electronic molecule with another hydrogen atom. An observant reader will notice, at this point, that we could, in the first place, attach the mesic hydrogen to one of the centres of the electronic molecule (see Fig. 2.1). To do so without the conversion electron we must have an accidental degeneracy [8]. The gain in energy from the mesomolecular formation must be absorbed by the internal degrees of freedom of the electronic molecule. Indeed, such a degeneracy has been observed [7]

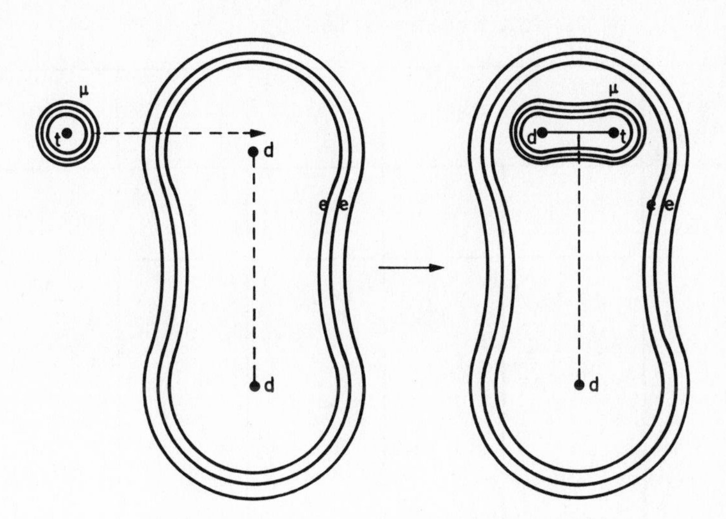

Figure 2.1 : Illustration of the resonant mesomolecule formation in the process $\mu t + (d_2ee) \rightarrow [(td\mu)dee]^*$.

and we shall speak of this process as the resonant mesomolecule formation. Naturally, the rates of the resonant process can be influenced by external parameters, such as temperature [8]. We will discuss this at length further below.

In order to achieve an accidental degeneracy one has to compare energies of the electronic molecular hydrogen of the order eV with those of mesonic atoms and molecules that are usually 200 times larger. However, the mesic molecules can have weakly bound excited states. In Table 3 we show the meso-molecular spectrum recently obtained by the Dubna group [10),14)] for the mesomolecular energies. J is the rotational quantum number and ν the vibrational quantum number. We notice that both in $dd\mu$ and $dt\mu$ weakly bound states of 2 and 0.9 eV have been predicted. This is a very important result, since it opens the possibility of the accidental degeneracy described above. The energies given in Table 3 are the binding energies

Table 3 : Binding energy $\epsilon_{J\nu}$ (eV) of μ molecules of the hydrogen isotopes

Angular momentum	J = 0		J = 1		J = 2	J = 3
Vibrational state	ν = 0	ν = 1	ν = 0	ν = 1	ν = 0	ν = 0
$pp\mu$	253.0	-	105.6	-	-	-
$pd\mu$	221.5	-	96.3	-	-	-
$pt\mu$	213.3	-	97.5	-	-	-
$dd\mu$	325.0	35.6	226.3	2.0	85.6	-
$dt\mu$	319.1	34.7	232.2	0.9	102.3	-
$tt\mu$	362.9	83.7	288.9	44.9	172.0	47.7

with respect to the mesic atom - that is, e.g., in the case of a $dt\mu$ molecule, the total binding of the μ is obtained adding 2711.18 eV from Table 1. From this example, we see that in order to establish that the mesomolecular bound state actually exists, one has to solve the three-body Coulomb problem with a relative precision better than 10^{-4} !

We note further that the weakly bound states happen to have allowed E1 transitions to the ground states of the respective mesomolecules. Consequently, within 10^{-10} sec, we are actually in the mesomolecular ground state and the nuclear fusion can follow.

In Table 4 we give the different possible fusion reactions that can proceed in mesomolecular states. The reactions of dd and dt are here of particular interest. They can proceed via strong interactions and are accompanied by the emission of fast neutrons. For this reason there is a significant recoil of helium as indicated in the last column of Table 4. Due to this recoil, muons can be shaken off the fusion product. We shall consider the probability W_s of the muon to stick to the recoiling helium further below in Section 5.

A compilation of the essential rates concerning the catalyzed fusion process is shown in Table 5. We see that in the dd and dt reactions the non-resonant molecular formation rate is very small. This has been the bottle-neck in the catalytic chain until now. We notice that when the resonant mechanism is included, there is still the bottle-neck for the $d+d$ reaction (the reader is warned that all numbers involving t are <u>theoretical predictions</u>). Without the resonant effect the chances for two fusions generated by one muon are highest in the $d+d$ case and should occur in less than ~10% of the events (in liquid hydrogen). This changes when we satisfy the resonance conditions in the dt reaction. Here we expect up to 100-200 fusions per muon under optimal conditions.

Table 4 : Fusion reactions, their Q values and recoil

Fusion reaction	Q value	Recoil/Nuc.
$p + p \rightarrow d + e^+ + \nu$	2.2 MeV	-
$p + d \rightarrow {}^3He + \gamma$	5.4 MeV	-
$t + p$ (50%)	4 MeV	-
$d + d \rightarrow {}^3He + n$ (50%)	3.3 MeV	0.27 MeV ($\frac{1}{12}$ E)
${}^4He + \gamma$ (~0%)	24 MeV	-
$d + t \rightarrow {}^4He + n$	17.6 MeV	0.88 MeV ($\frac{1}{20}$ E)
$p + t \rightarrow {}^4He + \gamma$	20 MeV	-
$t + t \rightarrow {}^4He + n + n$	10 MeV	-

Table 5 : Essential rates for catalyzed fusion (liquid hydrogen density)

Process	Non-resonant molecular rate $\Lambda_m \cdot \tau_\mu$	Total molecular rate $\Lambda_{mT} \cdot \tau_\mu$	Nuclear fusion rate $\Lambda_f \cdot \tau_\mu$	Fusion before sticking W_s^{-1}
$p\mu + p$	5	5	~ 0	-
$d\mu + p$	5 ~ 10	5 ~ 10	0.7	1
$d\mu + d$	0.1	1.8	~ 10^5	~ 8
$t\mu + d$	0.07	207	~ 10^6	~ 100
$t\mu + p$	1	1	10	1
$t\mu + t$	1.5	1.5	?	1

We shall now explain in the next sections how the critical numbers Λ_{mT} and W_S can be computed. We start with a brief discussion of the three-body Coulomb problem in the adiabatic basis.

3. - Mesomolecular spectrum

In this section we shall only outline the main aspects of the theoretical approach and describe the difficulties in computation of weakly bound mesomolecular states. As indicated in Section 2 we want to establish the binding energy of a muon in a mesomolecule up to 0.1 eV out of c. 3 keV. Thus we must reach a relative precision of better than 10^{-4}. On the other side we notice that since the muon is relatively heavy we must take account of the nuclear recoil due to the muon motion. Both these facts make the numerical work extremely difficult, while the formulation of the problem is relatively straightforward.

In the Born-Oppenheimer adiabatic approach the fast motion of the muon is treated as a separated two-centre problem for a fixed separation $|\vec{R}|$ between the nuclei that are assumed to oscillate slowly. Thus we solve the Schrödinger equation to obtain the adiabatic basis :

$$\left(-\frac{1}{2m_\mu^r}\Delta_{\vec{r}} - \frac{e_1 e}{r_1} - \frac{e_2 e}{r_2}\right)\phi_n(\vec{r};R) = \mathcal{E}_n(R)\phi_n(\vec{r};R) \tag{3.1}$$

where m_μ^r is the reduced mass of the muon with respect to the two centres $M_1 + M_2$: $1/m_\mu^r = 1/m_\mu + 1/(M_1+M_2)$. The quantities r_1 and r_2 are the distances of the muon from the centres 1 and 2, as shown in Fig. 3.1. In the adiabatic approximation the total wave function $\psi(\vec{r},\vec{R})$ of the three-body problem is expanded in terms of this two-centre basis

$$\Psi_i(\vec{r},\vec{R}) = \sum_n \phi_n(\vec{r};R)\,\chi_n^i(\vec{R}) \tag{3.2}$$

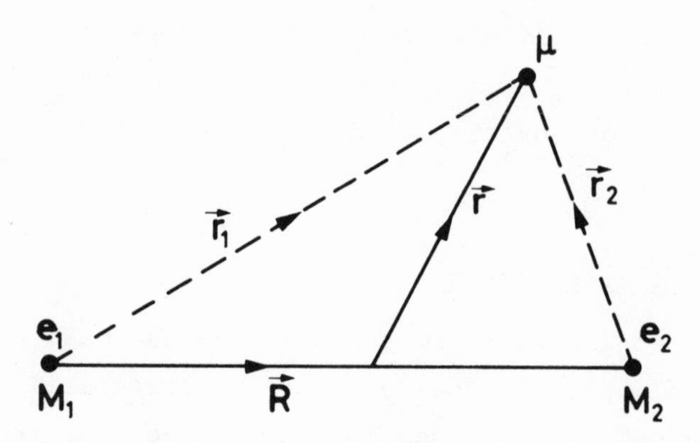

Figure 3.1 : The co-ordinates of the three-body Coulomb problem.

ψ_i satisfies the three-body Coulomb problem Hamiltonian. In order to obtain an equation for $\chi(R)$ we can consider the projection of the total three-body problem with the basis ϕ_n. We then find

$$\left(-\frac{1}{2\bar{M}}\Delta_{\vec{R}} + \frac{e_1 e_2}{R} + \mathcal{E}_n(R) - E_i\right)\chi_n^i(\vec{R}) = \tag{3.3}$$

$$= \frac{1}{2\bar{M}}\sum_k \int d^3r \left(\phi_n^* \Delta_R \phi_k\right)\chi_k^i + \frac{1}{\bar{M}}\sum_k \int d^3r \left(\phi_n^* \vec{\nabla}_R \phi_k\right)\cdot\vec{\nabla}_R \chi^i$$

where $1/M = 1/M_1 + 1/M_2$ and $\bar{M} = M m_\mu / m_\mu^r$.

Usually the dynamic coupling of the molecular motion of the nuclei to the muon wave function, the right-hand side of Eq. (3.3), is neglected and we have in the left-hand side of Eq. (3.3) the usual molecular problem. Another way of seeing this is to assume that m_μ/M is very small : this is the case for electronic atoms.

In particular when weakly bound molecular states are considered we have to include the effects of the right-hand side of Eq. (3.3). The diagonal $(n = k)$ terms can be easily included in the molecular potential on the left-hand side of

Eq. (3.3). The off diagonal terms demand a simultaneous solution of coupled channel equations, a task recently performed by the Dubna group [10),14)]. With the high accuracy required here, in particular, the coupling to the continuum is essential to obtain the proper energy shown in Table 3 for the $J = 1$, $\nu = 1$ states in dd and dt. The multichannel calculations increase the binding in both cases by ~1.3 eV (a 10^{-4} effect on the total energy). In the simple molecular picture the ddμ state could be weakly bound, while the dtμ state would be weakly unbound. The dynamic coupling makes also the (dtμ) state weakly bound and opens the way for new and exciting process of μ catalyzed dt fusion

In Table 3 we showed the difference between the molecular energy E_i and the atomic energy of the heaviest isotope given in Table 1. Thus Table 3 gives the additional binding due to the presence of a second centre ; the molecular quantum number in Table 3, angular momentum J and vibration ν, are summarized by the index i in Eq. (3.3) while n describes the muon configuration of the molecule.

As we have already emphasized, the absolute precision of both atomic and molecular energies must exceed 0.1 eV in order to obtain to within 10% the energies of weakly bound mesomolecular states. In order to appreciate the level of precision needed, let us consider as an example the effect of vacuum polarization. The vacuum polarization shifts for the 1s states in μp, μd and μt are [15)] -1.896, -2.126, -2.212 eV, respectively (the minus sign indicates increased binding). The molecular energies are affected also, since the form of the wave functions at the nuclear centre is different from the atomic case and because there is additional repulsion between the nuclei due to the internuclear vacuum polarization. Fortunately, these effects nearly cancel for the weakly bound molecular states : Melezhik and Ponomarev [15)] find 0.008 eV for the $(dd\mu)^{\nu=1}_{J=1}$ state and -0.003 eV for the $(dt\mu)^{\nu=1}_{J=1}$ state. We will discuss further below that even these small contributions will ultimately be measurable and are relevant for the μ catalyzed process. Naturally we must gain control of all other contributions to the molecular energy levels at this high relative precision of 10^{-6}-10^{-7} beforehand. It is worth mentioning that this precision has an "equivalent" of perhaps 10^{-10}-10^{-11} in electronic molecules, due to the increased importance of the recoil effect in mesic molecules.

4. - Resonant formation of mesomolecules

From the point of view of an electron a mesic hydrogen looks much like a quasi-neutron. Thus it can penetrate the electron cloud of the molecule and stick to one of the molecular centres. This process can proceed without the conversion (Auger emission) of an electron only when the mesomolecular energy gain - the energy given in Table 3 - is degenerate with vibrational or rotational states of the electronic molecule. We must also take under consideration the thermal energy ϵ_T at temperature T

$$\begin{aligned} \epsilon_T &= 3/2\, kT \\ &= .054\,\mathrm{eV} \quad \text{for } T = 400°K \end{aligned} \tag{4.1}$$

We recall that the thermal distribution

$$\gamma(\epsilon, \epsilon_T) = \left(\frac{27}{2\pi}\frac{\epsilon}{\epsilon_T}\right)^{1/2} \frac{1}{\epsilon_T} \exp(-3/2\,\epsilon/\epsilon_T) \tag{4.2}$$

is a weakly peaked distribution, thus the resonance condition on the energy of the electron molecule ϵ_r

$$\epsilon_r = \epsilon_T + \epsilon^{\mu}_{mol} \tag{4.3}$$

is not very pronounced - in the distribution (4.2) we have in the range of temperatures $30^{o}K < T < 1000^{o}K$ always an appreciable probability of finding a particular value of ϵ in the domain $0.005 < \epsilon < 0.1$ eV.

The resonance condition is shown in Fig. 4.1 with the special example of the $[(dt\mu)dee]$ molecule illustrated in part (b) of the Figure. As indicated there, the amount of ~1 eV mesomolecular binding can be absorbed by exciting the fourth vibrational state. In the case of the $[(dd\mu)dee]$

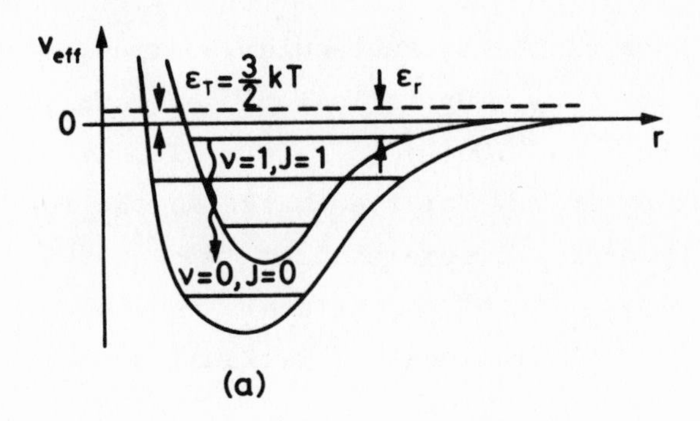

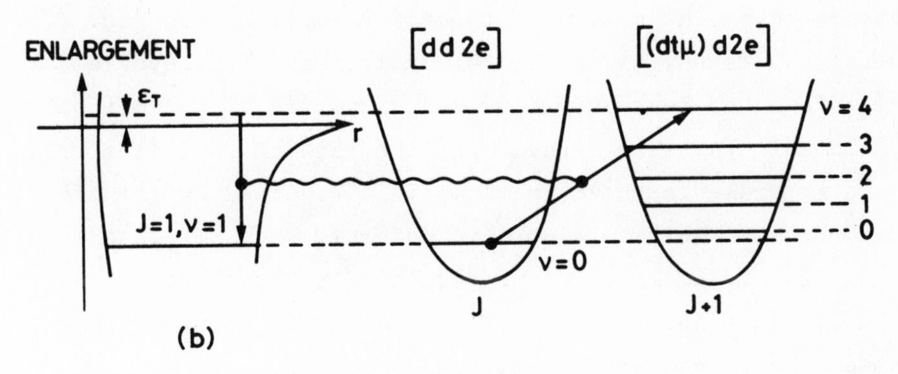

Figure 4.1 : a) Schematic level scheme of mesomolecules in the effective potential and the thermic energy.

b) Resonant molecular formation and the excitation of electromolecule.

molecule it is the eighth vibrational state that is excited by ~2 eV mesomolecular energy. Another component in the energy balance indicated in Fig. 4.1 is the rotational energy of the electronic molecule, before and after the attachment of the mesonic t or d to one of the molecular centres. This energy is

$$E^J = J(J+1)/2\Theta$$

$$\Theta = R^2\mu \tag{4.4}$$

Here μ is the reduced mass of the two molecular centres of the electromolecules and $R = 1.4$Å. Up to isotopic effects of the order of 10% this energy is $0.004\ \frac{1}{2}J(J+1)$ eV. We have to include a suitable amount of energy for the change in the rotational quantum number by one. We note that at $T \sim 400\text{-}500^\circ$K several rotational states $(J \leq 3)$ of the molecule can be excited thus making the set of rotational-vibrational states suitable for accidental degeneracy dense.

We expect that the molecular rates for the formation of $(dt\mu)$ and $(dd\mu)$ molecules will depend on temperature as illustrated by the resonance condition (4.3) and in Fig. 4.1. Indeed in a recent experiment of Bystritsky, Dzhelepov et al.[7] this effect has been investigated systematically. The collection of all available results for the rate of formation of the $(dd\mu)$ molecule is shown in Fig. 4.2. We see a systematic behaviour and the rate rises by factor 10 when T reaches 400°K. The solid line is the result of the theoretical study of Vinitsky, Ponomarev et al.[10] - a satisfactory agreement between theory and experiment can be achieved.

In view of these experimental and theoretical results we are led to the conclusion that the similar effect predicted for the $(dt\mu)$ molecule should also be present, leading to very high mesomolecular formation rates. The theoretical prediction here is based on the lower excitation of the electromolecule in order to absorb the lower mesomolecular binding - ~1 eV - of this mesomolecule. Gerstein and Ponomarev [9)] base their prediction of $\Lambda_{mT} \cdot \tau_{\mu} \approx 200$ on the excitation of the fourth vibrational state in the electromolecule. Indeed the success of the theory for the $(dd\mu)$ molecule as documented by Fig. 4.2 lends credibility to this prediction. On the other hand we should remember that a relative error of 10^{-5} in the determination of the total energy (c. 2710 eV) of the mesomolecule $(dt\mu)$ for $J = \nu = 1$ would significantly change the resonance condition and the predicted mesomolecule formation rate. In the detailed numerical work the difference between the $(dd\mu)$ and $(dt\mu)$ molecules is large - the latter being asymmetric, with the weak binding provided by the non-trivial terms in the right-hand side of Eq. (3.3) Furthermore the adiabatic

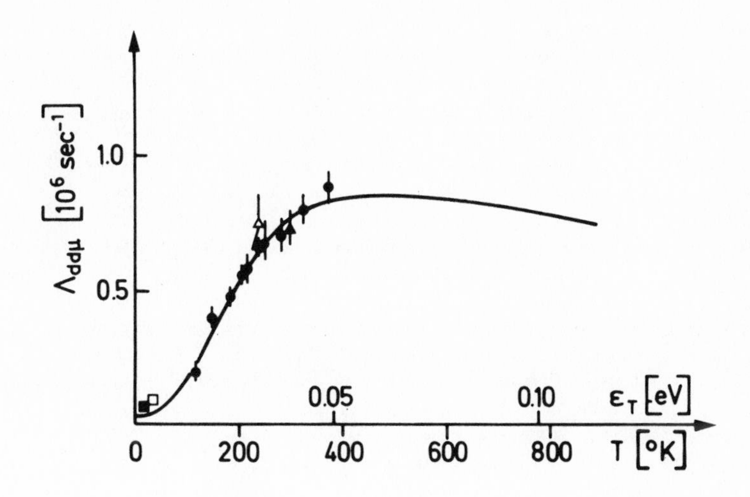

Fig. 4.2 : Collection of all data on the $dd\mu$ formation rate as function of temperature

approach is not quite correct, as will be described in the next section. This also influences the energies of the weakly bound states. We see that we must establish the total mesomolecular energy with a precision of 10^{-5}-10^{-6} to obtain a reliable prediction for Λ_{mT}. As we have already discussed, even the effect of the vacuum polarization has to be considered - and indeed the numbers given at the end of the previous section indicate that this effect changes the resonant temperature by ~5-10% Incidentally, the measurement of the resonance temperature can be viewed as the measurement of the molecular vacuum polarization effect [15].

5. - Muon sticking probability

The motion of the muon is faster than that of the oscillating nuclei - that is the basis for the adiabatic approach outlined in Section 3. Thus the muon cloud can adjust its wave function to each distance R between the oscillating nuclei. When we follow the mesomolecular energy in the ground state of (dt) as a function of R, we see that it evolves into the 1s state of helium in the united atom limit. Thus when fusion occurs in the vicinity of $R=0$, muon wave function resembles much the 1s wave function of helium. The process of fusion itself is sudden from the point of view of the muon. Let us now start the calculations assuming with Zel'dovich and Gershtein [6] that the initial state of the muon when fusion occurs is the 1s state in helium.

We can apply the sudden approximation to compute the probable fate of the muon after fusion. That is we consider probability amplitude

$$W_{fi} = \int d^3x \, \phi_f^{*He} \varphi_i \tag{5.1}$$

where φ_i is the initial muon wave function and ϕ_f^{He} is the final state of the muon in the presence of the fusion product, here indicated by the upper index He in a particular channel of the dd fusion and for all dt fusions.

We now concentrate on the fusion in the dtμ molecule, this being the only case where many fusions seem feasible per muon considering the numbers given in Table 5. In the process of the dt fusion 17.6 MeV energy is liberated of which a large part is carried away by the neutron, the remainder of c. 3.5 MeV being the recoil energy of the α particle (see Table 4). Thus with respect to the muon "frozen" in the orbital in the sense of the sudden approximation, the new nucleus moves with substantial momentum P_{rec}. This situation is illustrated in Fig. 5.1. Therefore, if we want to find the probability of sticking W_s, of the muon to the moving helium we must find its overlap with the moving bound muonic state of the fusion product. Thus :

$$W_s = \sum_n W_{ni} = \sum_n \int d^3x \, \Phi_n^{He}(\vec{x}, P_{rec}) \, \varphi_i(\vec{x}) \tag{5.2}$$

where

$$\Phi_n^{He}(\vec{x}, P_{rec}) = \varphi_n^{He}(\vec{x}) e^{i \vec{P}_{rec} \cdot \vec{x}} \tag{5.3}$$

and φ_n^{He} are the usual Schrödinger Coulomb bound states of He. As already discussed φ_i is taken to be the 1s state of He.

The largest contribution in Eq. (5.2) arises from the $n = 1s$ state. This has been computed [4] to be

$$W_s \approx (1+R)^{-4} \simeq .01$$
$$R = E_{rec}^{\mu} / 4\epsilon_0 \tag{5.4}$$

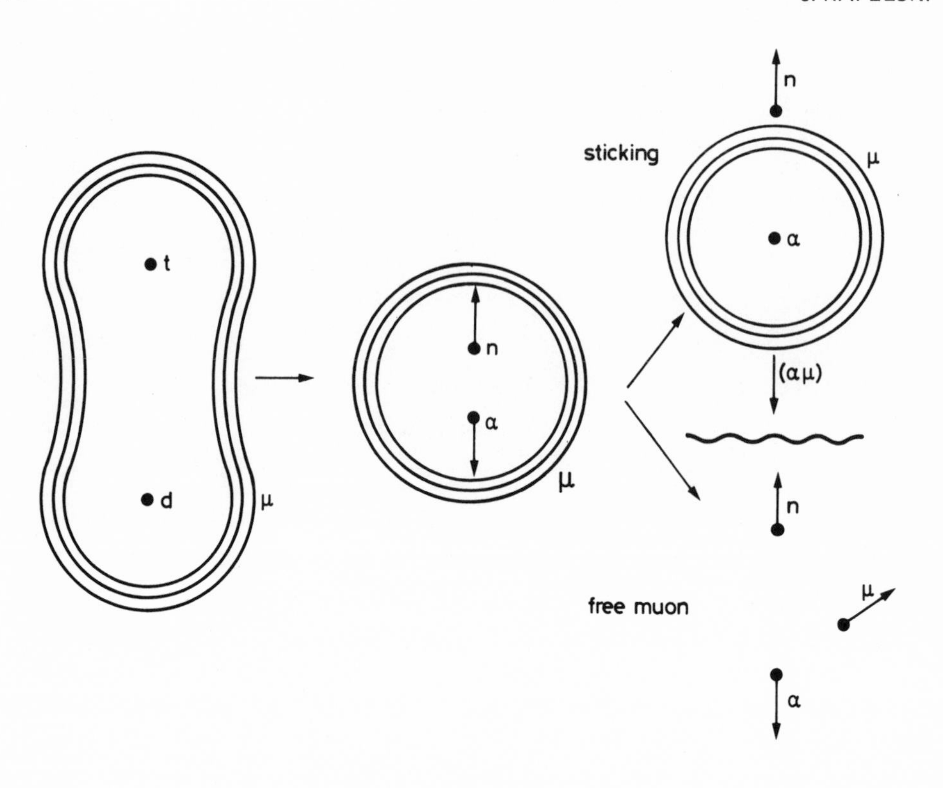

Figure 5.1 : Muon sticking in fusion.

where E^{μ}_{rec} is the muon kinetic energy when sticking and ϵ_o its binding energy in the 1s state of helium : 11.1 keV. In the case of the dt fusion $E^{\mu}_{rec} \sim 0.1$ MeV and we find the value indicated in Eq. (5.4). All other matrix elements in Eq. (5.2) are negligible compared to (5.4) (see below).

So far this calculation has neglected the non-adiabasy of the nuclear motion and the slight difference in the initial wave function as compared to 1s-μHe and further states in Eq. (5.2). Clearly, a more thorough study is needed here ; in particular the non-adiabasy effect will lead to a reduced population of the 1s-He state. This is due to the fact that the nuclear oscillations are only several times slower than muonic velocities. Thus during the nuclear motion leading to the united atom limit muons from the (μdt) molecule can populate the 2s or even 2p state in helium. For the symmetrical $(dd\mu)$ molecule only the 2s state needs to be considered, but it could be populated relatively more strongly. Incidentally the above-described non-adiabasy affects also slightly the energies of mesomolecules, as mentioned before.

It is less likely that more weakly bound muon in the 2s state will stick - it has larger chances to be shaken off - we find the sticking probability in the 2s state to be ~5 times smaller than in the 1s state. The probability that the muon will then continue in the 1s state dominates this contribution. Of this order of magnitude is also the sticking to the 2s state if the initial state is 1s. Thus we conclude that in most cases the sticking muon will be in the 1s state of the recoiling helium, although some population of the 2s state can be expected.

As we continue to follow the fate of the muon that sticks to the fusion product, we realize that the muon is not lost yet from the catalytic chain of reactions. The muonic helium will have appreciable kinetic energy of the order of 3.6 MeV - the muon's part being some 100 KeV. Thus there is enough energy to overcome the binding energy of muons in Helium of 11 keV in the 1s state, or less if transfer to a bound state of d or t occurs. Thus muons can be, in principle, reactivated.

First we observe that those muons that are initially in the (metastable) 2s state will either be directly transferred to the t or d 1s state or decay to the 1s state in consequence of the collisional quenching. As regards the transfer process we only recall that the 2s state of He and the 1s state of t are almost degenerate. However, it can be expected that the collision induced radiative decay to the 1s state of

He will dominate the transfer process in the collision of $(\mu He)^+$ with d^+ or t^+. Again this point deserves further attention.

Let us now consider what is known on the fate of the muon in the 1s state of He. In particular the process of direct ionization into the continuum has been considered by Jackson [5]. Using the conventional Bohr model he finds an activation probability of ~40% for the muon. He then corrects his estimate comparing with the (electronic) experimental He^+ ionization cross-sections in H_2 and finds 22%. He neglects the fact tha from the point of view of a muon many more channels are open - there is only one muon at a time. Thus direct transfer processes from μHe^+ to d or t, forbidden in the He^+ experimen can now proceed. It is very likely that the transfer processes are as important as the direct ionization contribution. To see this we recall that we are in the sudden limit of the Coulomb collision process. The velocity of the recoiling helium in initially c. 5 times higher than that of the muon in orbit. Therefore direct ionization can proceed only in relatively deep $\mu He^+ + d$ (or t) collisions, where the high frequencies are available. But in such collisions the muon may choose to go to the bound states of d (or t) and this at comparable strength. In liquid hydrogen many such collisions will occur, since the slowing down by electro-excitation of electrons is $\sim 10^{-9}$-10^{-10} sec while the collision time for Coulomb collisions is $\sim 10^{-11}$-10^{-12} sec. Therefore even a small probability for the transfer process can add up to a substantial contribution. We note that since in direct ionization more energy is required then to transfer the muon from the 1s state in He to the 1s state in t or d the direct transfer process could be more important than the ionization into continuum. Finally, there is the possibility that multistep processes will occur, that is that the muon will be excited to an intermediate state of helium in one collision and in a second collision it will be either transferred or ionized. However, since the radiation rate for dipole transitions is 10^{13}/sec, most of the intermediate states will have decayed before the second collision takes place. All these processes can be computed and the degree of difficulty does not exceed the one of similar calculations in atomic physics. Since the need for a more precise study of muon regeneration has arisen only recently, this problem has not been settled yet.

For completeness we would like to mention that in the dd fusion reaction an estimate of the sticking probability gives $W_{dd} \sim 0.1-0.2$, in view of the now much smaller recoil (see Table 4). For the same reason the reactivation probability is negligible. Nonetheless, as can be seen from Table 5, it is the mesomolecule formation rate that limits here the number of possible fusions.

6. - How to catalyze 100-200 fusions in dense hydrogen

We consider the fate of a muon stopped in hydrogen isotope mixture. Our assumption will be that the ratios of rates that follow from Tables 2, 5 are correct. In Fig. 6.1 we see a flow chart with the essential reaction channels concerning muon transfer and mesomolecule formation. The numbers show us rate ratios at equal relative isotope densities. They indicate up to which relative isotope density a particular reaction is dominant.

Let us begin at the point $p\mu$ in Fig. 6.1. We have here the competition between the $pp\mu$ molecule formation and the transfer to d or t.

Since once $pp\mu$ molecule is formed, the muon is definitively lost from the reaction chain, we conclude that the common relative heavy isotope concentration must be larger than c. 1%

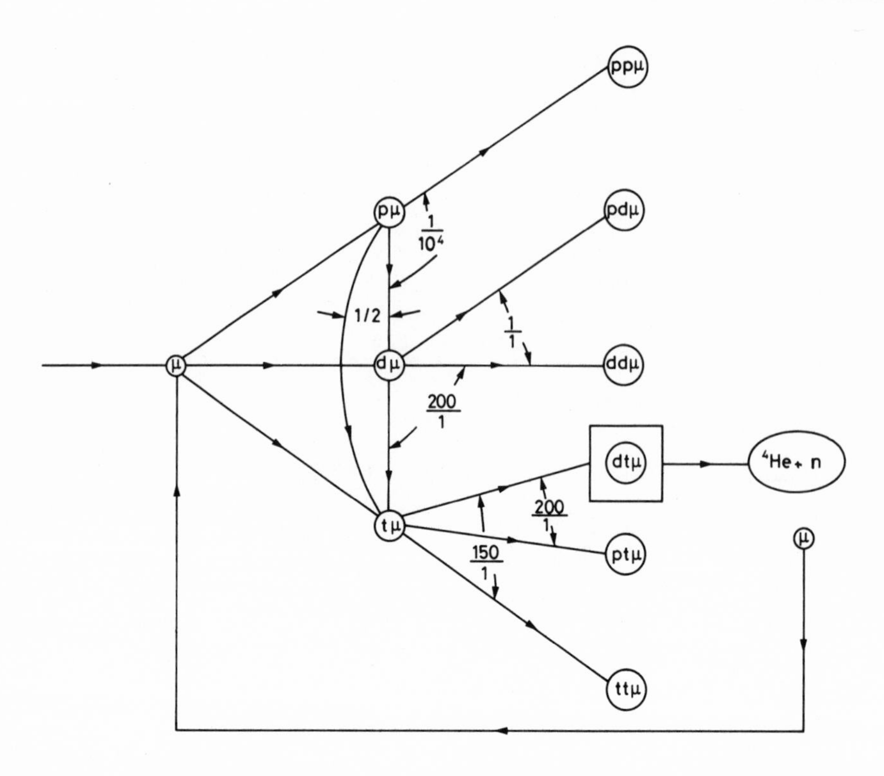

Figure 6.1 : Muon flow chart in hydrogen with the relative strength of reaction channels indicated at equal relative isotope density.

to allow that the ppμ molecule is avoided in 100-200 reactions. Since we wish a dt reaction we must have both isotopes present. Therefore let us turn now to the point dμ in Fig. 6.1. Now we must avoid both the dpμ and ddμ molecules. In order that the transfer of μ from d to t can proceed 200 times the relative concentration of t and p must be of the same order of magnitude ; otherwise the pdμ molecule is formed. Only now are we sure that all muons find their way to tritium. We can now turn our attention to the tμ point in Fig. 6.1. In order to avoid the ptμ molecule we must have as many d's present as p's. Consequently we arrive at the result that all isotopes must be present in about equal (order of magnitude) relative concentrations in order to sustain 100-200 (dtμ) mesomolecule formations per lifetime of the muon. In this optic the presence of p's seems not to be desirable. Naturally a completely different conclusion is reached when one wants to measure relative rates and uses competitive and known p processes as reference. Then a fraction of the $(dt\mu)$ formation per muon is sufficient, thus p can dominate the isotope mixture.

Now we continue our quest for 100-200 fusions and consider, in view of the above conclusions, the d-t isotope mixture as shown in Fig. 6.2. We wish to determine the best concentrations c_d and c_t of d and t, respectively, with the property :

$$c_d + c_t = 1 \tag{6.1}$$

Thus we must minimize the fraction of losses due to the formation of $dd\mu$ and $tt\mu$ molecules. We shall denote by R_d the reaction branching ratio at the point $d\mu$ in Fig. 6.2 :

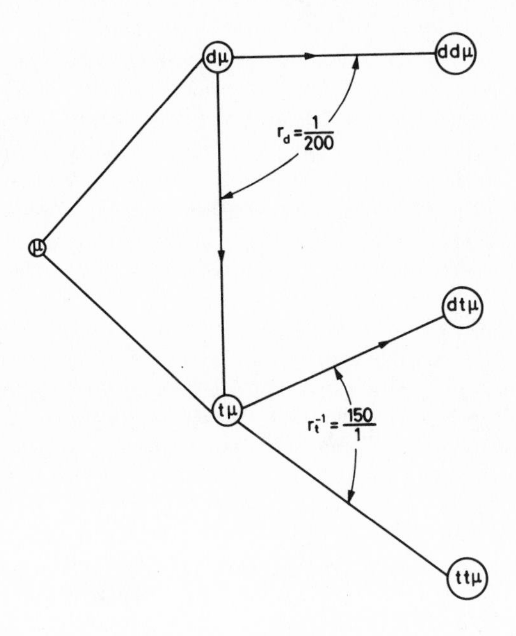

Figure 6.2 : Muon flow chart in deuterium-tritium.

$$R_d = \frac{\Lambda^m_{dd\mu} \cdot c_d}{\Lambda^T_{dt} \cdot c_t} = r_d \frac{c_d}{c_t} \tag{6.2}$$

where r_d is ~1/200 and the upper indices m and T refer to mesomolecules and transfer reactions, respectively. Similarly at the point $t\mu$ we introduce the ratio

$$R_t = \frac{\Lambda^m_{tt\mu} \cdot c_t}{\Lambda^m_{dt\mu} \cdot c_d} = r_t \frac{c_t}{c_d} \tag{6.3}$$

The actual prediction for r_t at the resonance of the $dt\mu$ ratio is ~1/150. We assume that the capture of the muon in d or t is directly proportional to their respective relative densities. Thus the probability of loss of the muon from the chain of catalyzed fusion reactions is given by

$$P_{Loss} = c_d \frac{R_d}{1+R_d} K_d + \left(c_t + \frac{c_d}{1+R_d}\right) \frac{R_t}{1+R_t} K_t \tag{6.4}$$

The factors K_i express the fact that the loss of the muon may not be necessarily permanent - in the $dd\mu$ case it corresponds to the sticking probability W_{dd} (see Section 5). It is perhaps of the same order of magnitude in the case of the $tt\mu$. Both K_d and K_t should be largely independent of c_d and c_t. Thus we can minimize Eq. (6.4) with respect to c_t upon elimination of c_d with the help of Eq. (6.1). We have, since for $c_d \sim c_t$, $R_d \sim R_t \ll 1$

$$P_{Loss} \approx \frac{(1-c_t)^2}{c_t} (r_d K_d) + \frac{c_t^2}{1-c_t} (r_t K_t) \tag{6.5}$$

and from $\partial P_{loss}/\partial c_t = 0$ we find (the inverse) relation

$$X = \frac{(1-c_t)^3}{c_t^3} \frac{c_t+1}{2-c_t} , \quad 0 < c_t < 1 \tag{6.6}$$

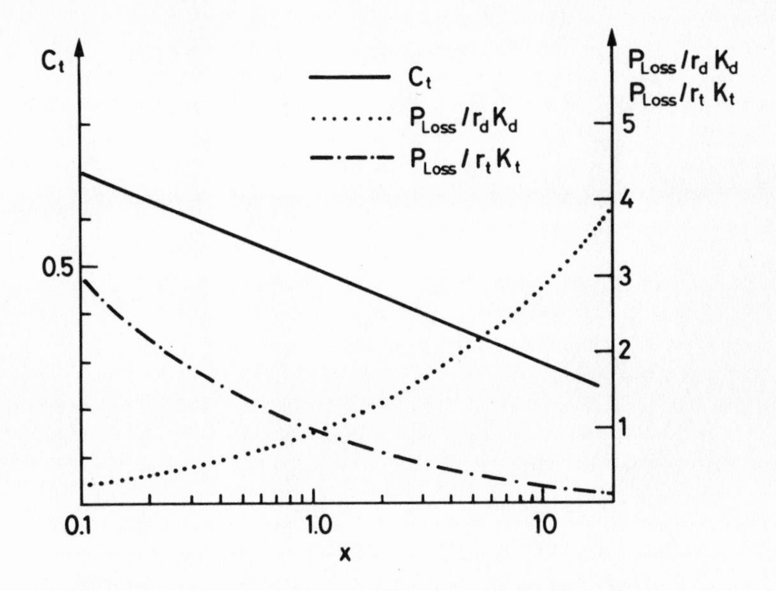

Figure 6.3 : Left scale, full line - optimum density for tritium as a function of X, the ratio of rates, Eq. (6.7). Right scale, dotted and dashed lines - reduced loss rates of muons in deuterium-tritium mixture.

where

$$X = \frac{r_t K_t}{r_d K_d} = \frac{\Lambda^m_{tt\mu}}{\Lambda^m_{dd\mu}} \cdot \frac{\Lambda^T_{dt}}{\Lambda^m_{dt\mu}} \cdot \frac{K_t}{K_d} \qquad (6.7)$$

In Fig. 6.3 (solid line, left scale) we show c_t as a function of X as defined by Eq. (6.6). As can easily be expected when the losses to $tt\mu$ and $dd\mu$ are equal and $X = 1$ we find $c_t = \frac{1}{2}$. As we can see in Fig. 6.3, a large change in X is reflected by a small change of optimum c_t. Since we expect $X \sim 1$, c_t will be ~ 0.5. This is, as one could expect in the first place, but we have also now a quantitative feeling about the dependence of c_t on the different rates. For $X = 1$, i.e., $c_t = \frac{1}{2}$, we can easily compute $P_{loss} = r_d K_d \sim 1/2000$. Thus we see that if the estimates of the different rates are correct, we could have more fusions (2000) than can be sustained at the present theoretical molecular formation rate $\Lambda^m_{dt\mu}$ during the lifetime of the muon. This situation is illustrated by the dotted line in Fig. 6.3, right scale, which describes P_{loss} at the extremum of Eq. (6.5). The inverse of the dotted values times $(r_d K_d)^{-1}$ is the number of fusions expected (up to the restriction imposed by lifetime of the muon). Thus we see that as X decreases below 1 the number of possible fusions increases (at constant $r_d K_d$). The opposite is true at constant $r_t K_t$ as shown in Fig. 6.3 by the dash-dotted curve (right scale).

Thus we see that competitive processes do not limit the number of fusions in a d-t mixture and that the actual limit is imposed at present by the mesomolecular formation rate as compared to the lifetime of the muon. As we have alluded to before, when this limit is overcome by some kind of activation of chemical rates by external parameters, it will be ultimately the slow-down time of the muon that will limit the number of fusions per muon to at most 10.000.

So far we have excluded from our discussion the possibilit that μd reacts directly with t to form a μdt molecule (an similarly with other systems). Here the energetics of the reaction changes, since we are gaining in addition to the energ shown in Table 3 the isotopic energy shown in Table 1, in the case, 48.04 eV. This is the reason why the transfer process is a much more likely event, except if there were a molecular resonant state for $J = 2$ or 3 at this energy - but there is no indication at present that this is the case.

7. - Outlook

We have seen that the prediction of the weakly bound mesomolecular $(dd\mu)$ and $(dt\mu)$ states has led to a breakthrough and has completely changed the relevance of muon catalyzed fusion among the mesomolecular reactions. In particular recent experimental evidence for temperature dependence of the mesomolecular formation rate of the $dd\mu$ molecule lends strong credibility to the theoretical work.

Many relevant rates seem not to be known that determine the fate of muons stopped in a deuterium-tritium mixture. There is in particular strong interest in the muon transfer rates between heavy hydrogen isotopes and in molecular formation rates that involve tritium. Apparently, the problem of the muon sticking to the fusion products, as well as the subsequent muon reactivation probability in the process of slowing down of muonic helium through collisions with heavy hydrogen isotopes, seems not to have been definitively settled.

The important fundamental aspect of this field is the test of our ability to understand the three-body Coulomb problem, including such an effect as the vacuum polarization.

We have so far been assuming that the resonant reaction proceeds in thermal equilibrium. However, as is now beginning to be known, the chemical reaction rates can be activated by external means such as high intensity lasers [16]. Another reason to think off-equilibrium is the fact that any muonic hydrogen is formed at finite - order eV - kinetic energy. Given the dense set of vibrational and rotational states of the electronic (dd) and (dt) molecules - resonant formation of a mesonic (dt) molecule could proceed even before thermal equilibrium has been reached - this particular aspect is absent if the dominant isotope is p. The non-equilibrium reactions do not seem to depend on the density of the hydrogenic target but on the number of collisions until thermalization, thus on the initial kinetic energy of the μt and the relative isotopic concentrations.

It goes without saying that even though the catalized fusions may lead to more thermal energy than the mass of the muon, the energy lost in the conventional way of making muons will probably outweigh the eventual fusion gains - even perhaps in most optimistic circumstances. However, one can think of the "cold" catalyzed fusion as a possible ignition point of the hot fusion process. Anyway an effort to reduce the energy cost for a muon could be of great practical importance concerning the relevance of the μ catalyzed fusion process.

In the last paragraph of this overview, let me remind the experimentalists reading this paper that the μ catalyzed fusion is still practically an open field. Clearly tritium is, in many respects, the name of the game - to the author's knowledge no experimental rates have been established involving t and μ up-to-day.

Acknowledgements

I would like to thank L.I. Ponomarev and H. Rafelski for many fruitful discussions that have stimulated many of the thoughts presented here as well as J. Duclos for his constant interest and encouragement. I also would like to thank G. Fiorentini for a careful reading of the manuscript and stimulating discussions that have clarified many points of this report and led to the discussion of the muon regeneration problem as presented in Section 5.

References

1. L.W. Alvarez, H. Bradner, F.S. Crawford, Jr., J.A. Crawford, P. Falk-Vairant, M.L. Good, J.D. Gow, A.H. Rosenfeld, F. Solmitz, M.L. Stevenson, H.K. Ticho and R.D. Tripp, Phys.Rev. 105:1127 (1957).
2. F.C. Frank, Nature 160:525 (1947).
3. A.D. Sakharov, Report of the Physics Institute, Academy of Sciences (1948).
4. Ya.B. Zel'dovich, Dokl.Akad.Nauk SSR 95:493 (1954).
5. J.D. Jackson, Phys.Rev. 106:330(1957).
6. Ya.B. Zel'dovich and S.S. Gershtein, [Soviet Uspekhi 3:593 (1961)], Usp.Fiz.Nauk 71:581 (1960).
7. V.M. Bystritsky, V.P. Dzhelepov, A.I. Rudenko, V.M. Suvorov, V.V. Filchenkov, N.N. Khovanskii and B.A. Khomenko, "Mesons in Matter", Proc.Intern.Symp. on Mesonic Chemistry and Mesic Molecular Process in Matter, Dubna (1977).
8. E.A. Vesman, Zh.Eksp.Theor.Fiz.Pisma 5:113 (1967) ; [Soviet Phys.JETP Letters 5:91 (1967)].
9. S.S. Gerstein and L.I. Ponomarev, Phys.Letters 72B:80 (1977) ; Erratum 76B:664 (1978).
10. S.I. Vinitsky, L.I. Ponomarev, I.V. Puzynin, T.P. Puzynina L.N. Somov and M.P. Faifman, "Resonance Formation of Hydrogen μ Mesomolecules", Dubna Preprint P4-10929 (1977) ; Zh.Eksp.Teor.Fiz. 74 (1978).
11. W.P.S. Tan Nature 263:656 (1976). E.P. Hincks, M.K. Sundaresan, P.J.S. Watson, Nature 269:584 (1977).
12. S.S. Gershtein and L.I. Ponomarev, "Mesomolecular Processes Induced by μ^- and π^- Mesons", in Myon Physics III - Chemistry and Solids, V.W. Hughes and C.S. Wu Editors, Academic Press, New York (1975), p. 141.
13. L.I. Ponomarev, "μ Atomic and μ Molecular Processes in Hydrogen Isotope Mixtures", SIN Preprint PR-77-011 (1977).
14. S.I. Vinitsky, L.I. Ponomarev, I.V. Puzynin, T.P. Puzynina and L.N. Somov, "The Calculation of the Energy Levels of the Hydrogen Isotope μ Molecules in the Adiabatic Representation", Dubna Preprint P4-10336 (1976).
15. V.S. Melezhik and L.I. Ponomarev, Phys.Letters 77B:217 (1978).
16. A.M. Ronn, "Laser Chemistry", in the May 1979 issue of "Scientific American", p. 103. See also references p. 150.

Part IV
Muon Spin Rotation

MUON DIFFUSION AND TRAPPING IN SOLIDS

A. M. STONEHAM

Theoretical Physics Division AERE Harwell

Didcot - Oxon OX11 ORA - UK

ABSTRACT

This paper surveys the theory underlying studies of muon motion in solids. It covers what is measured in muon experiments as well as the various forms of muon motion. The basic theory for metals is the quantum theory of diffusion, supplemented by the phenomena associated with trapping. In insulators and semiconductors there are extra effects associated with the several possible charge states and with ionisation-enhanced motion.

1. Introduction

What is actually measured in a muon experiment? Strictly, it is the muon polarisation at the time of decay. With proper timing circuitry and several detectors, one can measure

(A) The rate and sense of muon precession, which gives the local magnetic field, including any local, bulk internal and external fields, and including terms from hyperfine interactions with electrons;

(B) The decay of polarisation. Both longitudinal and transverse polarisation can be monitored. At high fields, longitudinal depolarisation monitors the transfer of muon Zeeman energy to other degrees of freedom, e.g. lattice vibrations.

After correcting for muon decay, the time dependence contains a term which oscillates at the precession frequency and which falls off with time as exp $[- f(t)]$. The relaxation function $f(t)$ depends on two factors:

(a) a mean-square dipole-dipole interaction with neighbouring magnetic moments, σ^2. From the dependence of σ^2 on the orientation of the external field one can tell which site the muon occupies;
(b) a correlation time τ_c such that the dipole-dipole interaction $h(t)$ is characterized by $\langle h(o)h(t)\rangle \equiv \sigma^2 \exp(-t/\tau_c)$. It is from τ_c that any information on diffusion is deduced. Clearly, only a restricted range of diffusion rates can be monitored. Correlation times significantly longer than the muon lifetime will be elusive: at the other extreme, fast rates are limited by the precession frequency and the time response of the counting. So, roughly, 10^{-6} sec $\gtrsim \tau \gtrsim 10^{-9}$ sec. is the working range. The correlation time does not always measure diffusion, nor necessarily muon motion. In some important cases muon motion is measured, and the type of motion are considered in §2.

An excellent summary of the experimental situation and an outline of the theory has been given recently by Seeger, in "Hydrogen in Metals" (ed. Alefeld & Völkl, Springer 1978).

2. Types of Muon Motion in Metals

The manner in which a muon moves through a metal depends on its own energy and on the temperature of the host. The motions may be local, so that they involve only a very restricted region of the crystal, or they may be delocalised motions which are restricted only by the muon lifetime and the crystal size.

Delocalised Motions When a muon enters a crystal, it has an energy far above the thermal and a kinetic energy much greater than any of the potential barriers set up by the host atoms. It exhibits a translational motion, whilst losing energy rapidly. If the host is at a high temperature, there may still be thermodynamically-significant of the (thermalised) muon showing the same sort of translational motion (Oates, Mainwood & Stoneham 1978).

At the normal temperatures of experiments two other possibilities are likely. One, the most probable, is that the muon will diffuse by an incoherent hopping motion. This diffusive motion will be discussed in the next section. The other is that the muon could propagate through the lattice by a coherent motion, rather like an electron in a semiconductor or metal. The coherent propagation is unlikely to be important unless the temperature is very low, unless the lattice distortion caused by the muon is modest, and unless the host crystal is remarkably free of defects.

Localised Motions The simplest motion is vibration. There is little doubt that the vibrational motion of hydrogen (and presumably of the muon) is highly anharmonic. This is shown both by the observed energies of excitation and by calculations of potential energy surfaces. One special form of anharmonicity is often described by the phrase "local tunnelling states". It refers to cases where the region in which the particle is localised has a potential energy surface with several minima separated by small barriers comparable with or less than the zero point energy. This particular type of tunnelling motion has been known for many years in molecules and for certain defects in solids. It was suggested for hydrogen in metals relatively recently (Stoneham 1972), and has become a particularly tempting explanation of anomalous elastic dipoles (Bucholz et al 1973), specific heats (Birnbaum & Flynn 1976), and isotope effects (Stoneham 1978). The tunnelling motion among some group of sites should not be confused with much earlier suggestions that the interstitial site with lowest free energy might change between low and high temperatures.

Clearly the localised and delocalised motions need not be independent. There may be a correlation of the two motions, and certainly one would expect the probability of a diffusive jump to depend on the degree of local vibrational excitation.

2.1 Muon Interactions in Solids

There are three principal types of interaction to concern us, namely (1) Chemical, (2) Elastic, and (3) Magnetic. The same interactions occur for muonium, though some details are altered. Chemical and Elastic interactions determine where a muon moves in a solid; the magnetic interactions determine what one can tell about what the muon is doing.

Chemically, the muon can be considered a proton and muonium as a hydrogen atom. Thus chemical bonding can be expected in any case hydrogen would bond, with small changes from isotope effects. Bond energies can be large, up to several eV; in metals, the screening by conduction electrons reduces the binding.

Elastic interactions occur because the muon distorts the lattice locally. A typical effect in a metal would be an expansion of order one atomic volume, with the neighbours moving out a few percent. The elastic interactions have three consequences: (i) Self-trapping: the local distortion caused by the muon itself may effectively immobilise the muon; (ii) Trapping, from elastic interaction with defects or impurities. Typical bindings might be 0.1-0.5 eV. If the muon causes local expansion, it will tend to prefer defects which cause local compression; (iii) the muon will have a lower energy in the site actually occupied than if it moved instantaneously to one of the adjacent sites which would have been equivalent in the perfect crystal. This underlies muon diffusion. It is clearly important to establish whether the muon is self-trapped in real solids. Both analogies with hydrogen and explicit calculations (Hodges and Trinkaus, Sol.St.Comm. 18 857 (1976); Leung, McMullen and Stott, J.Phys. F6 1063 (1976); Teichler Phys.Lett. 67A 313(1978) make it almost completely certain self-trapping will occur.

Magnetic interactions are very small compared with the others. They are determined by the dipole-dipole interactions between the muon moment $\underset{\sim}{\mu}_\mu$ in the lattice, with the general form:

$$\mathcal{H}_{dd} = [\underset{\sim}{\mu}_\mu \cdot \underset{\sim}{\mu}_L - 3(\underset{\sim}{\mu}_\mu \cdot \underset{\sim}{r})(\underset{\sim}{\mu}_L \cdot \underset{\sim}{r})/r^2]/r^3$$

where $\underset{\sim}{r}$ is the vector joining the two moments. Features to note are these: (a) The electron may not be localised. If so, the expectation value of $\mathcal{H}_{dd}$ is needed. (b) The electron may overlap the muon. If so, there is a "contact" interaction, proportional to $\underset{\sim}{\mu}_L \cdot \underset{\sim}{\mu}_\mu \, \delta(r)$. No new or subtle interactions are involved; the contact term is just a special manifestation of $\mathcal{H}_{dd}$. The constant of proportionality gives the electron density at the muon $|\psi_e(o)|^2$. (c) As a special case, in muonium the contact interaction has a large and well-defined value expressed as a field of 1585 gauss. (d) The electron moment can contain orbital angular momentum as well as the spin term. (e) The spins are quantised along the local magnetic field. As the field direction changes relative to the join of the two moments, $\underset{\sim}{r}$, so the second part of $\mathcal{H}_{dd}$ changes. It is through this that one can identify the site the muon occupies. Finally, (f), if the muon moves, $\mathcal{H}_{dd}$ determines the observable relaxation function, and this allows estimates of diffusion rates.

2.2 Muon Diffusion and its Measurement

Experimentally, there are two features which can be used to monitor muon diffusion. They are:

(I) Measurement of the correlation time of the dipole-dipole interactions. If the interaction is written h(t), then the correlation time τ_c determines how $\langle h(t)h(o)\rangle/\langle h(o)h(o)\rangle$ falls to zero at long times. The usual assumption is that $\exp(-t/\tau_c)$ fall-off occurs; this can give either Gaussian or experimental behaviour in the relaxation functions.

The correlation time, if simply interpreted, leads to a hopping time, and this is directly related to a diffusion rate. There are, of course several complications. One is not told simply if the muon motion is diffusion, rather than some other type of motion.

(II) What effects do impurities and defects have on other observed muon properties? If the muon diffuses very rapidly, only the traps will be seen, not the bulk of the host lattice. This puts a bound on the diffusion rate.

3. Quantum Theory of Muon Motion in Solids

3.1 Quantum and Classical Diffusion

In surveying muon motion in metals, I want to make two main points. The first is that quantum theory is necessary. The second is that one can construct a theory which gives expressions for diffusion rates from first principles, and that this theory predicts many qualitative features and some of the quantitative results obtained experimentally for the analogous problem of hydrogen in metals.

There are two obvious differences between classical and quantal models. One is the existence of zero-point motion, which is important here because it helps to determine the lattice distortion near a light interstitial. The second is that the energy levels of the interstitial are discrete: there is not a continuous distribution of levels up to some saddle point. But there are more profound differences which result from the recognition that it is the Schrödinger equation, not Newton's laws of motion, which determine microscopic behaviour. These differences are especially important in diffusion theory. First, the classical concept of the saddlepoint does not carry over into quantum theory, where the saddlepoint configuration does not approximate an eigenstate. Thus the saddlepoint in quantum theory loses the special position it occupies in rate theory. Secondly, as we discuss later, the nature of the local states between which transitions occur is different. Partly this is a problem of definition, and partly it comes from the problem of zero-point motion and its effects on the lattice. In particular, the initial and final states (and the saddlepoint, however defined) differ for different isotopes through the effects of zero-point motion. This difference is one reason why isotope-dependent activation energies should be expected, contrary to common mythology. The third point is that quantum effects enter into the dynamics of diffusion processes, as well as statistics. The correspondence principle defines a limit ($h \to 0$) in which the quantum theory reduces to the classical. If only statistics are important, h enters only in the form (h/KT); in this case quantum theory reduces to classical at high temperature. It is a myth that quantum theory and classical theory are always equivalent at high temperatures. There will surely be superficial similarities, but there will undoubtedly be differences in detail. It is useful to think of the classical description of motion over potential barriers as just one of several possible diffusion channels, rather than a unique limit. This is discussed in §3.4.

3.2 Formal Theory

First we consider the nature of the states between which transitions occur. In classical theory there are no problems. There is a deep potential well, and there are barriers so high that transitions between sites are prevented except during special fluctuations in quantum theory, for every state with an interstitial muon on one site, there is an equivalent state with the muon on another. From the symmetry of the total Hamiltonian it is easy to see that the exact eigenstates have the corresponding translational symmetry, and do not localise the muon on a particu lar site. However, one can define sensibly-exact stationary states, related to the exact ones as Wannier functions are related to Bloch functions in band theory. We write these states of the whole crystal as $|p,\nu\rangle$ where p labels the muon site (any other quantum numbers associated specifically with the muon can be included here) and ν labels other quantum numbers, such as the occupancies of the lattice phonon modes. The states are "sensibly exact" if their lifetime τ satisfies

$\tau \gg$ (1/frequencies of important phonons)
$\tau \gg$ (phonon lifetimes).

Since these states are not exact eigenstates, the total Hamiltonian, $\mathcal{H}$, induces transitions between them. These transitions give the atomic motion we want. The transition probabilities between specific states can be obtained from the time-dependent Schrödinger equation, and are written $w_{pp'}(\nu,\nu')$. However, we are not interested in the ν variables, which describe the detailed state of the lattice motion; we are concerned with the motion of the muon from site p to site p'. The relevant transition probability is given by

$$W_{pp} = \langle \sum_{\nu'} w_{pp} (\nu,\nu \rangle_T$$

which includes a thermal average over initial states ν and a sum over final states ν'.

We now ask: Is it obvious that the interstitial defect will diffuse by hopping from site to site, or will it propagate like a conduction electron in a metal? The answer is that both hopping and propagation can occur at appropriate temperatures, but that hopping occurs at the temperatures accessible to experiment.

At the lowest temperatures, the lattice will be in its lowest vibrational state, ν_o. The transitions which dominate are "diagonal", i.e. from $|p,\nu_o\rangle$ related by translation symmetry to the initial state. In this regime ($T < 10^{-5}$ ^oK typically) propagation occurs, not hopping.The transport rate decreases as temperature increases (or the effective mass of the interstitial increases with temperature). Even at very low temperatures, crystal defects and impurities can interrupt propagation by trapping and scattering.

At normal temperatures it is the "off-diagonal" transitions

which dominate: there is a redistribution of energy over the lattice modes in the transition $|p\nu\rangle \rightarrow |p'\nu'\rangle$. There are so many more final states accessible at higher temperatures that the diagonal transitions are completely overwhelmed. The coherence of the initial and final states is destroyed, and hopping motion dominates. Of course the coherence can be destroyed in other ways, such as by rapid transitions among states $|p\nu''\rangle$ at a given site. The essential point is that once a transition from one site to another is made, the probability that the next transition takes the system back to its initial state must be negligibly small.

3.3 Application to Light Interstitials in Metals:Basic Approximation

It proves necessary to make some simplifying approximations if we are to get quantitative results. The approximations are these

I. Born Oppenheimer approximation: that the electrons follow all heavier particles (the muon hydrogen and host metal) adiabatically. In consequence there is a unique potential energy for each lattice configuration.

II. Adiabatic Approximation: that the light interstitial follows the motion of the host nuclei adiabatically. This is equivalent to the requirement that the frequency of the local mode of the interstitial is much higher than the Debye frequency of the host. Since the interstitial can then follow the lattice motion, we may calculate an energy $E_{interstitial}(R)$ for each lattice configuration R. This energy acts as an extra potential energy in determining the motion of the lattice atoms. It includes the zero-point energy of the interstitial. Capture of an electron to form muonium will, of course, affect the potential energy surface.

Note that this present adiabatic assumption is the opposite one to that which is commonly made, where the lattice distortion is determined by the instantaneous position of the interstitial. Such an assumption would be reasonable for a heavy interstitial, but it is quite inappropriate to assume the slowly-moving lattice ions can follow the rapid motion of the light hydrogen interstitial.

III. Linear approximation: that the term $E_{interstitial}(R)$ is linear in the displacements of the host lattice atoms

$$E_{interstitial}(R) = E_o(R_o) + (R - R_o)E_1(R_o).$$

In consequence the atoms of the host are simply displaced; the frequencies and eigenvectors of the modes are not altered. The distortion of the lattice when the interstitial is on a given site is called the "self-trapping" distortion, since it reduces the mobility of the interstitial. There is an associated energy, the self-trapping energy

$$1/2\ \Sigma_q\ M_{host}\ \omega_q^2 |\underset{\sim}{Q}_q|^2$$

which is the reduction in the sum of $E_{interstitial}$ and elastic strain energy on distortion of the lattice. The importance of this distortion, and the analogy with the electronic polaron, was noted by Schaumann et al. Note that the displacements $\underset{\sim}{Q}_q$ also determine the volume of solution, δV. This will be important later.

It is perfectly possible to go beyond the Linear Approximation, but the extra complexity is rarely justified.

IV. Condon approximation: that the transition matrix element should be independent of the host lattice configuration, R

$$<\phi_{muon}(r_\mu;R)|\mathcal{H}|\phi_{muon}(r_\mu;R)> = J = \text{constant}.$$

This seems a good approximation for Direct Processes, where the barrier to motion comes from the self-trapping displacements alone. Other cases will be described later which go beyond this approximation.

V. The Harmonic Approximation. This is much less important than in classical theory, and enters in a different way. It is unnecessary to assume that the muon-host atom interaction is harmonic. Only the harmonicity of host atom motion is important, and this enters through the overlap of the initial and final host lattice states. These states are different because the interstitial causes self-trapping distortions about two different sites in these states.

3.4 Transition Rates

The transition rate can be obtained by straightforward, if complicated, application of quantum mechanics. The result can be seen more directly by simpler physical arguments. These are based on the idea that there is some range of lattice configurations which permit particularly fast motion from one site to another. The rate then depends on the probability of the desired configuration being achieved. In the high temperature limit of a single oscillator we distinguish two cases:

i) The probability that displacement $\underset{\sim}{Q}$ lies between $\underset{\sim}{Q}_o$ and $\underset{\sim}{Q}_o + d\underset{\sim}{Q}$

$$W_I \sim T^{-1/2} \exp\{-E(\underset{\sim}{Q}_o)/kT\}\, dQ$$

where $E(Q_o)$ is $\frac{1}{2} M\omega^2 Q_o^2$.

ii) The probability that displacement Q exceeds Q_o

$$W_{II} \sim T^{1/2} \exp\{-E(Q_o)/kT\}$$

At low temperatures, a variety of quantum effects become important, both in the statistics and dynamics of the problem. Naturally the constants of proportionality depend on the details of the system. But the high temperature behaviour is often governed by probabilities of the types given above.

We first give examples of type (i), where specific configurations $\underset{\sim}{Q}_o < \underset{\sim}{Q} < \underset{\sim}{Q}_o + d\underset{\sim}{Q}$ are involved.

a) In transitions between electronic states whose energy surface cross at certain values $\underset{\sim}{Q} \sim Q_o$, the non-radiative transitions are dominated by the configurations near the crossing point. Such transitions occur, for example, in the excited state of the F centre, and the existence or non-existence of luminescence is determined by them.

b) The motion of hydrogen in bcc metals in the Flynn-Stoneham model is by transitions of type (i). When the lattice phonon energies are much less than the hydrogen local mode energy, the hydrogen moves much faster than the slow-moving lattice atoms. The important transitions are those between the lowest hydrogen levels in the initial and final interstices. But, because of the much faster hydrogen motion, these can only occur when the lattice configuration $\underset{\sim}{Q}_o$ is such that the initial and final hydrogen states have the same energy. As Flynn has observed these states must be degenerate to within the tunnelling energy, so that $d\underset{\sim}{Q}$ in W_1 is proportional to Ω_T, the tunnelling frequency for this "coincidence" state where the energies are the same.

Since the transition probability is also linear in Ω_T when Q is in the correct regime, the expected high-temperature rate varies at $\Omega_T{}^2 T^{-1/2} \exp\left\{-E(\underset{\sim}{Q}_o)/kT\right\}$, as predicted in the more detailed calculations.

c) In small-polaron motion, as in the motion of V_k centres in ionic crystals, the behaviour is exactly analogous to that of proton motion described above. The main differences stem from the fact that a self-trapped hole, rather than hydrogen, moves, and that the self-trapping distortions are primarily lattice polarization rather than dilatation.

The best-known example of a type (ii) process is classical diffusion. In this picture the diffusing particle must have a displacement of the reaction-coordinate which is sufficiently large to take the particle up to saddle-point of the transition, or beyond. Here $\underset{\sim}{Q}_o$ corresponds to the saddle-point. Another example occurs in the Flynn-Stoneham model of hydrogen diffusion in fcc metals where the Condon approximation fails. The tunnelling matrix element Ω_T depends strongly on the positions of the atoms straddling the jump-path (there is some analogy too with classical arguments based on atom-size and misfit, but the correspondence is not exact). When these atoms are sufficiently displaced $(\underset{\sim}{Q} > \underset{\sim}{Q}_o)$ the value Ω_T becomes large enough that transitions in this regime dominate.

3.5 Results Using the Condon Approximation: Direct Processes

The most transparent results are those at high temperatures (in practice for temperatures near the Debye temperature or above) when the hopping rate becomes

$$W = \frac{\pi}{4\hbar^2 E_a kT} |J|^2 \exp(-E_a/kT).$$

This has obvious similarities to the classical expressions. The "activation energy" E_a is given by

$$E_a = 1/2 \ \Sigma_q M_{host} \omega_q^2 |\Delta Q_q/2|^2$$

where ΔQ_q is the change in the displacement of the mode in the jump. E_a is not the potential barrier to the motion of the interstitial. It is a lattice activation energy, the energy needed to take the lattice from its initial self-trapped configuration to the "coincidence" geometry, where the interstitial has the same energy on either the initial or the final site. Two other features should be noted. First only modes whose displacements are altered in the transitions contribute to E_a. These we call "antisymmetric" modes, from their symmetry with respect to the jump plane. Secondly, E_a is a strain energy, and this leads to the Zener relations.

At low temperatures (in practice below about 10% of the Debye temperature) the hopping rate tends to a T^7 dependence (the exact power depends on some of the details of the model). There is no abrupt change, and the parameters are usually such that a clear division into low-and high-temperature regimes is not possible.

We discuss the values of the parameters and the detailed temperature dependence later. The activation energy E_a can be obtained entirely in terms of independently-obtainable parameters.

A further complication is that there are excited states of the muon within each interstice, and these may be involved in diffusion. The basic formulae were given by Flynn & Stoneham. The idea has been taken further by workers at Sandia recently, who note both effects on the rate and the isotope effect.

3.6 Beyond the Condon Approximation: Lattice-Activated Processes

In the direct processes, the only self-trapping distortion provides the barrier to motion. This is not always valid, as can be seen from Fig. 2. Case (a) corresponds to a direct process. In case (b), however, there are two atoms in the jump plane which hinder the motion of interstitial. The important point is that the transition matrix element will be sensitive to (symmetric) motion in the jump plane of these atoms. The Condon approximation will cease to be valid.

A simple model which shows the important effects assumes the transition matrix element is zero unless the two atoms of interest are separated by more than some critical distance, when the matrix element has a constant value J_o. The theory goes through without difficulty since this symmetric mode (whose mean displacement does not change in the transition) is independent of the asymmetric modes contributing to E_a. The final result is

$$W = \frac{1}{4\hbar} |J_o|^2 \frac{1}{\sqrt{E_a E_s}} \exp\left[-(E_a + E_s)/kT\right]$$

There is an extra term in the activation energy, and the temperature dependence of the pre-exponential factor has vanished. E_s is the elastic energy to produce the critical distortion giving J_o.

It appears that for interstitials on octahedral sites of bcc metals, where there are no atoms straddling the jump path, the Condon approximation holds and the self-trapping term dominates in the activation energy. For interstitials in fcc metals there are atoms straddling the jump path. The Condon approximation is not valid, and E_s forms an important path of the activation energy. The Sandia workers have also argued for corrections to the Condon approximation, though in a different form. They note that J may differ between the unrelaxed lattice geometry and the geometry appropriate to the "coincidence" state where the muon has the same energy on either site. This correction affects the absolute rate and isotope dependence, rather than the Flynn-Stoneham correction which increases the activation energy. As a further comment on the point made by the Sandia workers, the "tunnelling states" models for hydrogen and muon diffusion only need tunnelling in the coincidence state. Thus tunnelling states could be important in diffusion yet unobservable in neutron scattering.

3.7 Quantum Theory of Diffusion: Summary of Theory

Bringing together the basic ideas, one can envisage three main regimes of muon motion:

I. A propagating regime where, even in the lowest vibrational state the muon may propagate through the lattice rather like an electron. This regime is very sensitive to lattice strain, and will normally be suppressed by the dislocations and other defects present. Even in a perfect crystal the regime occurs only at very low temperatures when there is strong coupling, as for hydrogen in metals.

II. A hopping regime where the activation energy is associated with the lattice distortion produced by the particle and not by a potential energy barrier in the energy surface for particle motion.

III. A "classical" regime in which diffusion is dominated by either excitation to states above the potential energy barriers or to states below the barriers for which tunnelling is sufficiently rapid.
I know of no system where all three regimes are unambiguously observed. For hydrogen in metals II is by far the most important, and I am not aware of clear evidence for I or III. For hydrogen in insulators or oxides the situation is less clear.

The important features of the hopping motion can be identified from three simple ideas. The first is that the muon is much lighter than the host lattice atoms and moves much more rapidly than they do. In a hopping jump from one site to another, the host atoms hardly move at all, and little if any energy is exchanged between the muon and the host. Thus the host atoms must be in some instantaneous configuration such that the total energy is the same whichever of the two sites the muon occupies. The second idea is that, when the energies do indeed match, the probability of the jump occurring depends on a transition matrix element (J(Q) which may depend on the precise atomic positions. The third point is that, when the muon is localised in one interstice for even a short time, the lattice distorts around it to give the "self-trapping" distortion discussed in the next section. The activation energy for the hopping motion then contains two principal contributions: the lattice strain energy E_a needed to distort the crystal from its self-trapping configuration to one in which the total energy is the same whichever of the two sites the muon occupies, and any further lattice strain energy E_s needed to give the optimum value of the transition matrix element J(Q). This second contribution is omitted in standard small polaron theory but, as Flynn & Stoneham (1970) argued, gives a major contribution in fcc hosts.

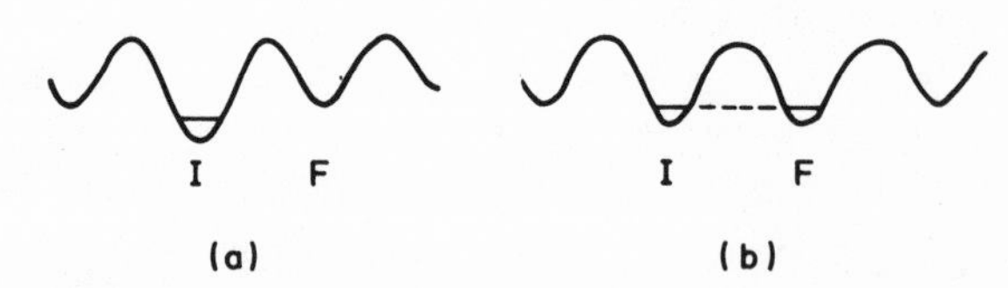

Figure 1. Illustration of the diffusion limited only by the self-trapping distortion. In (a) the interstitial has a lower energy in the central site than in the adjacent ones and cannot tunnel; in (b) a fluctuation makes two sites energetically equivalent. The activation energy is the energy needed to produce the fluctuation.

4. Comparison of the Quantum Theory with Experiment

So far the arguments for quantum theories, as opposed to classical, and the basic formalism have been presented. The real test of the theory is its success in practice. The comparison will concentrate on the theory of Flynn and Stoneham since they attempt to use reasonable physical assumptions and to make predictions which can be compared with experiment. Most other workers treat just simple model systems or use a phenomenological analysis. The analysis here obviously depends a lot on data for hydrogen in metals, where data are more complete.

4.1 Hydrogen in Metals: General Aspects

A) Distinction Between Motion in bcc and fcc Hosts

The distinction was described in Section 3.5. In essence, the jump probability in an fcc host is sensitive to the motion of the two host atoms which lie in the jump plane; there are no corresponding atoms in bcc hosts. The motion in fcc lattices is thus "lattice activated", and that in bcc hosts "direct". Other things being equal, a higher activation energy is expected and observed (see (C) below) for fcc hosts.

B) Relation Between Activation Energy and Volume Change

The linear approximation leads to the relation

$$E_a \sim (\Delta V/V)^2 .$$

For a given host, this result does not depend in the use of continuum elasticity. Results are available for carbon, nitrogen and hydrogen in bcc Fe, and show the following ratios:

$$\underset{\sim}{Q}_C : \underset{\sim}{Q}_N : \underset{\sim}{Q}_H \quad = \quad 1.00{:}0.081.0.08$$

$$(\Delta V)_C^2 : (\ V)_N^2 : (\ V)_H^2 = 1.00{:}0.88.0.11$$

Other theories, in terms of transitions through excited states in an anharmonic potential well, do not predict a simple relation between E_a and ΔV.

C) Absolute Magnitudes of Activation Energies

The final expression for the activation energy, E_a, involves no free parameters. For bcc host it has the form:

$$E_a = \frac{M\omega_D^2 d^2}{360} \left(\frac{1+\nu}{1-\nu}\right)^2 \left(\frac{\Delta V}{\Omega}\right)^2 \Phi$$

Here M is the host atom mass, d the jump distance, ω_D the Debye frequency, ν Poisson's ratio and $(\Delta V/\Omega)$ the volume change per interstitial in terms of the atomic volume. Φ is a complicated but well-defined function which appears because of the tetragonal symmetry of the interstitial sites. Predictions for bcc hosts are in the range 0.05 - 0.15 eV, depending on the precise assumptions made. There are uncertainties in the various parameters - notably $(\Delta V/\Omega)$, but also in which of the Debye temperatures is appropriate - which probably make the calculated E_a unreliable to as much as 50%. However, the general agreement with experiment is most satisfactory. For the fcc host Pd, the qualitative predictions of (A) are confirmed: the observed Q $\sim$ 0.26eV for Pd:H is much larger

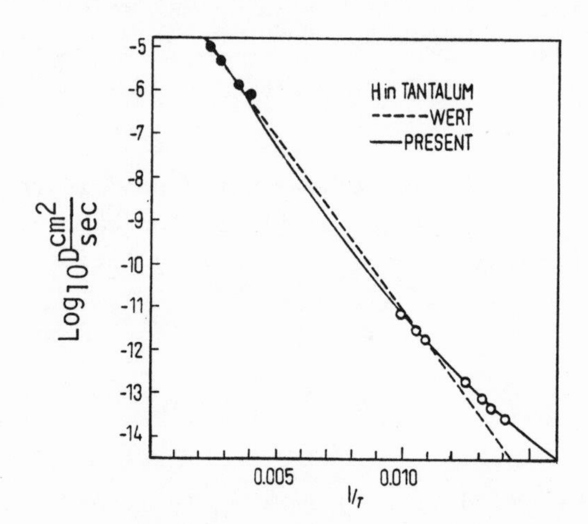

Figure 2. Temperature dependence predicted for Ta. Predictions of the Plynn-Stoneham theory compared with experiment. --- Wert (1970). —— Stoneham (1971). ● Merisov et al. (1966). o Cannelli and Verdini (1966).

than predicted $E_a \sim 0.05$eV; the difference is a measure of the energy E_s.

D) Temperature Dependence of the Diffusion Rate

The results in 2 give the diffusion rate analytically in the low and high temperature limits. Results at intermediate temperatures have also been obtained. They show (i) that the high temperature asymptote, which is very similar to the Arrhenius expression, holds to much lower temperatures than one might expect; the range of temperature is largest when the ratio $(E_a/\hbar\omega_D)$ is least; (ii) that the low temperature asymptote holds only at very low temperatures, usually below those met in practice, and that deviations from the low-temperature asymptote at higher temperatures may be of either sign; and (iii) that an excellent fit to the results for Ta:H and for H trapped by O in Nb may be obtained.

E) Zener Relations

In 1952 Zener observed an empirical relationship between the motion entropy S_m and motion energy for interstitials in bcc lattices. E_a should be regarded as a Gibbs function for migration; it is both temperature and pressure dependent. The motion entropy, S_m, and motion volume, V_m, may be obtained from E_a (or $E_a + E_s$ where relevant) by the standard thermodynamic formula

$$S_m = -(\partial E_a/\partial T)_p,$$

$$V_m = (\partial E_a/\partial P)_T$$

Since E_a and E_s are elastic strain energies, the motion entropy is obtained from the temperature dependence of the appropriate elastic constants (principally shear constants) in the Debye approximation. This explains the empirical relation between Q and S_m pointed out originally by Zener. Our equations for E_a and the self-trapping energy show that the motion entropy may be related rather generally to the lattice mode frequencies. A similar explanation has been given by Flynn in 1968 for the Zener relation in solvent diffusion. We note, however, that when energy contributions other than elastic strain energy are important the Zener relation may not hold. Thus for direct transitions, in which $|J_{pp'}|^2$ may be sensitive to temperature and pressure, the Zener relation need not hold. In lattice-activated processes, the pressure and temperature dependence is dominated by the dependence on the strain energies E_s and E_a; the relation should then be valid.

4.2 The Isotope Effect

The isotope effect is very complicated. Experimental data are reviewed by Alefeld & Wipf (1975) and the theory is discussed in detail by Stoneham (1978). For present purposes, the differences in muon and hydrogen masses are important for these reasons:

(1) The self-trapping distortion may be modified by the different zero point motion and altered importance of local tunnelling states;

(2) The transition matrix element may be altered, both for the same reasons as in (1) and because of a different degree of breakdown of the Condon approximation (dependence of J(Q) on Q);

(3) The altered importance of excited states, both localised and delocalised. These are amplified in the following table (see Stoneham 1978).

SCHEMATIC DIAGRAMS OF OBSERVED

ISOTOPE EFFECTS

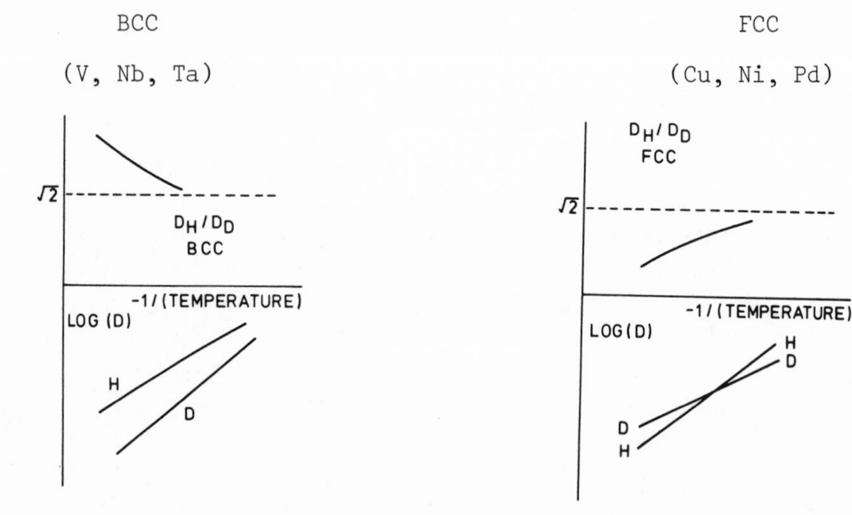

Contribution to the isotope effect

Here a positive effect is one which increases the rate for lighter masses, i.e. the lighter the mass, the smaller the activation energy or the larger the pre-exponential factor.

Effect	Pre-exponential factor	Activation energy	Comment
1. Classical: Conventional:	+ $(m_{eff}^{-1/2})$	nil	m_{eff} is a weighted average of the host and particle masses. All interactions assumed harmonic.
Probability of successful jump	–	nil	Classical theory is usually developed for harmonic systems,and assumes a jump will occur whenever a saddle point is crossed in the right direction. Molecular dynamics calculations (which usually include anharmonicity) stress how hard it is to justify this view, for there are often "jumps" in which the diffusing particle moves straight back to the site it has vacated. Bennett has suggested another criterion, as follows. Suppose one considers the motion from A to B via a saddle-point S. Then a successful jump is one in which the system passes from A through X (between A and S) and then through Y (between S and B) without making any subsequent intermediate crossings of X.

Contribution to the isotope effect (Continued)

Effect	Pre-exponential factor	Activation energy	Comment
2. Quantal			
(a) zero-point motion polaron effect		-	There is zero-point motion which causes a distortion of the surrounding lattice. This "polaron" effect alters the activation energy, increasing if for lighter isotopes.
(b) tunnelling matrix elements purely overlap effect: dependence on geometry:	+	nil	The overlap of initial and final state wavefunctions, and is usually bigger for the larger zero-point motions of lighter atoms. In some cases the matrix element is very small unless atoms straddling the jump path move by more than some critical amount. The critical movement leads to a term in the activation energy corresponding to the strain energy needed to make the movement. Since the critical amount is isotope-dependent, the activation energy will change with isotope. Probably the contribution to E_a will be less for lighter isotopes, simply because the larger zero-point motion makes smaller critical displacements possible.

Contribution to the isotope effect (Continued)

Effect	Pre-exponential factor	Activation energy	Comment
(c) excited states delocalized:		+	If the excited state is delocalized so that its energy depends little on the isotope mass, then the activation energy (the difference between ground and excited state energies here) will be less for lighter masses because the zero-point energy $\frac{1}{2}\hbar\omega$ raises the ground state energy the light mass relative to the heavy mass.
harmonic oscillator:		-	If the levels are like those of a harmonic oscillator, the larger frequency ω for the light mass raises the excited state energies and increases the activation energy.

Contribution to the isotope effect (Continued)

Effect	Pre-exponential factor	Activation energy	Comment
(d) local tunnelling states		(reduce isotope dependence)	Local tunnelling states. One effect which proves especially important for II in bcc metals like V, Nb and Ta comes from the possibility that light atoms can tunnel between interstitial sites within one interstice, as well as jumping between interstices. This phenomenom has as its main effect the increase of the transition matrix element and a diminution of the polaron isotope effects. However, the situation is complicated because the precise form of the potential energy surface is affected by the local lattice distortions.

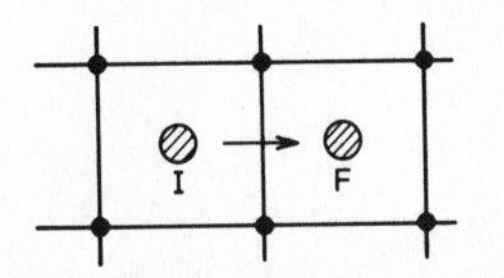

I:
SIMPLE INTERSTITIAL:

NO TUNNELLING

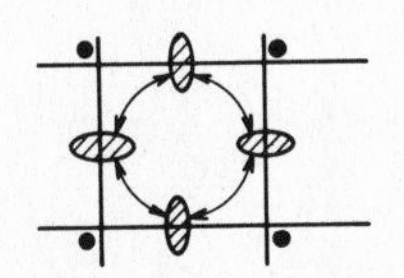

II:
TUNNELLING IN INITIAL STATE.
SELF - TRAPPING DISTORTION SHOWN.

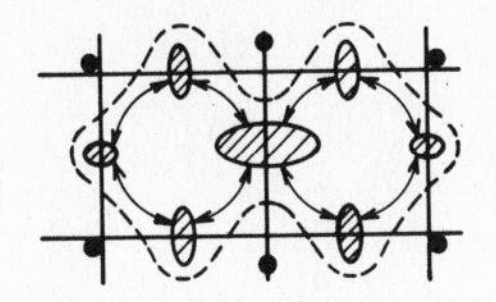

III:
TUNNELLING IN COINCIDENCE STATE.
CASES I OR II CAN APPLY FOR INITIAL STATES.

● METAL ATOM

REGION WHERE INTERSTITIAL PROBABILITY DENSITY EXCEEDS SOME SPECIFIED VALUE

----- LOWER DENSITY CONTOUR

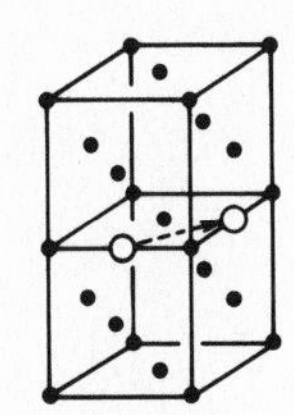

fcc LATTICE
● METAL
○ OCTAHEDRAL INTERSTITIAL

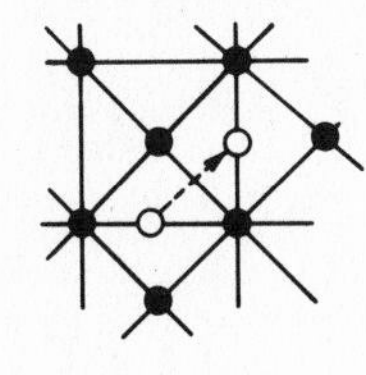

fcc (100) PLANE

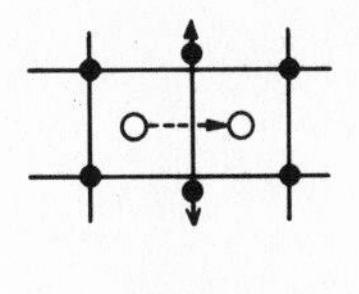

2 - DIMENSIONAL SQUARE LATTICE: BREAKDOWN OF CONDON APPROXIMATION

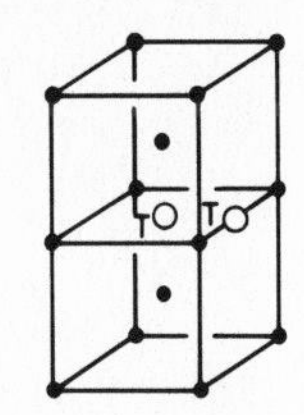

bcc LATTICE
● METAL
○ OCTAHEDRAL SITE

T TETRAHEDRAL SITE

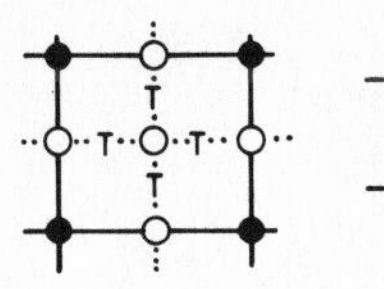

bcc LATTICE (100) PLANE

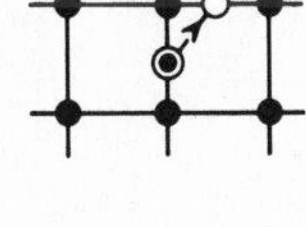

2- DIMENSIONAL SQUARE LATTICE: CONDON APPROXIMATION ADEQUATE

5. Muon Trapping

Several groups have found evidence that muons can probe defect and impurity properties in metals. Since the defects include deformation-induced dislocations and damage from neutron irradiation, and since the impurities include both alloying metals and trace impurities like oxygen, the muon promises to be a very versatile tool. Many gaps remain. To understand these, one must look at the principal processes involved. The first is the slowing-down process, in which the muon comes to rest. Usually the muon will end in an interstitial site, self-trapped by the lattice distortion it causes, and vibrating like any normal chemical impurity. The second process is diffusion. Presumably this is described by the quantum theory developed for ligh interstitials by Flynn and Stoneham; the features of importance here are that the rate should increase with temperature, as a power law at low temperatures and thermally-activated at higher temperatures. One might expect μ^+ diffusion to resemble diffusion of H^+, D^+ or T^+, with a suitable isotope effect, though present data are too limited to check in detail. The third process is defect trapping, due to the strain fields of imperfections or to their perturbation of the electronic structure. At high temperatures, release from traps occurs There is also a fourth process, independent of the solid: the muon decays with a characteristic time of 2.2 μsec, and this imposes limits.

5.1 Trapping Mechanisms

Chemical and elastic interactions determine where muons go in solids. It is thus these features of impurities and defects which determine trapping. Very weak traps simply moderate observed diffusion; very deep traps may eliminate all but local motion.

5.1.1 Weak Traps

Several mechanisms can give small variations of muon energies from site to site. If the muon is hopping, as in Regime II, there is a small extra term in the activation energy (Flynn and Stoneham).

At lower temperatures in the same regime, where two-phonon processes would be expected to give a characteristic T^7 behaviour, there should be an extra component linear in temperature. This has been discussed as a possible explanation of the observed rate in vanadium (Fiory et al 1978).

The main source of the fluctuations in energy from site-to-site will be the random strain fields due to distant defects. The theory of these strains is reviewed by Stoneham (1969). Alloys of like elements (perhaps of Na and K) could also provide small chemical fluctuations. One interesting suggestion by Professor Crowe concerns the effects of isotopes: are there any significant differences between metals with just one isotope (e.g. Al) and ones with a mixture of isotopes (e.g. Cu)? The energy fluctuations involved can be worked out as follows. In the metal without the muon present, the isotopes modify the zero-point energy. Light isotopes vibrate with larger amplitude and cause local expansion, with displacements U_o. If a muon is put in, it exerts forces F on the lattice atoms. The terms $\underline{F}.\underline{U}_o$ give one contribution to the energy. Another possible contribution is the relaxation energy from the self-trapping distortion. However, Hughes ((1966); see also Stoneham 1975 p.163) has shown this second term has no host isotope dependence. Thus only the $\underline{F}.\underline{u}_o$ term gives fluctuations: each light isotope can be considered as a centre of dilatation (and heavy isotopes with corresponding a local contraction), and can affect muon motion just like the other random strain fields. The effects are weak, however; even at sites adjacent to the odd isotope, the interaction energy will be only of order 10^{-4} eV.

5.1.2 Deeper Traps

Broadly speaking, one expects positive muons to be trapped easily where hydrogen and/or positrons are trapped. Thus vacancies, dislocations, voids, and chemical impurities like oxygen are all likely traps. The precise association may be complicated. For example, one expects μ^+ to trap in a vacancy. Yet in some damaged metals (Al, Cu, Ag), Ligeon and his colleagues find hydrogen settles in the adjacent site, i.e. as an impurity interstitial-vacancy pair. Presumably the chemical and elastic interactions are working in opposite directions.

5.2 Effects on Muon Properties

If there is only a single type of trap, one can outline the way the depolarisation rate should vary with temperature. Four main regimes are expected. In I, at the lowest temperatures, the muon does not diffuse significantly; it merely comes to rest, and measures an average property of the solid. In II the muon diffuses faster as the temperature rises, and the rate at which it samples different internal magnetic fields affects the depolarisation rate and can be measured. This regime ends at a temperature fixed by the trap density, and in region III, the muon monitors properties of the trapping site. Finally, in region IV, release from traps and motional-narrowing occur. Obviously, behaviour gets more complicated when several distinct traps are involved, but the principles are the same. Many workers have obtained results fitting this qualitative description of the temperature dependence. The general theory has been given by several workers, including Kehr, Honig and Richter (μSR Newletter =20 p794).

For this picture to be useful, it is important that the muon can indeed diffuse to the traps of interest. The muon lifetime sets a limit here. In most metals studied (V, Nb, Ta, Cu) at accessible temperatures the muon hopping time between diffusive jumps is in the range 10^{-5} - 10^{-7} sec. corresponding to a mere 0.2 to 20 jumps in a lifetime. Diffusion from a random stopping site to any trap present in low concentration is negligible. And indeed

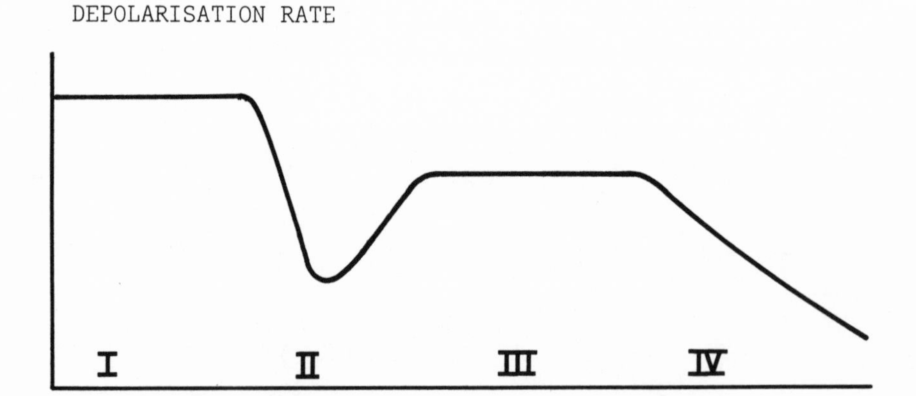

the irradiation of Cu and the doping of Nb by Ta have little observable effect on muon properties. Two metals are different, and appear to show very fast diffusion: Cr, with a hopping time around 10^{-11} sec. and Al, where motional narrowing appears to be complete even at the lowest temperatures, with times like 10^{-10} sec. at 10°K and 10^{-12} sec. at 50°K indicated.

Several groups have studied trapping in Al, including workers from seven American, one Swedish and three German Laboratories. Two results are particularly interesting. One was reported by Seeger at the recent Rutherford Laboratory meeting, and is from Stuttgart workers, who looked at neutron-irradiated Al. The point here is that the basic defects, and hence the traps, can be modified by annealing. Reversible changes with temperature reflect the muon motion, whereas irreversible ones indicate changes in defect structure and can be related to previous damage work. Two traps were detected in Al: one was shallow, possibly an interstitial cluster; the other trap was deeper and is not yet identified. The second result comes from a collaboration among five of the American laboratories (Phys.Rev.Lett. 41 1558 (1978)). The experiments showed the effects of Cu impurity and of deformation. The dislocations act as deep traps, with a binding of around 0.2 eV, and also seem to show the rapid pipe diffusion known from conventional diffusion studies. The Cu appears to give rise to some four different shallow traps (binding the muon by less than 1/50 eV) even at the low concentrations used (420 and 1300 ppm). The authors argue these traps are microclusters of Cu, and that the clusters are present at levels far above random. If so, the muon technique is giving new information about alloy structure in circumstances hard to match otherwise.

5.3 Muons as monitors of materials properties

If the interpretations are right, muon spin resonance adds another tool to the study of alloy structure, complimentary to positron annihilation, neutron scattering and various forms of electron microscopy. Clearly muon experiments need a large accelerator, and cannot be as portable as techniques like positron annihilation; applications to macroscopic defect structure, as in non-destructive testing, will have to be reserved for special cases. Open questions remain too. Are there really so many microclusters in Al:Cu? Are there systems where such muon results can be checked by other methods, like magnetic or Mössbauer techniques? If these microclusters are real, do they affect mechanical properties? Another question is in how many systems can the muon probe defects and impurities? Diffusion rates mentioned earlier suggested that the success with Al was the exception, not the rule. This need not be so. Slow diffusion only prevents the muon probing a trap if the stopping site is random within the crystal. However, a slowly-moving muon may be stopped preferentially near defects, by its interaction with their strain field or otherwise. This may be why results on vanadium are very sensitive to interstitial oxygen. In such cases the muon will probe the defects, but will give little information about the perfect host: one will see only the regimes III and IV of §5.2.

6. Muons in Insulators and Semiconductors

6.1 General Differences from Metals

For rather general reasons there are important differences between insulating or semiconducting hosts and metals. First, the muon may exist in one of several charge states. Secondly, the muon may form local chemical bonds with one or more neighbours. Thirdly, the ionisation which is produced by the muon may produce observable effects: luminescence, ionisation-enhanced diffusion, and possibly defect production analogous to photochemical damage.

Some of these features can be illustrated by what is known hydrogen impurity. In the alkali halides, interstitial hydrogen is seen in both the H_i^o (U_2 centre) and H_i^- (U_1 centre) forms. Beck, Meier and Schenk (Z.Phys. B22 109 (1975)) have identified the muon analogue of the neutral defect (μ_i^o). The localised chemical bonds involving OH in oxides and CH in carbides are also found frequently. The hydroxyl is observed in different types of oxides varying from the ionic, like MgO, through silica to oxide glasses, and it would be surprising if (Oμ) was not to be found. Hydrogen is rather elusive in semiconductors. It is not observed in crystalline Si, for reasons still uncertain, though it may be that H_2 is formed. Some forms of amorphous Si, however, clearly contain massive amounts of hydrogen saturating dangling bonds, and these would repay studies with muons. In SiC an especially interesting form of hydrogen has been seen by Choyke, Patrick and Dean, involving CH formation at a Silicon vacancy.

In any muon experiment there will probably be a large amount of ionisation produced in the wake of the muon and as a consequence of any other radiation present. Since there is typically only a single muon in the system at any time it will not be possible to detect luminescence or damage associated directly with the muon. However, the muon itself may be affected by the ionisation. One possibility is that vacancies are produced and that the muon is trapped in such a vacancy. This is most likely in cases like the alkali halides where anion vacancies and interstitials can be produced by non-radiative electron-hole recombination (see, e.g. Itoh 1976) and where the corresponding substitutional hydrogen centre (U centre, H_s^-) is known. The second possibility is ionisation-enhanced diffusion of the muon. This phenomenon is well known in semiconductors (see e.g. the references in Stoneham, Tasker and Catlow 1978). Enhanced motion of hydrogen in MgO in the presence of ionisation has also been observed (Chen, Abraham and Tohver 1976). There are three general types of ionisation-enhanced motion. In Local Heating models the ionisation leads to vibrational excitation which enhances diffusion. This should be very effective for muons, since vibrational excitation of the local mode associated with muon itself should speed diffusion jumps. In Local Excitation models, electronic excitation occurs to excited states for which the potential energy surface allows more rapid diffusion. This may operate in the alkali halides, where hydrogen centres can be affected strongly by optical excitation. The third model, due to Bourgoin and Corbett (1972), could operate if the muon occupied different sites in different charge states (e.g. different interstitial sites for $[\mu^+e]$ and $[\mu^+ee]$). Since the motion proceeds by successive capture of electrons and holes, one would assume the mechanism more effective for semiconductors with mobile holes than for polar crystals with self-trapped holes. Ionisation-enhanced diffusion by any of these mechanisms may be identified by the low (possibly zero) activation energy expected. There are thus special problems with non-metals. Ionisation-enhanced diffusion in semiconductors may ensure that the muon can always reach a defect; recombination-produced vacancies and interstitials in ionic crystals may ensure there are always defects to be reached.

6.2 Muons in Silicon

Two forms of muonium are seen in silicon, with properties contrasted in the following table. Note the hyperfine structure referred to has the form

$$A_{\parallel} S_{Ze} S_{Z\mu} + A_{\perp}(S_{xe} S_{x\mu} + S_{ye} S_{y\mu})$$

in both cases; isotropy implies $A_{\parallel} = A_{\perp}$.

	Normal Muonium	"Anomalous" Muonium
1.	Less common form	"Common" form
2.	Isotropic hfs	Anisotropic hfs$_7$ (111) symmetry.
3.	High hfs $A_{\parallel} + 2A_{\perp} \sim 45\%$ of free muonium value	Low hfs $A_{\parallel} + 2A_{\perp} \sim 2\%$ of free muonium value
4.	Seen in p-type Si below 80°K Not seen in n-type Si	Seen in p-type Si below 80°K Seen in n-type at low temperatures
5.	References: Adrianov et al., Sov.Phys. JETP 31, 1019 (1970) Brewer et al.,Phys.Rev.Lett. 31 143 (1973).	Reference: Patterson et al., Phys.Rev.Lett. 40 1347 (1978)

Models for the two forms are still primitive. A high hyperfine structure tends to go with a compact, tightly-bound centre, and normal muonium could be like a hydrogen atom at a tetrahedral interstitial site. Against this, the state of hydrogen itself has not been identified - indeed, it could easily be present in molecular form. The anomalous muonium shows parallels with results for Si implanted with deuterium (Picraux and Vook, Phys.Rev. B11 2066 (1978)) where the D also has (111) symmetry, apparently lying about 1.6Å from an Si atom along the joint to the tetrahedral site. Any model should recognise that there are always many impurities present (notably O, C and H; electrically-active impurities are better controlled) and that the charge state is not established.

6.3 Muons in Oxides

Most measurements have been on magnetic oxides, and the aim has been to identify the sites occupied and the internal fields at them. Ambiguities remain which are not directly relevant to the present discussion, so I merely cite the work:

α-Fe_2O_3	Graf et al., Sol.St.Comm. 25 1079 (1978)
Cr_2O_3	Rueg et al., Rorschach abstract 2.8 1978

Other work on powdered SiO_2 (G.M. Marshall et al., Phys.Lett.65A 351 (1978)) is of interest because it shows the muonium tends to emerge into the inter-particle voids. Just as positrons tend to associate with open regions, so, it seems, does the muon.

Only one measurement of diffusion is given. Hayano, Vemura, Imazato, Nishida, Nagamine, Yamazaki and Yasuoka (Phys.Rev.Lett. 41 421 (1978)) find the muon diffusion in MnO characterised by an activation energy of 72 meV. Whilst there are uncertainties, this value is very close to what one would expect for the small-polaron type of hopping motion. The calculation for MnO (Catlow, Mackrodt, Norgett and Stoneham,Phil.Mag. 35 177 (1977)) gives 340 meV for the intrinsic small polaron, compared with the experimental 300 meV. Since the interstitial muon jump distance is probably about $\sqrt{2}$ times the OH separation in OH^-, i.e. about half that for the intrinsic small polaron, and since the activation energy is roughly quadratic in the jump distance, the same theory gives almost exactly the value 72 meV observed. By way of contrast, hydrogen diffusion in oxides may involve motion of the (OH^-) species; activation energies are typically 2-3 eV.

Appendix

History of the Quantum Theory of Light Interstitial Diffusion

I have included this Appendix because there are three separate strands of theory which have sometimes been confused. The three strands are:

(A) Small polaron theory, developing from the various studies of an electron strongly coupled to lattice distortions. The small polaron theory is also closely related to the theory of non-radiative transitions.

(B) Quantum theory of Light Interstitial diffusion in a form parallel to small-polaron theory. Thus the interstitial is strongly coupled to lattice distortions, and the purely mathematical formalism resembles that of small polaron theory. The differences from small polaron theory are (i) which adiabatic approximations are made, (ii) validity of the Condon approximation, (iii) effects of specific lattice geometry, (iv) quantitative estimates of activation energies, volumes of solution, etc., (v) analysis of isotope effects and (vi) discussions of the several possible interstitial sites, of tunnelling states and of vibrational excited states in a given interstice, etc.

(C) Other quantum theories of diffusion, including (i) theory of quantum crystals like He, (ii) generalisations of classical models with quantum statistics, and (iii) models based on band theory.

Some of the main references are now listed. I have not covered the very useful work on trapping by other impurities, since quantum effects are not involved to any great extent. The formal theory of muons diffusing in the presence of traps has been given by Kehr et al., μSR Newsletter # 20 p.794 (1978).

(A) Small polaron theory

The original concept was first described in:

(A1) L.Landau, Phys.Zeits. d. Sowjetunion 3 664 (1933).

However, the first analyses of small polaron motion were the discussions in:

(A2) J.Yamashita and T.Kurasawa, J.Phys.Soc. Japan 15 802 (1960), J.Phys.Chem.Sol. 5 34 (1958).

(A3) T.Holstein, Ann.Phys. (NY) 8 325, 343 (1959),

with many later studies following on from Holstein's work, notably by L.Friedman and by D.Emin. The work by Kagan and Klinger up to 1974 appears to describe small polaron theory only, and indeed their later papers:

(A4) Y.Kagan and M.I.Klinger, J.Phys. C7 2791 (1974)
M.I.Klinger J.Phys. C8 2343 (1975)
M.I.Klinger J.Phys. C11 914 (1978)

are basically small polaron theory, though the authors do describe "defecton" diffusion. This is obviously intended to cover light interstitial motion, though the solid-state physics is not discussed.

The connections between non-radiative transitions and small polaron theory are discussed by

(A5) A.M.Stoneham Phil.Mag. 36 983 (1977)
T.Holstein Phil.Mag. 37 399 (1978)

There are only a few detailed quantitative calculations of small polaron properties:

(A6) K.S.Song, J.Phys.Chem.Sol. 31 1389 (1970) and Sol.St.Comm. 9 1263 (1971).
M.J.Norgett and A.M.Stoneham J.Phys. C6 238 (1973).
R.Monnier, K.S.Song and A.M.Stoneham J.Phys. C10 4441 (1977).
C.R.A.Catlow, W.C.Mackrodt, M.J.Norgett and A.M.Stoneham Phil.Mag. 35 177 (1977)

These parallel calculations of muon activation energies.

(B) Light Interstitial Diffusion (Small polaron approach)

To my knowledge, the first paper to discuss this in a way analogous to small polaron theory is (B1) which, for reasons outlined in the introduction, is fundamentally different from the Class A papers:

(B1) C.P.Flynn and A.M.Stoneham, Phys.Rev. B1 3966 (1970)

Specific aspects were then followed up in these papers:

(B2) A.M.Stoneham, Ber der Bunsenges 76 816 (1972) (Reviews relating quantal J Collective Phenomena 2 9 (1975) and classical theories)
C.P.Flynn Comments on Solid State Physics 3 159 (1971).

(B3) A.M.Stoneham J.Phys. F2 417 (1972) (Temperature dependence at intermediate temperatures).

(B4) A.M.Stoneham & C.P.Flynn J.Phys. F3 505 (1973) (Electro and Thermomigration)

(B5) A.M.Stoneham J.Nucl.Mat. 69/70 109 (1978) (Isotope effect together with other work (not listed) on potential energy surfaces).

On specific aspects (i) paper (B5) discusses the isotope effect, and the papers (B2) review many other quantal and classical theories.

(ii) the idea of tunnelling states is very old, but its first application to H in metals was in my Jülich paper (B2); the most important applications are probably

(B7) H.K.Birnham & C.P.Flynn, Phys.Rev.Lett. 37 25 (1976)
J.Bucholz, J.Volkl & G. Alefeld, Phys.Rev.Lett. 30 318 (1973)

(iii) The recent work (eg by the Sandia group at the Rorschach muon meeting) is basically based on (B1), without new qualitative features, though with the important addition of more detailed quantitative calculations.

Note that the Kagan-Klinger work falls into class A; also, they pay attention to weak coupling, and it is usually considered that H or μ^+ in metals fall into the strong-coupling regime.

(C) Other Quantum Theories

I have already compared these in papers (B2) and (to a lesser extent) (B5); these papers also discuss the pioneering work and review of J.A.Sussmann, Ann.Physique (Paris) 6 135 (1971).

One reference should be added, since Kagan and Klinger refer

to it, namely A.F.Andreev and I.M.Lifshitz, Sov.Phys. JETP 29 1107 (69).
This is aimed at the quantum crystal case (eg solid He); when it refers to defects, it has in mind ^3He in ^4He, for example.

HYDROGEN DIFFUSION AND TRAPPING IN BCC AND FCC METALS

D. Richter*

Brookhaven National Laboratory
Upton, New York 11973 U.S.A.

*On leave from IFF-KFA, Jülich, West Germany.

INTRODUCTION

The behavior of hydrogen in metals is a subject of intense current interest (1,2,3) both for scientific as well as technological reasons. Recently, a number of practical applications like hydrogen storage in the form of metal hydrides or hydrogen as a working fluid in refrigeration devices have been considered. In particular, the properties of tritium in the construction materials as well as in the lithium blanket of future fusion reactors are investigated currently. This lecture restricts itself to the fundamental aspects of the metal-hydrogen systems. Here the large number of anomalous properties are the reason for continuous scientific effort. For instance, the time scale of hydrogen motion is extremely short. The characteristic frequencies of the localized modes of hydrogen in Ta, Nb, or V are in the order of 10^{-14} sec (energies between 0.1...0.2 eV); the jump frequencies for H-diffusion at elevated temperatures in those systems are between 10^{+12} to 10^{+13} sec^{-1}. They are comparable with the correlation times for diffusion in liquids and more than ten orders of magnitude larger than e.g. the jump times for nitrogen in Nb. Out of the large number of experimental data this paper will survey only some recent results on representative fcc and bcc metals for dilute H solutions. The first part of this lecture deals with the nature of the elementary step in H-diffusion. Here the temperature and isotope dependence of the H-diffusion coefficient gives hints to the mechanism involved. The experimental results are discussed in terms of semiclassical and quantum mechanical diffusion theories.

Quasielastic neutron scattering reveals microscopic details of both the time and space development of the diffusive process on an atomic scale. After outlining the method on the example PdH_x in the second part of this lecture, new results on the jump geometry in bcc metals with special emphasis on the anomalous behavior at higher temperatures, where correlated jump processes are important, are presented. An examination of the Debye-Waller-factor shows a point-like localized proton at all temperatures. In part 3 we deal with the problem of H-vibrations in bcc metals and their possible conjunction to the fast diffusion mechanism. In particular, the dispersive step in the TA acoustic branch in $NbH_{0.15}$ (4) and the recently discovered energetically low flat phonon modes in $NbD_{0.85}$ (5) will be discussed (6).

The influence of substitutional and interstitial impurities on the hydrogen diffusion properties is treated in part 4 of this lecture. The trapping capability of point defects has been concluded firstly from increasing H-solubility with increasing amounts of impurities. Resistivity, internal friction and Gorsky-effect experiments reveal further evidence for the trapping process. Again, quasielastic neutron scattering gives insight into the microscopic details of the diffusion process in the presence of trapping impurities. Results on NbH_x doped with nitrogen impurities are interpreted in terms of a two-state model which includes multiple trapping and detrapping processes (7). In a second step the data are compared with numerical calculations using a realistic potential. The effect of substitutional impurities is demonstrated in the example of vanadium impurities in Nb. NMR and preliminary neutron results are presented.

Finally, the dynamical behavior of the hydrogen in the trapped state is surveyed. Internal friction experiments on H trapped on O-impurities at low temperatures seem to be in accordance with predictions of the tunneling hopping theory (8). Specific heat experiments on NbN_xH_y-systems show a Schottky-type anomaly at about 1 K with a strong isotope effect (9). These results are explained in terms of tunneling states of the proton in the neighborhood of the impurity.

TEMPERATURE AND ISOTOPE DEPENDENCE OF THE H-DIFFUSION COEFFICIENT IN BCC V, Nb, AND Ta AND FCC Pd, Cu, AND Ni

The experimental methods to obtain the hydrogen diffusion coefficient can be divided into two groups according as it is measured by studying the relaxation of a nonequilibrium distribution to equilibrium or it is investigated under equilibrium conditions. The first group contains the commonly used macroscopic methods like Gorsky-effect, permeation-, resistivity relaxation-, heat of transport-, outgassing experiments, etc. These methods have been reviewed elsewhere (3,10) and will not be described here. The second group comprises microscopic experiments like nuclear magnetic resonance, Mössbauer effect and quasielastic neutron scattering (QNS). The large number of experimental results obtained by the various methods have been compiled recently and are presented in Table 1 for our examples. In all cases, the data are represented in the form of an Arrhenius relation. $D = D_o \exp -E_a/kT$ where E_a is the activation energy for the diffusive process. In the case of Pd and Ni, the various methods reveal remarkably consistent values for the diffusion constant. This can be attributed to the relatively well-defined surface conditions which lead to reliable results also for the different permeation techniques. In the case of the other materials, a good consistency of the data only exists if the results of surface independent methods are compared. For the case of H in Nb, Fig. 1 presents selected data obtained by the different methods (12-17). Some recent permeation results on Pd-plated Nb samples are included (18). Fig. 1 shows clearly the anomalous temperature dependence of the hydrogen diffusion constant in Nb which exhibits two different activation energies in the considered temperature region. An even more pronounced break in the temperature dependence of the diffusion coefficient has been reported recently for H in Ta (19). Whereas in Nb for deuterium no change of slope has been observed so far, there are experimental indications for this effect also for deuterium in Ta.

In vanadium, however, where the activation energy for diffusion has the lowest value of all three bcc-metals, no change of slope has been observed so far. Concerning the isotope dependence of diffusion coefficient which primarily gives hints to the nature of the diffusion mechanism we have the following situation for the 3 bcc-metals: H diffuses always faster than D and T in the whole temperature range investigated so far. The preexponential factors in the high temperature range are nearly independent of the isotope and if one corrects for the lattice constant a the preexponential $1/\tau_o = D_o a^2/48$ is also independent of the material. For the activation energies $E_H < E_D < E_T$ holds. The ratio of the diffusion coefficients of hydrogen and deuterium are in general far away from the classical prediction of the Vineyard theory (20): $D_H/D_D = \sqrt{m_D/m_H}$.

As an example for the temperature and isotope dependence of the diffusion coefficient in an fcc metal Fig. 2 presents the results for Pd. In contrast to the behavior in the bcc metals, where the jump rates increase with decreasing mass of the isotope (22), here D diffuses faster than H whereas the values

Table 1. Activation energies E and preexponential factors D_o of the Arrhenius relation for the diffusion coefficient in the various metals.

System	E_a(meV)	D_ox10^4 (cm^2/sec)	System	E_a(meV)	D_ox10^3 (cm^2/sec)
Nb-H					
T>270K(11)	106	5.0	Pd-H(11)	230	2.9
T<250K	68	0.9	Pd-D(22)	206	1.7
Nb-D (11)	127	5.2	Pd-T(23)	276	11
Nb-T(11)	135	4.5	Cu-H(21)	403	11.3
Ta-H			Cu-D(21)	382	7.3
T>270K(11)	140	4.4	Cu-T(21)	378	6.1
T<200K(11)	40	0.02	Ni-H(21)	409	7.0
Ta-D(11)	160	4.6	Ni-D(21)	401	5.3
V-H(11)	45	3.1	Ni-T(21)	395	4.3

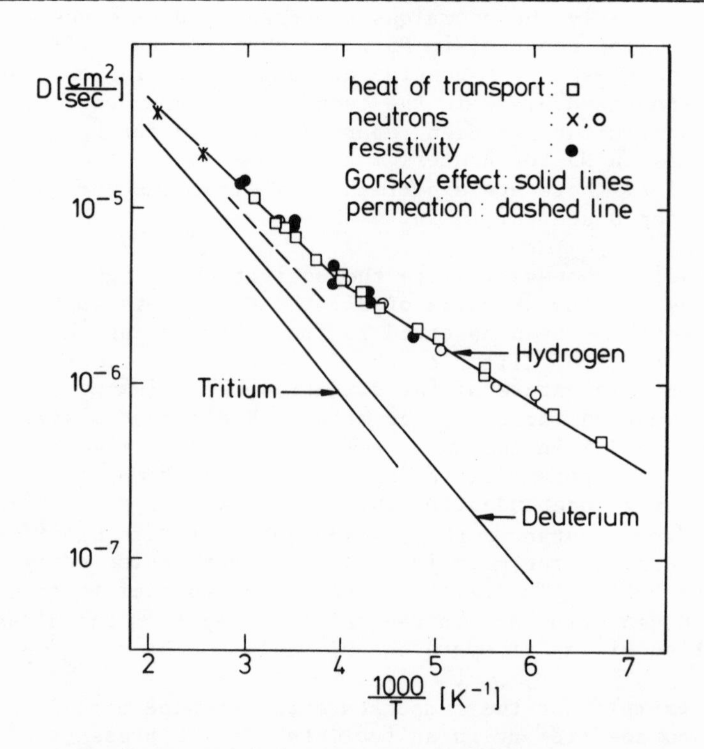

Fig. 1. Diffusion coefficient for H, D, T in Nb. Heat of transport (14), Neutrons (16,17), resistivity (15), Gorsky effect (12,13) permeation (18).

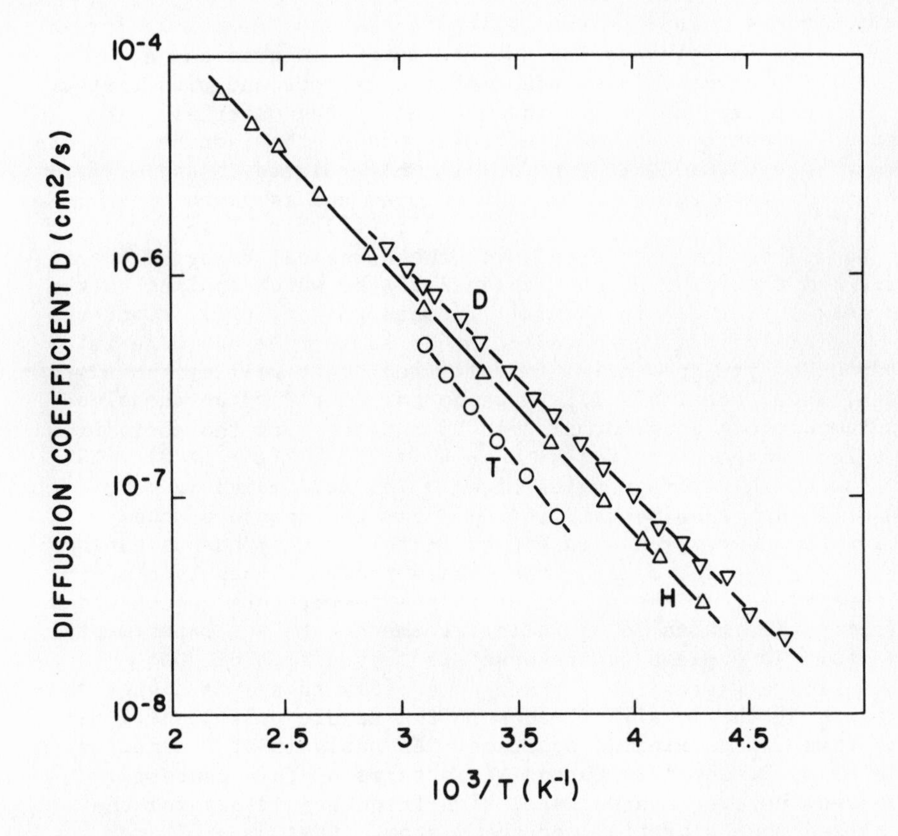

Fig. 2. Diffusion coefficient of H-isotopes in Pd, H and D (22), T (23).

for T are below those of H (23). An extrapolation of the high temperature data of Katz et al. (21) on Ni and Cu would result in a similar reversed isotope effect for the diffusion coefficients at lower temperatures yielding $D_H < D_D < D_T$. For all three, fcc metals the preexponentials D_o for H and D behave close to $\sqrt{m_D/m_H}$. This holds also for T in the case of Ni and Cu; i.e. D_{oH}: D_{oD}: D_{oT} = 1: $1/\sqrt{2}$: $1/\sqrt{3}$. Contrary to the bcc metals the activation energies decrease in going from H to T causing the reversed isotope effect at lower temperatures. In order to obtain evidence whether the hydrogen diffusion is described more likely by a classical over-barrier jump process (20) or by a quantum mechanically tunneling hopping process (24, 25) which are both reasonable for the investigated temperature and concentration region we shall compare the experimental results with predictions of both theories. The classical rate theory modified with respect to the discreteness of the hydrogen excitations $\hbar\omega$ leads to the following results (21): (i) for kT $<< \hbar\omega$ which is relevant for the bcc metals it predicts a universal prefactor independent of the isotope and with respect to the jump frequency also independent of the material. The activation energy is predicted to depend on the isotope. In its simplest form the isotope effect is only related to the different oscillation energies of the isotopes and is given by $E_D - E_H = 1/2\ (\hbar\omega_D - \hbar\omega_H) = 1/2\ \hbar\omega\ (1 - 1/\sqrt{2})$; (ii) for $kT >> \hbar\omega$ the theory becomes identical with the classical Vineyard theory. (iii) For the intermediate region $2kT \backsim \hbar\omega$ which applies to the fcc metals we refer to the full formula in Ref. (21). Concerning the bcc metals, the prediction of an isotope and material independent preexponential is fulfilled quite well by the experimental data (see Table I), if we do not consider the anomalous low temperature behavior of H in Nb and Ta. For the activation energies, we get e.g. for Nb $E_D - E_H = 1/2\ \hbar\omega\ (1 - 1/\sqrt{2}) = 23$ meV (using the average value of $\hbar\omega = 155$ meV) which is very close to the experimental finding. For the change of the activation energy for H in Nb and Ta this theory has no explanation. For the fcc metals, the theory predicts roughly a $m^{-1/2}$ isotope effect in the prefactor in the temperature region under investigation which is again in agreement with the experimental results. To explain the reversed isotope effect of the activation energies, it was suggested (21) to assume higher frequencies of the localized modes in the saddle point configuration than in the minimum position. An analysis of the diffusion data of H, D, and T in Cu and Ni in terms of this concept revealed, however, unrealistic high-frequency values for the localized mode vibrations of the H atom. (Ni: $\hbar\omega = 330$ meV in the minimum and $\hbar\omega = 470$ meV in the saddlepoint configuration).

The tunneling hopping theory in its simplest form (24) predicts the following behavior: (i) For $T << \theta_D$ (θ_D: Debye temperature) the diffusion coefficient should be proportional to J^2T^7, where J is the tunneling matrix element between adjacent sites and T is the temperature. For $T \simeq \theta_D$ the theory predicts an Arrhenius behavior similar to the classical theory. The parameters, however, have another meaning. The prefactor is governed again by the tunneling matrix element J whereas the activation energy E_a is given by the energy necessary to distort the lattice in such a way that the diffusing particle has the same energy in the initial and final position. Again, we have to examine the experimental data with respect to predictions of the theory. First, we can state that for long-range diffusion of H and its isotopes a T^7 behavior has never been observed which is not very surprising concerning the investigated temperature region. The absence of an isotope effect in the prefactor found for the bcc metals above 250 K, however, is in contrast to the predictions of the theory, where the prefactor is dominated by a strongly isotope dependent tunneling matrix element. The observed isotope effect in the activation energy $E_H < E_D$ again is in conflict with theory. For H and D in Nb the double-force tensors are equal (26) and no isotope effect should be expected. With respect to the double force tensors of H and D in Ta, a reversed isotope effect should have been observed (26). Thus, the tunneling hopping theory at least in its simplest form does not describe the experimental results for Nb and Ta above 250 K. An extension of the theory (25) using the concept of extended H wave functions in order to account for the outlined discrepancies is in disagreement with proton form factor measurements by neutron scattering (27). However, a recent reexamination of the small polaron tunneling hopping theory by Emin et al. (28) casts new light on the situation. This approach avoids previously made simplifications like a perturbation treatment of J, termed nonadiabatic regime, the so-called "Condon-approximation" which neglects the influence of the actual position of the host atoms on J, and finally, it includes excited states of the diffusing particle. Explicit calculations for the diffusion of the H isotopes including the positive muon in Nb revealed results as shown in Fig. 3. The following features should be noted: (i) the high-temperature activation energy of diffusion increases with increasing mass of the diffusing species; (ii) the adiabatic character of the high-temperature diffusion behavior manifests itself in the isotope independent prefactor; (iii) at lower temperatures, the contribution of nonadiabatic transfers which require a lower activation energy increases. This results in a change of slope in the Arrhenius plot of the H diffusion constant. Thus, the results are in very good agreement with the experimental finding and are the only theoretical calculations which can account for the change of slope observed for the H-diffusion constant.

Employing tunneling hopping theory beyond the Condon approximation and using pseudopotential results to describe the potential for H in Cu, very recently Teichler (29) could reproduce the experimental results for H-diffusion in this fcc-metal theoretically. For the more complicated transition metals Ni and Pd no theoretical calculations are available so far.

QUASIELASTIC NEUTRON SCATTERING STUDIES OF HYDROGEN DIFFUSION IN Pd, Nb, Ta, AND V

For the study of the diffusion of H in metals QNS is a unique tool (30) which allows the simultaneous investigation of the time and space development of the diffusive process on a microscopic scale. In the case of hydrogen which scatters predominantly incoherent such an experiment reveals information on the behavior of single protons in space and time. Following the

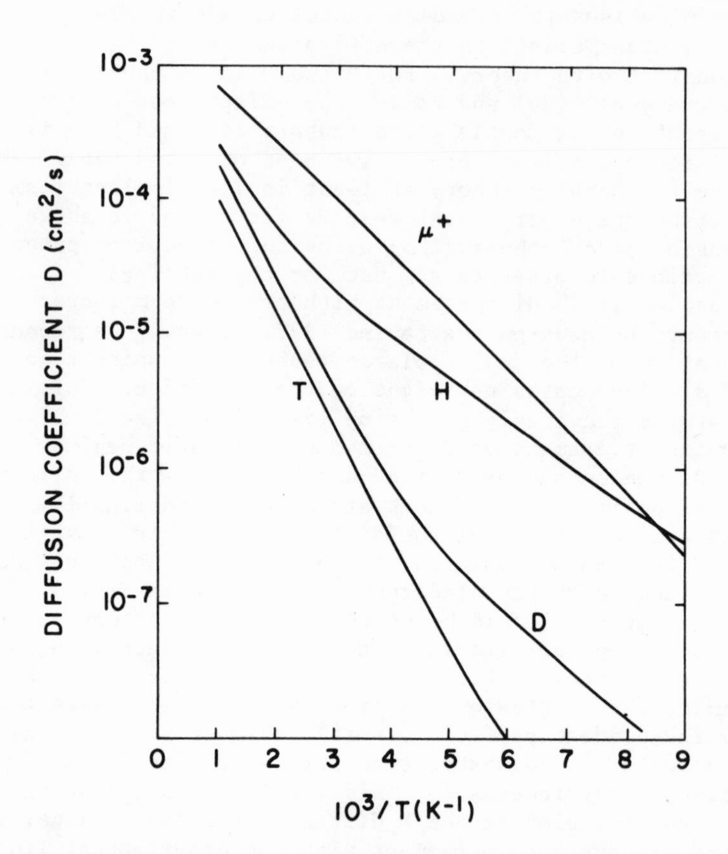

Fig. 3. Calculated diffusion coefficients for the H isotopes in Nb (28).

concept of van Hove (31) the observed double differential cross section $\partial^2\sigma/\partial\omega\partial\Omega$ is proportional to the Fourier transform of the self-correlation function $G_s(\underline{r},t)$ of the proton

$$\frac{\partial^2\sigma}{\partial\omega\partial\Omega} \sim S_{inc}(\underline{Q},\omega) = \frac{1}{2\pi} \int_{-\infty}^{+\infty} dt \int_{-\infty}^{+\infty} d^3r \; e^{i(\underline{Q}\underline{r}-\omega t)} \; G_s(\underline{r},t) \tag{1}$$

where S_{inc} $(\underline{Q},\omega)$ is the so-called incoherent scattering law, $\underline{Q} = \hbar\underline{K}_f - \hbar\underline{K}_i$ is the momentum transfer and $\hbar\omega = E_f - E_i$ is the energy transfer at the sample. In the classical limit $G_s(\underline{r},t)$ can be interpreted as the conditional probability to find a proton at a time t at a site $\underline{r}$ if it has been at $\underline{r} = 0$ for t=0. The results of QNS experiments on H-diffusion are commonly interpreted in terms of the Chudley-Elliott model(CE) (32) which introduces the following assumptions:

(i) the jump time from site to site is small compared to the mean rest time of hydrogen on its interstitial site.

(ii) there is no correlation between vibration and jump processes.

Under these circumstances the self-correlation function can be obtained by solving a master equation:

$$\frac{\partial}{\partial t} G_s(\underline{r},t) = \frac{1}{z\tau} \sum_{i=1}^{z} (G_s(\underline{r}+\underline{S}_i,t) - G_s(\underline{r},t) \tag{2}$$

where $\underline{S}_i$ are the jump vectors to the accessible neighboring sites, z is the number of these sites and τ the mean rest time at a certain site. Eq. (2) can be integrated by a Fourier transformation yielding in the case of Bravais lattices:

$$\frac{\partial}{\partial t} G_s(\underline{Q},t) = -\frac{1}{z\tau} \sum_{i=1}^{z} (1-e^{i\underline{Q}\underline{S}_i})G_s(\underline{Q},t) \tag{3}$$

$$= \frac{1}{\tau} f(\underline{Q}) \, G_s(\underline{Q},t)$$

$$G_s(\underline{Q},t) = e^{-\frac{t}{\tau}f(\underline{Q})} \tag{4}$$

$G_s(\underline{Q},t)$ fulfills already the appropriate initial condition. Finally, Fourier transformation with respect to the time yields

$$\mathrm{Sinc}(\underline{Q},\omega) = \frac{1}{\pi} \frac{f(\underline{Q})/t}{\left(\frac{f(\underline{Q})}{\tau}\right)^2 + \omega^2} \tag{5}$$

The intensity of the quasielastic line can be described by a the Debye-Waller - or form factor exp - $1/3Q^2 \langle u^2\rangle$, where $\langle u^2\rangle$ is the mean-square displacement of the proton at its site. Using the diffusion of H in Pd as an example we outline briefly what kind of information can be obtained from an investigation of the $\underset{\sim}{Q}$ and ω-dependence of the quasielastic scattering law. Fig. 4 shows the two possible interstitial sites for hydrogen in fcc Pd together with the jump vectors to nearest-neighbor sites.

Table II summarizes some characteristics for the two possible jump mechanisms.

For equal diffusion coefficients we have $\tau_o = 2\tau_t$. For both models the functions $f(\underset{\sim}{Q})$ can be easily calculated yielding:

$$f_o(\underset{\sim}{Q}) = \frac{1}{6}\{6 - \cos\frac{a}{2}(Q_x+Q_y) - \cos\frac{a}{2}(Q_x-Q_y) - \cos\frac{a}{2}(Q_x+Q_z) - \cos\frac{a}{2}(Q_x-Q_z) - \cos\frac{a}{2}(Q_y+Q_z) - \cos\frac{a}{2}(Q_y-Q_z)\} \qquad (6)$$

$$f_t(\underset{\sim}{Q}) = \frac{1}{3}\{3 - \cos\frac{a}{2}Q_x - \cos\frac{a}{2}Q_y - \cos\frac{a}{2}Q_z\} \qquad (7)$$

For small momentum transfers $\hbar\underset{\sim}{Q}$ the width $\Gamma(\underset{\sim}{Q})=\tau^{-1}\, f(\underset{\sim}{Q})$ of the Lorentzian is independent of the jump model $\Gamma(\underset{\sim}{Q}) = \hbar DQ^2$. It reveals the macroscopic diffusion coefficient measured over microscopic distances. The $\underset{\sim}{Q}$ dependence of line width at large $\underset{\sim}{Q}$'s is determined by the geometrical details of the jump mechanism. For fcc-Pd Fig. 5 shows the theoretical $\underset{\sim}{Q}$ dependence of line width for the models in question. The experimental points (33) are clearly in favor for H jumps between nearest-neighbor octahedral sites. With respect to the high barrier and the large spacing between the interstitial sites this result is quite reasonable.

For the bcc metals with their complicated hydrogen sublattice (6 sites per unit cell)(Fig. 6),the situation is much more complicated. Besides the complication due to the non-Bravais-H-sublattice, there is experimental evidence that the diffusion process itself is not as simple as for the fcc metals. Already earlier results on V (34) Ta (35) and Nb (36) showed systematic deviations from the predictions of a simple nearest-neighbor jump model. Also, anomalies in the intensity of the quasielastic line have been reported (34,37). These anomalies have been attributed to the occurrence of further neighbor jumps and/or to jumps between different types of institial sites

Table II. Characteristics of tetrahedral and octahedral jump mechanism in fcc metals

	Octahedral model	tetrahedral model
Number of nearest neighbors	12	6
Jump direction	(110)	(100)
Jump length	$a/2\sqrt{2}$	$a/2$
Diffusion coefficient $D = \ell^2/6\tau$	$a^2/12\tau_o$	$a^2/24\tau_t$

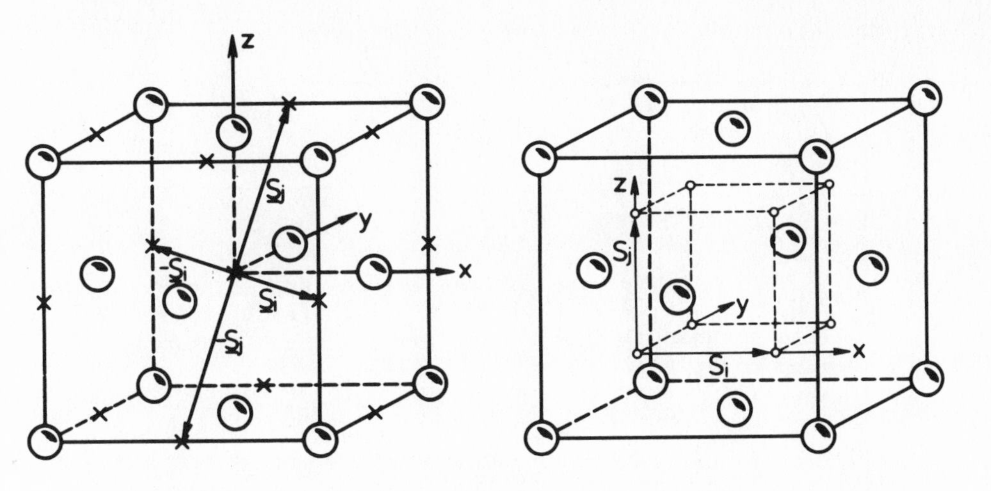

Fig. 4. Possible H-interstitial sites in fcc-Pd, O Pd atoms, x octahedral sites, O tetrahedral sites, <u>Si</u> H jump vectors.

(34,35) and to the influence of a finite time of flight (38) but no conclusive picture has been reached.

Very recently, Lottner et al. (27,39) have reexamined the problem for Ta, Nb, and V. Quasielastic neutron scattering experiments where performed at $NbH_{0.02}$, $TaH_{0.13}$ and $VH_{0.07}$ single crystals at temperatures between 290 and 760 K for Q values between 0.3 and 2.5 $Å^{-1}$. The data were analyzed in terms of 4 different models always assuming that H jumps occur between tetrahedral sites.

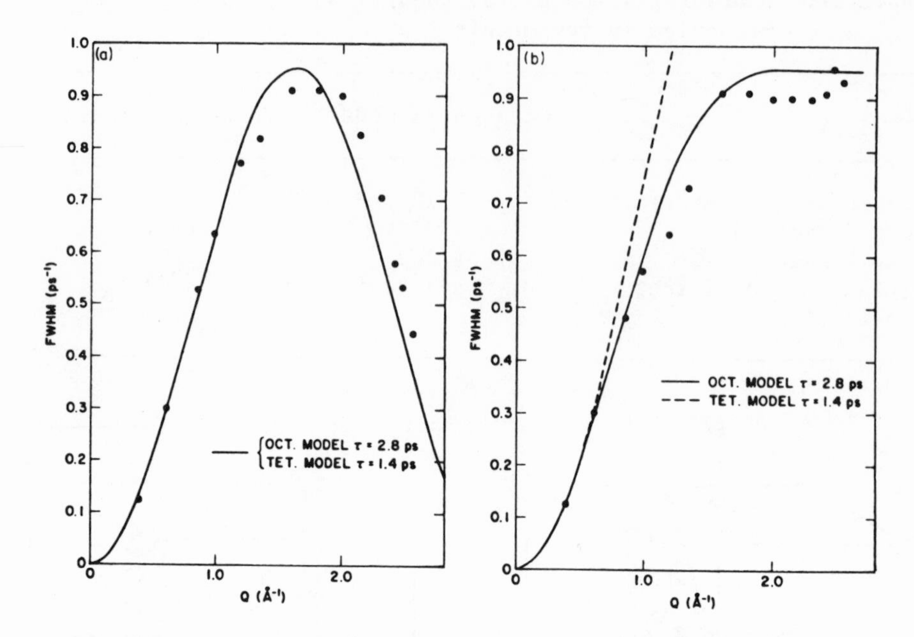

Fig. 5. Experimental results for the quasielastic line width in PdH_x for a (100) (left) and a (110) direction; solid line octahedral jump model, dashed line tetrahedral jump model (33).

Model (1): Hydrogen jumps occur only between nearest neighbor sites with jump vectors in (110) direction and a jump rate $1/\tau_1$.

Model (2): In extension to model (1) also jumps to second nearest neighbors are included. They occur in (100) direction across the cube face center with a jump rate $1/\tau_2$.

Model (3): Assumes that in addition to nearest-neighbor jumps correlated double jumps are possible. The corresponding jump rate is $1/\tau_2$.

Model (4): Generalizes model (3). It considers the H alternatively in a mobile "state" (life time τ), where it can perform repeated jumps to nearest neighbors with a jump rate $1/\tau_1$, and in an immobile self-trapped "state" (life time τ_t). The exchange between both "states" is described by transition rates given by the inverse life times. This concept is formal identical to the treatment of H diffusion in the presence of trapping impurities (40) which will be discussed later on.

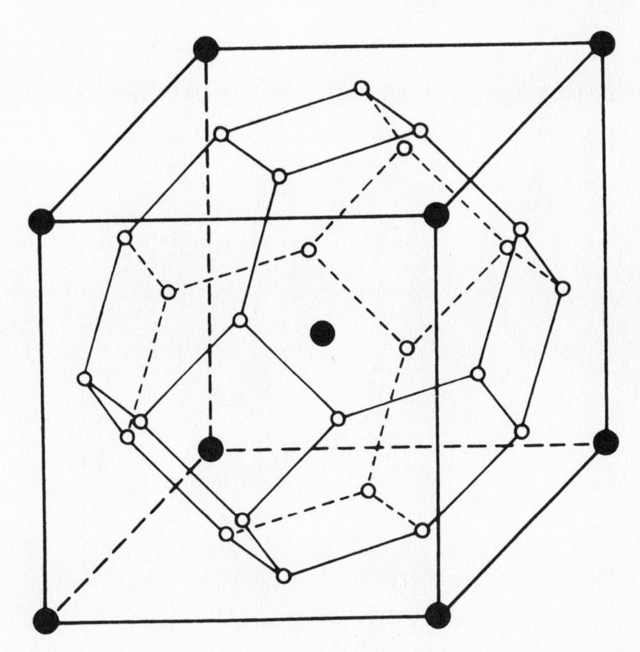

Fig. 6. Hydrogen sublattice for tetrahedral interstitial sites in bcc models.

Model (1) includes 2 independent parameters, the jump rate $1/\tau_1$ and the quasielastic intensity I(Q), models (2) and (3) have 3 parameters and finally (4) is a four-parameter model. Fig. 7 shows the results of a data analysis applying model (1). While at room temperature, all spectra yield the same $1/\tau_1$, at higher temperatures severe deviations appear. The decrease of $1/\tau_1$ with increasing Q can be understood qualitatively as an increase of the effective jump length. Fig. 8 presents results obtained by an analysis of the high temperature data with the models (2) to (4). The jump rates calculated by a simultaneous fit of the spectra measured at different Q values for one crystal orientation ϕ are plotted vs ϕ. For a correct description of the data the obtained jump rates should not depend on the crystal direction. Fig. 8 makes it clear that the extension of model (1) to jumps to next-nearest neighbors does not solve the problem whereas the assumption of correlated jumps leads to a satisfactory agreement between theory and experiment. Similarly, also for H in Ta and V, the simple jump model (1) fails to explain the data at elevated temperatures, whereas model (3) seems to lead to an adequate description of the experimental results also for Ta and V. Model (4) has not been used for the data analysis for Ta and V. However, the large ratio τ_1/τ_2 found for the contribution of double jumps makes the application of model (4) desirable.

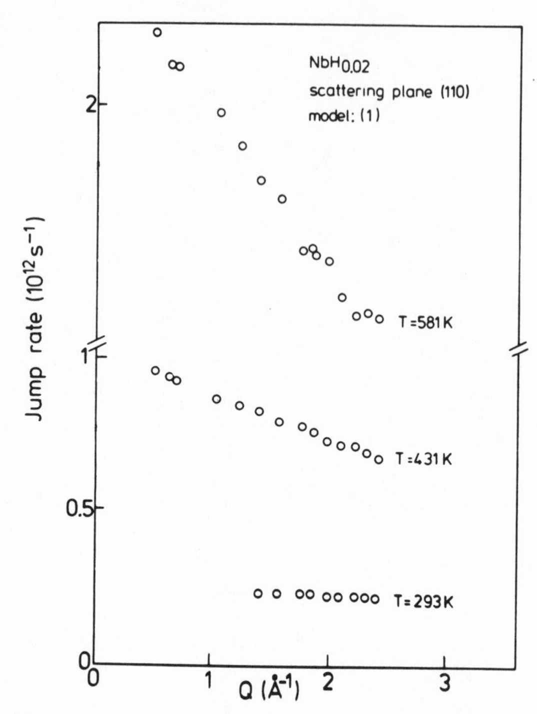

Fig. 7. Jump rate $1/\tau_1$ of model (1) as a function of $|Q|$ determined from the measurements at 293 K, 431 K and 581 K (39).

The authors do not specify the physical origin of the occurrence of correlated jumps. Here the relation between the jump rate and lattice relaxation time τ_r, the time the lattice needs to dissipate the energy necessary to produce the jump, may be important. For the case of electronic small polaron hopping Emin has shown (41,42) that for times short compared to τ_r the energy required for a successive jump is only 1/3 of the activation energy in the relaxed lattice. However, the involved electronic hopping rates were in the order of 10^{+14} sec^{-1}. In view of the much smaller jump rates of the proton 10^{+12} - 10^{+13} sec^{-1} on the one hand and the differences in the coupling to the lattice on the other hand, it is not clear, whether this concept is applicable. Quantitative calculations are necessary.

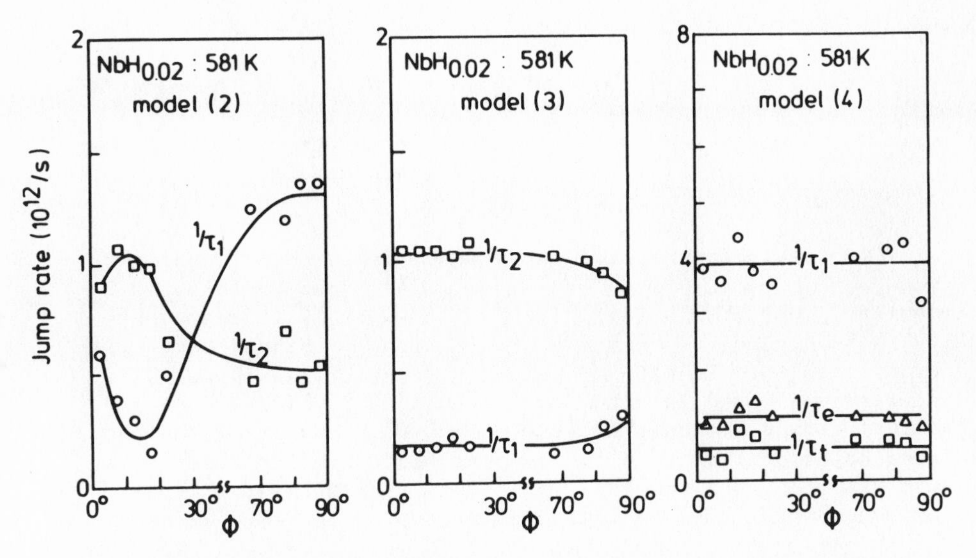

Fig. 8. Jump rates as a function of sample orientation ϕ obtained by a simultaneous fit of the scattering law for the models (2), (3), (4), to the QNS result at 581 K (39).

The Q dependence of the quasielastic intensity evaluated from this experiment follows a normal Debye-Waller factor (DWF) behavior with mean-square amplitudes for the H-motion of $1/3\langle u^2\rangle = 0.02 - 0.04$ Å^2 which is near to the value expected from harmonic calculations (27). For V an anomalous decrease of the intensity has been observed at larger Q values and high temperature (T = 763 K) which was attributed to non-negligible jump times (27,39). The normal DWF evaluated for Nb and Ta even at high temperatures is in contrast to earlier reports of an anomalous behavior, which was mainly caused by errors in the integration of the quasielastic spectrum neglecting wing contributions. The observation of a normal DWF shows that the proton is well localized at its interstitial site contrary to earlier speculations about an extended proton wave function (25).

PHONONS AND H-DIFFUSION

During its rest time at a particular interstitial site the proton exhibits vibrational motion around its equilibrium position. Two types of motion have to be distinguished:

(i) localized vibrations of the dilute H-atom against its metal neighbor with frequencies typically a few times higher than the cutoff frequency of the host vibrations and

(ii) acoustic vibrations where the H follows the motions of the host atoms. These vibrational properties have been surveyed recently (43). Here we shall concentrate on features in the vibrational spectra which might be correlated to the diffusive motion of the proton.

Ultrasonic and neutron experiments reveal consistently an increase of the bulk modulus $(C_{11} + 2C_{12})/3$ and the shear modulus C_{44} in Nb, Ta (44, 51, 45). However, with respect to the shear modulus $C' = (C_{11} - C_{12})/2$ both techniques yield different results. In $TaD_{0.02}$ a 12% increase of the phonon energies was found for the corresponding $(\xi,\xi,0)$ T_1A branch, whereas from ultrasonic experiments $(\xi \simeq 0)$ a change of the phonon frequencies of less than 1% was observed. Similar observations were also reported for $TaH_{0.18}$ (44,45). From this, at small wave vectors, a H caused dispersion step in the $(\xi,\xi,0)T_1A$ branch has to be assumed. In order to investigate whether this dispersion step is connected with the diffusive jump motion of the H-atom, the phonon frequencies in the region of the resonance condition $\omega\tau \sim 1$, were measured for $NbH_{0.15}$ at 350 C where this condition is reached at $q \simeq 0.08$ Å^{-1} (4). The observed relative frequency shift $\Delta\nu/\nu = (\nu_{NbH_{0.15}} - \nu_{Nb})/\nu_{Nb}$ as a function of the reduced wave vector ξ is shown in Fig. 9. For $\xi > 0.1$ a constant value of $\Delta\nu/\nu = -0.7\%$ is observed, for $\xi < 0.1$ $\Delta\nu/\nu$ decreases, and reaches $\Delta\nu/\nu \sim -6\%$ at $\xi \simeq 0.03$.

One is tempted to explain this behavior in terms of a relaxation process. The H-atom sitting on a tetrahedral site in the bcc lattice causes a tetragonal distortion field. A jump to a neighboring site turns this field by an angle of 90°. Such a rotation of the displacement field couples with the $(\xi,\xi,0)$ T_1A mode in question. The sound velocity changes then from a relaxed behavior at low frequencies to unrelaxed behavior at higher frequencies. The solid line in Fig. 9 represents the theoretical curve for such a relaxation process.

$$\Delta\nu/\nu = A/(1 + (vq\tau)^2) + C \tag{8}$$

where v is the corresponding sound velocity. The position of the dispersion step is directly related to the jump time of the proton and its shape is well described by the above formula. However, this explanation contradicts other observations:

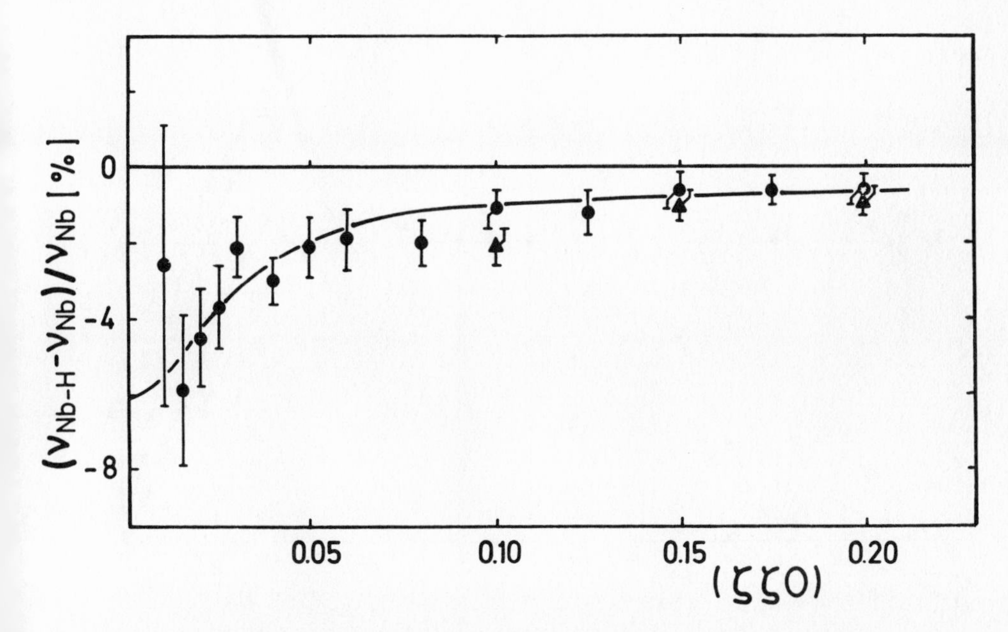

Fig. 9. Relative shift $\Delta\nu/\nu = (\nu_{NbH}-\nu_{Nb})/\nu_{Nb}$ for TA_1-phonons in (110) direction in $NbH_{0.15}$ at 350 C. The solid line represents the calculations based on Eq. (8) 4.

(i) diffuse neutron scattering experiments on NbD_x which probe the elastic distortion fields caused by the D reveal an essentially cubic displacement field (46).

(ii) internal friction experiments on bcc hydrides show the absence of the Snoek effect thereby supporting the view of a cubic displacement field (47).

(iii) the lack of a 1/T dependence of the step amplitude (44) points also in the direction that the relaxation process cannot be a normal Snoek effect. In conclusion, it is not understood how the long wavelength acoustic phonons can couple to the diffusive jump of the proton.

The acoustical and optical phonons in the bcc-H systems have been investigated thoroughly (43). Very recently, however, Shapiro et al. (5) reported the observation of new excitations in $NbD_{0.85}$ and $NbH_{0.82}$. Fig. 10 shows the dispersion curves measured along the three symmetry directions (1,0,0), (1,1,0) and (111) at T = 160 C in the liquid-like α'-phase of $NbD_{0.85}$. Included are also the dispersion curves in pure Nb (48) and pre-

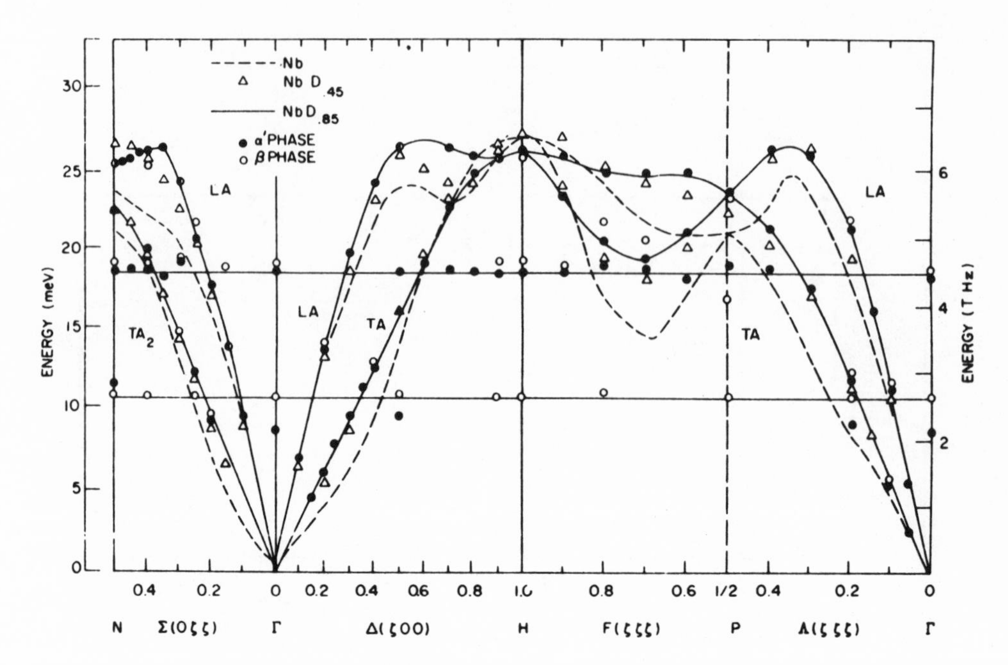

Fig. 10. Dispersion curve for $NbD_{0.85}$ in the α'-phase at T=160C and in the β phase at T=20°C(5) compound with results of $NbD_{0.45}$ (49) and those of pure Nb (48).

vious results on $NbD_{0.45}$ (49). The new features are the dispersionless excitations at 18.4 meV and at 10.8 meV. The positions of these frequencies do not depend on temperature nor do they change with isotope mass. The line width of the excitations broadens considerably in going from the ordered β phase to the α'-phase. In addition, the NbH system shows broader lines than the NbD sample. The authors argue against an explanation in terms of a density of state effect - in particular, there is no structure of the density of states in the 10 meV region - and against a defect mode; but they do not have a clearcut interpretation for the observed feature.

Thereafter, Lottner et al. (6) reported the observation of a strong peak at 15.7 meV in $NbH_{0.05}$ (Fig. 11). It appears at the position of the peak in the density of state for the transverse acoustic modes. The absence of a peak corresponding to the longitudinal cutoff frequencies (∿23 meV) is against an explanation simply in terms of protons mirroring the acoustic phonons of the lattice. The authors propose an interpretation of this feature as a resonant-like mode of the protons within the acoustic modes of the Nb host. The local density of state

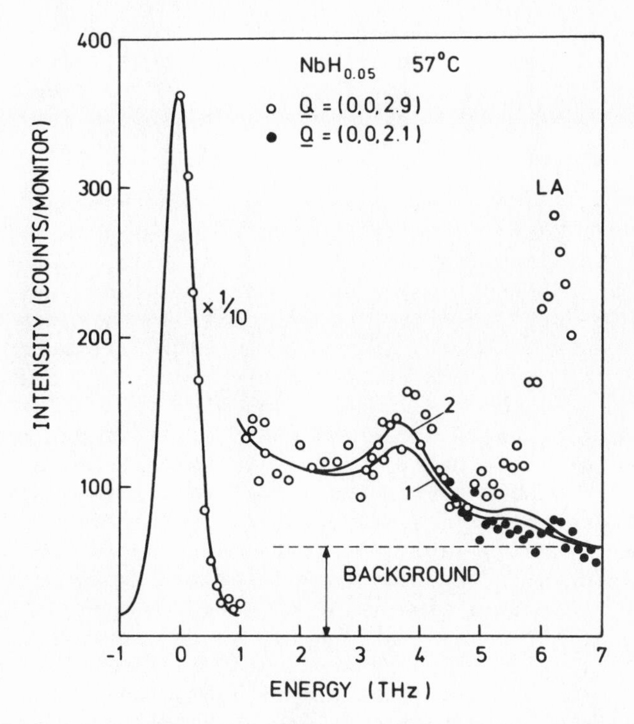

Fig. 11. Energy spectrum for $NbH_{0.05}$ curve 1:1-parameter model, curve 2:6 parameter model. The peak LA origins from the host lattice phonon (6).

for the H in the low frequency limit was calculated by standard Green-function techniques. Already the most simple approach using only one longitudinal coupling constant between the H and its nearest neighbors reveals a pronounced peak at 15.7 meV (curve 1 in Fig. 11). The H-atom vibrates with enhanced amplitude in phase with the surrounding host lattice atoms avoiding thereby a strain of the strong spring to the next-nearest neighbors. In a more sophisticated model, which describes more experimental data like the absence of Snoek effect and the change of the elastic constants, the calculations were repeated using now 6 parameters (curve 2 in Fig. 11). Although qualitatively the picture does not change, quantitatively, the intensity is even more enhanced. Because of the strong coupling of this resonant motion to the host lattice, only a very small isotope effect is expected (50). Whether the flat new modes, found by Shapiro et al. (5) in the high concentration regime, can be interpreted in the same way, is an open question.

THE INFLUENCE OF DEFECTS ON H-DIFFUSION

Numerous investigations have demonstrated that the physical properties of H dissolved in bcc-transition metals are strongly changed by the presence of small amounts of impurities. Vapor pressure data show deviations from Sieverts law at low concentrations which increase by introducing oxygen impurities into Ta or due to cold working of V (51,52). Solubility data reveal a shift of the phase boundary between the α- and β-phase in the presence of O-, N or C-impurities (53). Applying very low cooling rates in Ta-H the phase boundary of the pure material is observed (54) whereas in Nb-N (55) a shift independent of the cooling rate is found. Resistivity experiments on Nb-H doped with N (55) yield a reduced residual H-resistivity at temperatures well above the phase boundary to the β-phase. Furthermore, Gorsky-effect measurements in NbN_xH_y samples (56) showed a strong decrease of the H-diffusion coefficient in particular at lower temperatures. Finally, in the presence of O- and N-impurities in the Nb-H system additional relaxation peaks in internal friction experiments are observed (57). All these features are naturally explained in terms of H trapping at the impurities. In their vicinity they lower the ground state energy for the H. This gives rise to deviations from the Sieverts law at low concentrations due to an increase of the enthalpie of solution. If the binding energy at the trapping center is larger than the enthalpie of formation for the ordered β-phase, the solubility in the dilute phase increases. Resistivity measurements suggest that in NbN each N-impurity is capable of keeping one H from precipitating into the β-phase, whereas in TaH oxygen impurities do not prevent precipitation. Trapping implies the formation of impurity hydrogen pairs. They are assumed to scatter conduction electrons with a smaller probability than the two single scattering centers and they give rise to relaxation processes in internal fraction experiments. Their results will be treated in more detail in the next chapter. Such trapping processes also slow down the long range diffusion. The significance of the trapping processes will increase with increasing ratio of binding energy and thermal energy kT.

On a microsopic level, the influence of interstitial impurities on the H-diffusion process was studied recently by quasielastic neutron scattering on Nb-H samples doped with nitrogen impurities (7). The trapping capability of substitutional defects was investigated for the Nb-V-H system by means of NMR (58). For this system also preliminary neutron scattering results are available (59). Both will be discussed in the following section.

For low concentrations of trapping impurities, where trapping regions and regions of undisturbed host material are present, relatively simple arguments can be given, in order to explain what kind of information can be obtained from a quasielastic neutron scattering experiment. For small momentum transfers, the scattering process averages over large volumes in space (of the order $(2\pi/Q)^3$). Therefore, a long section of the diffusive path of the proton will be probed by the neutron wave packet. This path consists out of periods of undisturbed diffusion as well as of portions where the proton is trapped. Under these circumstances, the scattering law is expected to be a single Lorentzian whose width is given by the effective or macroscopic H-diffusion coefficient D_{eff} in the system. At large Q's, however, the average occurs over short distances and the scattering law depends on the single diffusive step. In this case, the scattering law contains information about the mean trapping time and on the fraction of protons being trapped.

Quantitatively, the scattering law has been calculated in terms of a phenomenological two-state random walk model (RWM) thereby approximating the complicated structure of the trapping region around the impurity by a single escape rate $1/\tau_o$. This approach was confirmed later using the average T-matrix approximation in the limit of dilute concentration of traps (60). We shall survey briefly the RWM approach. Here it is assumed that the proton diffuses in a crystal with randomly distributed traps. The proton diffuses e.g. through the undisturbed parts of the lattice for an average time τ_1 exhibiting jumps with the jump rate of the undisturbed lattice $1/\tau$. Thereafter, it is trapped at the impurity for an average time τ_o. Thus $1/\tau_1$ is the trapping rate and $1/\tau_o$ the escape rate. The self-correlation function is easily calculated using the self-correlation functions of the proton in the undisturbed lattice and in the trapped state. The mathematical procedure is outlined in detail in Ref. (7). The result for the incoherent scattering law is a superposition of two Lorentzians with width Λ_1 and Λ_2 and weights R_1 and R_2. $\Lambda(\underline{Q})$ is the line width in the undisturbed lattice.

$$S_{inc}(\underline{Q},\omega) = R_1 \frac{\Lambda_1/\pi}{\Lambda_1^2+\omega^2} + R_2 \frac{\Lambda_2/\pi}{\Lambda_2^2+\omega^2} \tag{9}$$

$$\Lambda_{1/2} = \frac{1}{2}\left(\tau_o^{-1} + \tau_1^{-1} + \Lambda(\underline{Q}) \pm \left(\left(\tau_o^{-1} + \tau_1^{-1} + \Lambda(\underline{Q})\right)^2 - 4\Lambda(\underline{Q})/\tau_o\right)^{\frac{1}{2}}\right)$$

$$R_1 = \frac{1}{2} + \frac{1}{2}(\Lambda(Q)\;\frac{\tau_1-\tau_o}{t_1+t_o} - \tau_1^{-1} - \tau_o^{-1})/(\tau_o^{-1} + \tau_o^{-1} + \Lambda(Q))^2 - 4\Lambda(Q)/\tau_o)^{\frac{1}{2}} \tag{10}$$

$$R_2 = 1 - R_1 \tag{11}$$

The typical properties of the scattering law are shown in Fig. 12, where the widths Λ_1 and Λ_2 of the two components are plotted together with their weights vs. DQ^2. Included also is $\Lambda(Q)$ of the undisturbed lattice. For the sake of simplicity the calculations are made for a simple cubic lattice. Using the average T-matrix approach, explicit expressions for $1/\tau_1$ and $1/\tau_o$ were found (60). The leading term of the escape rate $1/\tau_o$ is independent of the trap concentration whereas the trapping rate $1/\tau_1$ is proportional to the concentration as expected.

Quasielastic neutron scattering experiments were carried out on Nb samples doped with N and H. The N-concentrations were $C_N^1 = 0.7\%$ and $C_N^2 = 0.4\%$. The corresponding H concentrations were $C_H^1 = 0.4\%$ and $C_H^2 = 0.3\%$. The experiments have been performed in a temperature range $180 \leq T \leq 373$ K for $0.1 \leq Q$ 1.9 $\mathring{A}^{-1}$. The measurements were made, using the high-resolution back-scattering spectrometer at the ILL Grenoble. The energy resolution (FWHM) was between 0.7 and 1.5 μeV the scanning range $\Delta E = \pm$ 5.3 eV. As a consequence of the small value of ΔE, at larger Q's only the narrow component could be measured. The integrated intensity of the quasielastic line, as observed within ΔE, as a function of Q and the temperature is shown in Fig. 13. The intensity has been normalized to the total intensity I obtained at 73 K. For small Q nearly the full intensity is concentrated within the energy window. This demonstrates the existence of only one line which falls entirely within the instrumental energy range. However, at larger Q's, only a fraction of the intensity appears in the integration window which decreases with increasing temperatures. This behavior reflects the reduction of the trapped fraction of protons with the temperature. The experimental spectra for $Q \leq 0.9$ $\mathring{A}^{-1}$ below) were fitted with the scattering law of the RWM, τ_o and τ_1 being the only disposable parameters. The result is presented in Fig. 14. Compared to typical jump rates τ^{-1} of 10^{+11} jump/sec τ_o and τ_1 are two orders of magnitude smaller. At higher temperatures $\tau_1 > \tau_o$ holds; the protons are mainly in undisturbed regions. At lower temperatures, we have $\tau_1 < \tau_o$, the protons are predominantly trapped. τ_o as a local property of the traps does not depend on the N-concentration justifying the assumption of well-separated trapping and undisturbed regions in the lattice. From the activation energy of $1/\tau_o$ (166

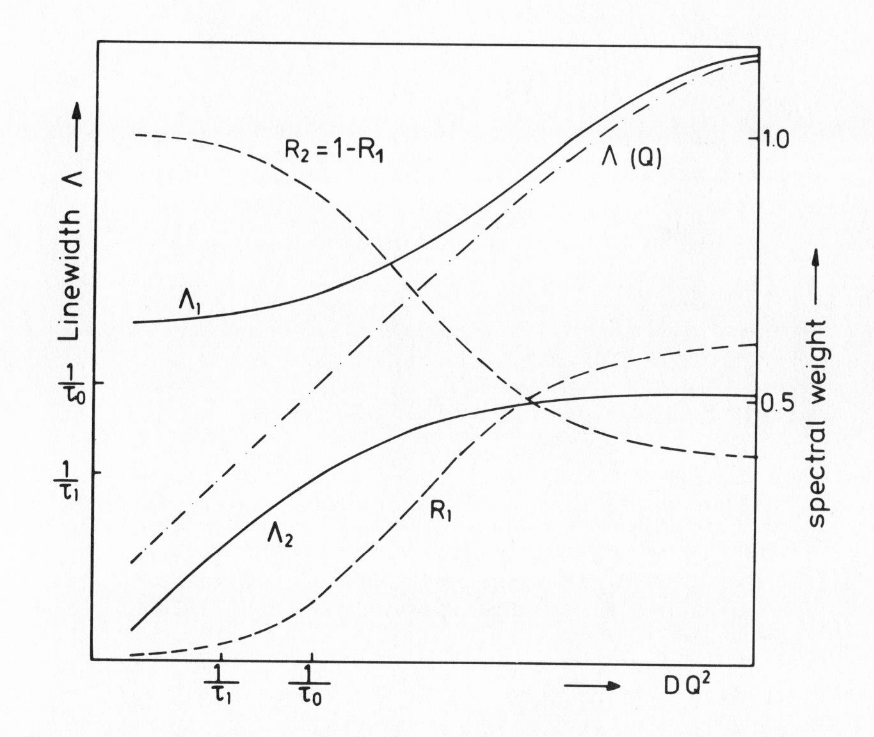

Fig. 12. Line width of the two components Λ_1 and Λ_2, (solid lines) and their spectral weights (dashed lines) R_1, R_2 in the scattering law for the RWM description of diffusion in the presence of traps. Dashed dotted line: line width for an undisturbed lattice (7).

meV) a binding energy of approximately 100 meV for the N-H pair can be deduced. The capture rate $1/\tau_1$ depends on the N-concentration. The ratio of $1/\tau_1$ for the two N-concentrations of 1.8 ± 0.2 compared to the ratio of N-concentrations of 1.9 ± 0.2 is in agreement with theory (60). According to the average T-matrix approximation as well as to results of simple reaction theory for trapping (61) which yields

$$1/\tau_1 = 4\pi R_t D c_t \tag{12}$$

where R_t is the trapping radius and C_t is the trap concentration, the activation energy for $1/\tau_1$ should agree with the activation energy for the self-diffusion coefficient D. The resulting higher values (94 meV) can be explained by the assumption that the nitrogen impurities are saturated as soon as they have trapped a single hydrogen (see also Ref. 55). After correction for saturation effects Eq. (12) allows the evaluation

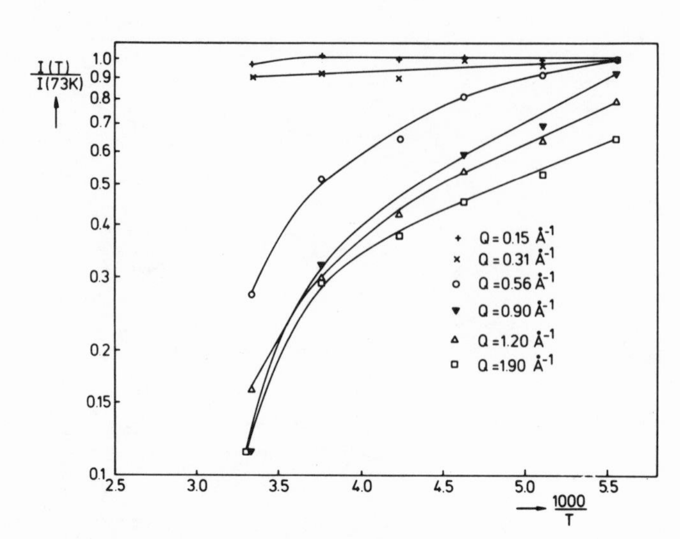

Fig. 13. Intensity of the observed quasielastic spectrum at the $NbN_{0.007}H_{0.004}$-sample within the energy window $\pm\Delta E$ as a function of inverse temperature for different Q. The intensity is normalized to the total intensity determined at 73 K (7).

of the trapping radius R_t. Values in the order of 5 Å result which are in good agreement with calculations using anisotropic continuum theory. Furthermore, the mean-free path between two trapping events can be calculated. It decreases with increasing temperature demonstrating the saturation effects of the traps. Its absolute value is slightly larger than the mean distance between the nitrogen impurities. Summarizing, the simple two-state RWM represents the experimental results below Q = 0.9 $Å^{-1}$ very well. It has been shown that it is also applicable to muon diffusion in the presence of traps (62). Also problems of chemical reactions implying hydrogen atoms might be treated with such a model.

At large Q values a direct determination of the escape rate $1/\tau_0$ should be possible. Fig. 15 presents the width of the narrow components of the quasielastic line for two larger Q values. The activation energy of an Arrhenius law for Γ (T) at Q ≃ 1.9 $Å^{-1}$ is only 115 meV, compared with the expected value of ∽ 170 meV. Thus the connection between $\Gamma(T)$ and the escape process is more intricate than expected. It is obvious that the detailed structure of the disturbed region around a nitrogen interstitial affects the scattering law at larger Q. Thus, a model has been investigated which takes into account the lattice distortions created by the nitrogen interstitials (63). The nitrogen atoms were located at tetrahedral sites in a niobium lattice. The

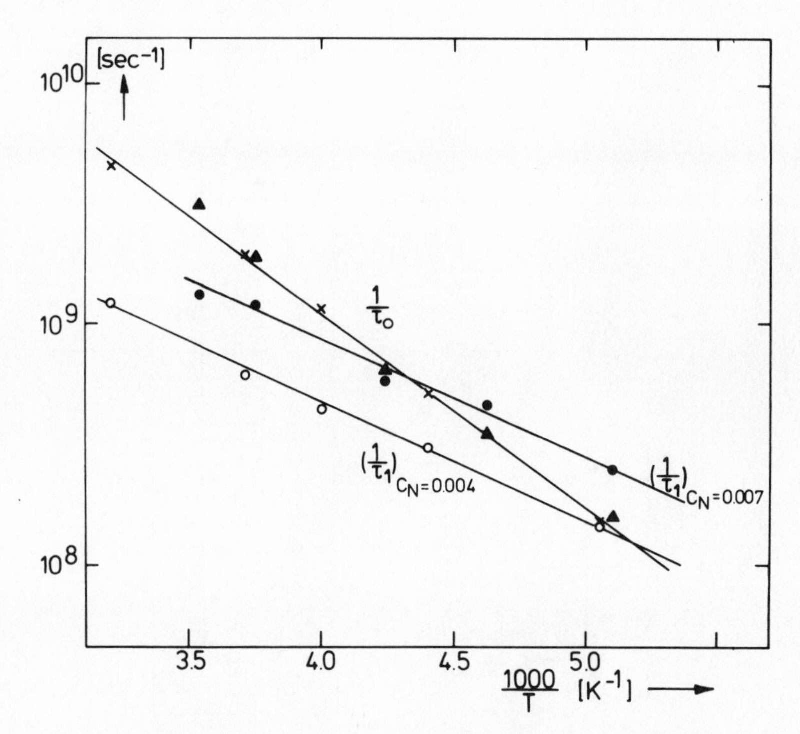

Fig. 14. Escape rate $1/\tau_o$ and trapping rate $1/\tau_1$ as a function of inverse temperature and nitrogen temperature. For $1/\tau_o$: Δ: $C_N = 0.7\%$, x = $C_N = 0.4\%$ (7).

lattice distortion was calculated from the Kanzaki forces of the nitrogen atoms, fitted to lattice expansion and Snoek effect measurements, and from the niobium force constants, determined from phonon spectra. The equilibrium energies were determined from the lattice distortion and the Kanzaki forces of a proton. The potential maxima were assumed to be equal. Electronic contributions to the hydrogen-nitrogen interaction were approximated by a hard-core potential. The self-correlation function for diffusion of a proton $G_s(\underline{r},t)$ was calculated by numerically solving the rate equations. A cube with 16 000 niobium atoms and 128 nitrogen atoms was used. The results for D_{eff} and the width Γ of the narrow component for large Q are shown in Fig. 5 as two lines. One obtains good agreement between the calculations and the experimental points for D_{eff} and Γ. Especially the activation energy of Γ was found to be similar to the experimental one. This relatively low activation energy can be understood from the combined effect of different small jump rates near the trap (64).

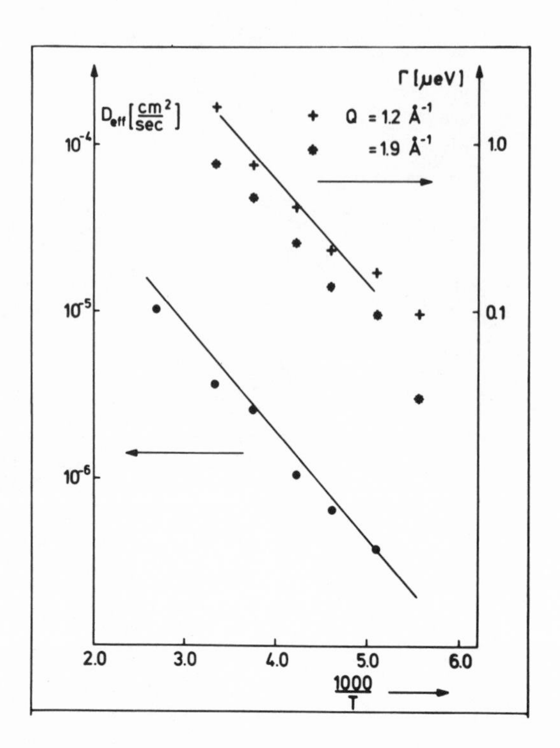

Fig. 15. Effective diffusion constant (lower left part) and width Γ of the quasielastic spectrum for two Q-values (upper right part) as a function of inverse temperature. The lines are the result of the numerical calculations (63).

The influence of substitutional defects on H-diffusion in Nb has been investigated recently by NMR (58) and later by quasielastic neutron scattering (59). The NMR experiments have been performed on $Nb_{1-x} V_xH_y$ - powder samples (x = 0.05 and 0.08) with varying H concentrations C_H. Investigated were: (i) the line width Γ_{Nb} of the Nb^{93} absorption line as a function of H concentration. The observed line width decreases sharply with increasing H-concentration. At the hydrogen concentration, where C_H reaches the vanadium concentration C_v, a break in the concentration dependence of Γ_{Nb} occurs which leads to a weak dependence of Γ_{Nb} or C_H for $C_H > C_v$. (ii) the hydrogen line width Γ_H as a function of temperature. In the temperature region $150 < T < 200$ K Γ_H decreases to its high temperature value which does not change up to 310 K, the highest temperature investigated. This result does not depend on the H-concentration. (iii) the Knight shift at the V nucleus as a function of H-concentration and temperature. Fig. 16 shows the observed sharp

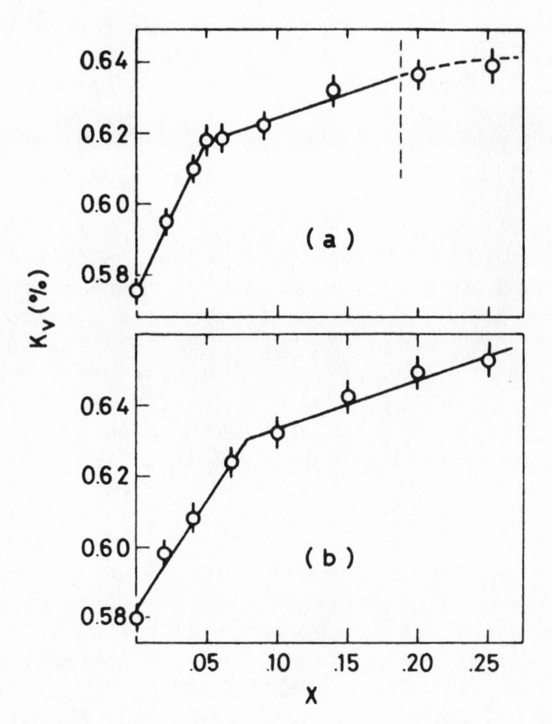

Fig. 16. ^{51}V Knight shift as a function of H-concentration in (a) $Nb_{0.095}V_{0.05}H_x$ and (b) $Nb_{0.92}V_{0.08}H_x$ systems at room temperature. The dotted line shows the solubility limit of hydrogen in the $Nb_{0.55}V_{0.05}$-alloy (58).

increase in the Knight shift K_V for $C_H/C_V < 1$. At $C_V = C_H$ again a break in the concentration dependence is observed. For $C_H > C_V$ K_V depends only weakly on C_H. Both the Knight shifts K_V in a H-free $Nb_{1-x}V_x$ sample and in $Nb_{1-x}V_xH_y$ increase with temperature. However, the rate of increase observed for the H-containing sample is much weaker than in the H-free sample.

All these results have been interpreted in terms of H-trapping at the V substitutional atom. The large Nb line width in $Nb_{1-x}V_x$ can be understood as a result of second order quadrupolar interaction. The V-atoms introduce internal strains which create an electric field gradient acting on the Nb nuclei. The sharp decrease of line width with increasing H concentration was attributed to a relaxation of the internal strains by the dissolved H atoms. In this picture V-H pair formation compensates for the lattice deformation caused by the small V-atom (atomic radius r_V 8.5% smaller than r_{Nb}). The observed complete motional narrowing of the proton line for temperatures T > 200 K independent of C_H led to the assumption of high H mobility near the V impurity. Fig. 6 shows that around a substitutional V-atom, which can be assumed to sit in the center of the cube, are 24 equivalent tetrahedral H sites. Rapid motion between these sites does not require a dissociation of the pair and can account for a similar line narrowing as observed for free H atoms.

Already earlier Knight-shift experiments (65) on VH_x compounds have shown an increase of K_v with increasing H concentration. This has been interpreted in terms of a H-atom donating its electron to the conduction band. Thereby the increase of K_v is due to the decrease of the negative core polarization contribution, K_c. K_c is related to the d-density of states at the Fermi energy, which decreases with increasing H concentration. The rate of increase in K_v above $C_H/C_v = 1$ is in accordance with the results found in the earlier measure measurements on VH_x. For $C_H/C_v < 1$, however, the increase in K_v is considerably larger than at higher concentrations. This has been attributed to a local change in the electronic structure at the V-nucleus due to V-H pair formation. Assuming that the electron of the hydrogen forming the pair contributes only locally to the change of the electron density in the d-states, a consistent description of the results on VH_y and $Nb_{1-x}V_xH_y$ evolves. In this picture a V-H pair gives rise to a local H concentration of 25% (1 H atom/4 surrounding atoms). The observed Knight shift corresponds to the value obtained in $VH_{0.25}$. The temperature dependence of the Knight shift provides information on the thermal occupation of the trapping sites. The weaker increase of the Knight shift with temperature compared to H-free $Nb_{1-x}V_x$ is then a consequence of the partial decomposition of the V-H pairs with increasing temperature. This tends to diminish the temperature effect on K_v. A quantitative evaluation of the binding energy yields $E_b = 90 \pm 50$ meV.

Again, quasielastic neutron scattering could provide direct insight into the trapping mechanism as well as in the dynamics of the protons in the trapped state. Preliminary results on $NbV_{0.002}H_{0.004}$ have been reported (59). The quasielastic spectra were measured in the temperature region $178K \leq T \leq 247K$ for Q's $0.13 \leq Q \leq 1.8$ $Å^{-1}$. Fig. 17 shows the width $\bar{\Gamma}$ of the spectra for the two smallest Q-values as a function of inverse temperature. Assuming that in this Q-range the line width is still proportional to the effective H-diffusion coefficient of the system, D_{eff} was evaluated. For comparison, the temperature dependence of the line width for a pure Nb sample (16) is indicated as a dashed line. In the next step of data refinement, all spectra with $Q \leq 1$ $Å^{-1}$ (see above) were fitted

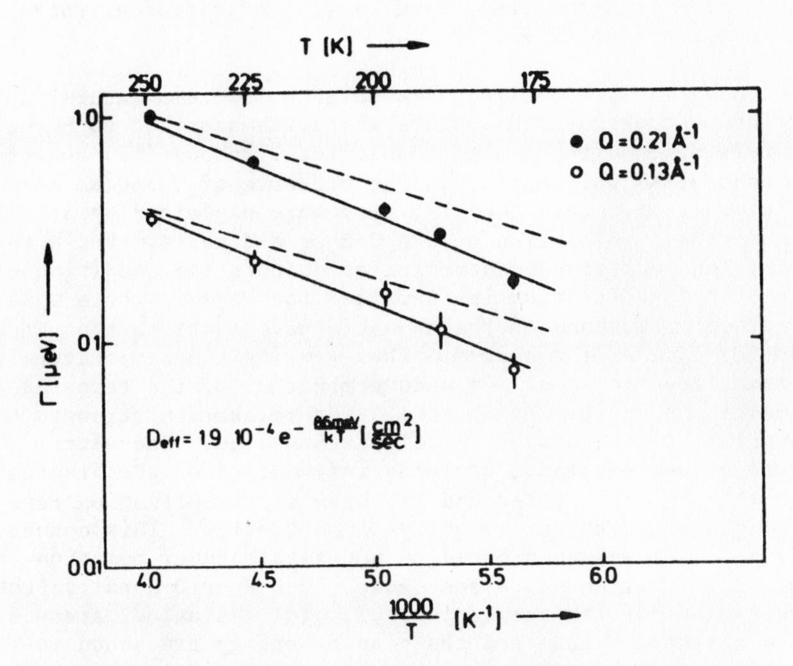

Fig. 17. Quasielastic line width of H in $NbV_{0.002}H_{0.004}$ at small Q's. Dashed line expected line width for a pure NbH-sample (59).

simultaneously with the scattering law of the RWM (Eq. (8)). Thereby an exponential temperature dependence of the two rates τ_o^{-1} and τ_1^{-1} was assumed. The activation energy for τ_1^{-1} was taken equal to the energy of migration in the undisturbed lattice (68 meV). Under these circumstances, the fitting process yields an activation energy for τ_o^{-1} of $E_a = 163 \pm 10$ meV. From E_a a binding energy $E_b = 95 \pm 10$ meV can be derived. This value is in agreement with the Knight shift result. The evaluated prefactor for $1/\tau_1$ of $\tau^{-1} = (3.6 \pm 0.3) \times 10^{-10}$ sec corresponds to a trapping radius of $R_t = 3.2$ Å. Due to the very low impurity concentration, however, this experiment did not yield information on the local dynamics at the trap.

DYNAMICS OF THE H-ATOM TRAPPED AT AN IMPURITY ATOM

In order to explain the observed concentration independent motional narrowing of the NMR-proton line in $Nb_{1-x}V_xH_y$, rapid H-motion around the substitutional V-atom had to be assumed (58). However, nothing is known about details of this motion. In contrast to the situation at substitutional impurities a larger number of experiments on the dynamics of the proton in the vicinity of an interstitial impurity like oxygen or nitrogen in Nb are available (8,57,9).

Already early experiments revealed a low temperature internal friction peak in Nb-H systems which was ascribed to H and D Snoek relaxations (66). Later Baker and Birnbaum (57) showed that these peaks only appear in the presence of O and N impurities in Nb. Consequently, they were explained by a reorientational relaxation of the O-H or N-H pairs. For a temperature independent concentration of defects the condition of maximum internal friction is fulfilled for $\omega\tau = 1$, where ω is the incident frequency of the acoustic wave and τ is the relaxation time of the system. Thus varying the temperature at different frequencies allows a determination of the temperature dependence of the relaxation time. The relaxation strength Δ is proportional to the number of relaxing species. Therefore, the temperature dependence of Δ yields information on the binding energy at the trap. Baker and Birnbaum found relaxation rates τ which changed with the impurity concentration. This concentration dependence was attributed to a possible super position of different relaxation processes caused by N-H or O-H pairs (the O/N ratio varied with sample purity). The evaluated parameters for the relaxation rate and the binding energy are shown in Table III. More recently these experiments have been repeated and extended on better characterized samples using H as well as D (67). Single crystalline UHV purified Nb-samples were doped with controlled amounts of O and H or D. The experiments were carried out in the temperature range 80-350 K and in the frequency range 30-210 MHZ. The longitudinal wave propagation direction was (100). The most interesting feature of these measurements is the pronounced isotope effect in the H-relaxation as well as in the binding energy of the O-H pair. The resulting parameters for the relaxation rate and the binding energy are shown in Table III.

A detailed site occupancy of the O-H pair has not yet been established (channeling experiments on NbO_xH_y samples are underway (68). However, the coupling of the longitudinal acoustic wave in (100) direction with the strain field of the pair requires a (100) pair symmetry. Octahedral sites for the O as well as for the H have been proposed but also tetrahedral site occupancy is possible for this symmetry. On the other hand, also a (111) pair symmetry for the O-H pair has been reported (69). Various mechanisms for the reorientation process, which involve partial dissociation of the pair, have been discussed. But, as long as the actual position of the proton is not known those discussions remain speculative. Also, the large isotope effect in the relaxation rate as well as in the binding energy is not understood. From the absence of an isotope effect in the double-force tensor for H and D in Nb similar binding energies for both species are expected in the framework of elastic theory. The ratio of H and D jump frequencies is far away from the prediction of the classical jump theory $\tau_D/\tau_H = \sqrt{M_D/M_H} = \sqrt{2}$ ($\tau_D/\tau_H(260K) = 2.3$; $\tau_D/\tau_H(100K) \simeq 200$). Also the tunneling hopping theory in its simplest form (24) cannot account for the observed isotope dependence. It would predict a large isotope effect in the prefactor due to the smaller tunneling matrix element for D and no isotope effect in the activation energy due to the absence of isotope effect in the double-force tensor for D and H.

Table III. Relaxation rates $\tau^{-1} = \tau_o^{-1} \exp(-E/kT)$ and binding energies for O-H and O-D pairs in Nb.

Defect	$\tau_o^{-1} (s^{-1})$	E(meV)	E_B(meV)
O-H (57) (low purity)	$(7.3 \pm 5)\ 10^{11}$	170 ± 25	90 ± 25
O-H (7) (int. purity)	$(8 \pm 10) \pm 10^{12}$	165 ± 25	---
O-H (8)	$1.3^{+1.2}_{-0.4}\ 10^{12}$	160 ± 10	90 ± 10
O-H (8)	$2.5^{+2.5}_{-0.8}\ 10^{12}$	200 ± 20	130 ± 20

Recently, the internal friction measurements on O-H(N-H) pairs in Nb were extended to low temperatures and long relaxation rates (8). Strain relaxation experiments allowed to monitor jump rates down to 10^{-3} Hz. The experiments were performed on U.H.V. purified polycrystalline Nb samples doped with controlled amounts of O or N. The observed relaxation frequencies were equal for H-N and H-O clusters. Fig. 18 presents the observed relaxation frequencies as a function of inverse temperature. Included are results from earlier experiments at higher temperatures. Fig. 18 clearly shows strong deviations from an Arrhenius behavior at low temperatures. The authors attributed this behavior to a tunneling hopping process of the proton around the impurity. The data were fitted with the expression of the small polaron theory for the jump rate in Debye approximation (25) (dashed line in Fig. 18). It depends on three

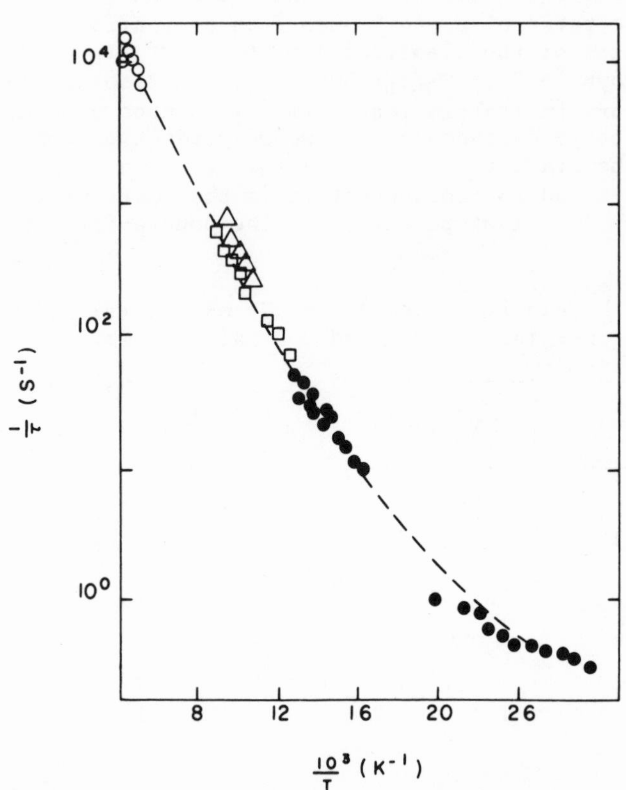

Fig. 18. The logarithm of the hopping frequency of H around a O impurity vs 1/T. O Mattas (67) |__| Schiller (69), o Chen (8), Δ Canelli (66).

independent parameters: the distortion energy necessary to equalize adjacent potential wells, the Debye temperature θ_D and the tunneling matrix element J. They were evaluated to: E_a = 211 meV, J = 53.4 meV and θ_D = 263.6 K. However, considering the presumable multistep reorientation mechanism involving partial dissociation in a highly distorted lattice near the impurity (67), the application of the tunneling hopping theory in its simplest form seems not to be adequate for the problem, as already demonstrated by its failure to explain the isotope effect for the reorientation rate. Therefore the resulting parameters might not have a direct physical meaning, e.g. a value in the order of 200 meV for E_a would imply an extraordinary deep potential well due to polaron formation. This has to be compared with an expected value of 26 meV calculated by Wagner and Horner (70) for pure Nb on the basis of elastic lattice theory.

A few years ago, Sellers et al. (71) reported the observation of a low temperature specific heat anomaly in Nb. The strong isotope effect made it clear, that the observed feature was caused by the H or D atoms. Birnbaum et al. (72) ascribed this phenomenon to proton tunneling among tetrahedral and triangular interstices on a cubic face within a polaron like distorted lattice region. Very recently the relation between the specific heat anomaly and the presence of N-impurities in Nb has been investigated systematically (9). The specific heat of UHV purified Nb, Nb doped with N or H and Nb doped with N and H or D was measured in the temperature region 0.07 - 1.5 K. Fig. 19 shows the obtained specific heat data together with the results reported by Sellers et al. They clearly demonstrate (i) that pure Nb as well as Nb doped either with H or with N exhibit essentially the same low temperature behavior which is very close to the expected calculated curve*; (ii) that samples containing both N and H or D show a large isotope dependent specific heat anomaly. Thus, the anomaly is related to the presence of both the N-impurity and the H-atom which at the low temperatures in question is trapped at the impurity. From the reported low residual resistivity ratio (100) it can be concluded that Sellers et al. already observed the same phenomenon. Because nothing is known about the geometry of the N-H pair, however, it is impossible to calculate the specific heat anomaly and the underlying system of tunneling levels. Also a correlation between the specific heat anomaly and the low temperature relaxation mechanism observed with internal friction is not obvious. The time scale of motion for both processes differs by more than 10 orders of magnitude (10^{-3} Hz and 35 K $10^{+11}s^{-1}$ at 1 K). Therefore, even a knowledge of the sites involved in the relaxation mechanism would not provide necessarily the geometry of the tunneling process. In this connection the authors speculate about (i) additional low energy sites around a nitrogen atom which are populated at low temperatures and (ii) that the reorientation in internal friction might occur between systems of tunnel split sites. The relatively low size of the anomaly (entropy increase due to the specific heat anomaly is only 0.055 k_B and 0.087 k_B per trapped H and D respectively) was attributed to stress induced interaction effects among neighboring N-H pairs. On the of basis elastic continuum theory the potential disturbance due to the neighboring defects was estimated to ≃ 1 meV in the average which is larger than the thermal energy of the anomaly. Their influence on the experimental result, therefore, is quite understandable. These blocking effects are also the reason that neutron scattering experiments, performed to reveal detailed microscopic information on the tunneling process, have been unsuccessful so far (73).

*for pure Nb.

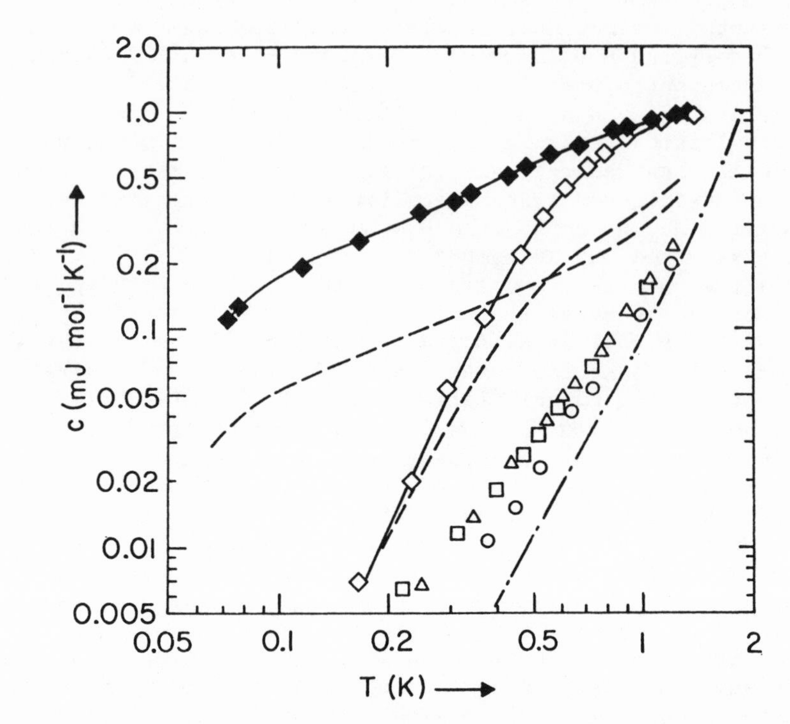

Fig. 19. Specific heat of Nb-samples partially doped with H,D and N in a log-log plot. O pure Nb, |__| $NbN_{0.003}$, Δ $NbH_{0.002}$ Δ $NbN_{0.006}$, Δ $NbN_{0.003}$ $D_{0.002}$. The solid lines are guide lines for the eye. The dashed line represents the results of Sellers et al. (71). The dashed pointed line gives the calculated specific heat for pure Nb (9).

ACKNOWLEDGMENT

The author gratefully acknowledges stimulating discussions with V. Lottner and K. W. Kehr (KFA-Jülich), S. M. Shapiro (BNL, Upton, NY) and T. Springer (ILL, Grenoble). Work at Brookhaven was supported by the Division of Basic Energy Sciences, Department of Energy,under Contract No. EY-76-C-02-0016.

Note added in proof

A reinvestigation of the "new excitations" in $NbD_{0.85}$ has shown that the 10.8 meV feature is a false peak. The 18.9 meV remains real.

REFERENCES

1. Hydrogen in Metals. Basic Properties (Eds. J. Völkl and G. Alefeld), Topics in Applied Physics Vol. 28, Springer Berlin, Heidelberg, New York 1978.
2. T. Springer in Topics in Current Physics (Eds. S. W. Lovesey and T. Springer) Vol. 3, Springer Berlin, Heidelberg, New York 1977.
3. Diffusion in Solids: Recent Developments (Eds. A. S. Nowick and J. J. Burton) Academic Press, New York 1975.
4. A. Magerl, W. D. Teuchert, R. Scherm, IAEA Report, IAEA SM-219/21 (1978).
5. S. M. Shapiro, Y. Noda, T. O. Brun, J. Miller, H. Birnbaum, T. Kajitani, Phys. Rev. Lett. 41, 1051 (1978).
6. V. Lottner, H. R. Schober, W. J. Fitzgerald, Phys. Rev. Letts. 42, 1162 (1979).
7. D. Richter, T. Springer, Phys. Rev. B 18, 126 (1978).
8. C. G. Chen, H. K. Birnbaum, phys. stat. sol. (a) 36, 687 (1976).
9. C. Morkel, H. Wipf, K. Neumaier, Phys. Rev. Lett. 40, 947 (1978).
10. H. K. Birnbaum, C. A. Wert, Ber. Bunsenges. Physik. Chem. 76, 806 (1972).
11. J. Völkl, G. Alefeld, chapter 12 in Ref. (1).
12. G. Schaumann, J. Völkl, G. Alefeld, phys. stat. sol. 42, 401 (1970).
13. G. Matusiewicz, H. K. Birnbaum, J. Phys. F 7, 2285 (1977).
14. H. Wipf, G. Alefeld, phys. stat. sol. (a) 23, 175 (1974).
15. D. G. Westlake, S. T. Ockers, D. W. Regan, J. Less Common Metals 49, 341 (1976).
16. D. Richter, G. Alefeld, A. Heidemann, N. Wakabayashi, J. Phys. F 7, 569 (1977).
17. W. Gissler, G. Alefeld, T. Springer, J. Phys. Chem. Sol. 31, 2361 (1970).
18. N. Boes, H. Züchner, Z. Naturforsch. 31a, 760 (1976).
19. J. Völkl, H. C. Bauer, U. Freudenberg, M. Kokkinides, G. Lang, K. A. Steinhauser, G. Alefeld, ICFUAS-6-485 (1977).
20. G. H. Vineyard, J. Phys. Chem. Sol. 3, 121 (1957).
21. L. Katz, M. Guiman, R. J. Borg, Phys. Rev. B 4, 330 (1971).
22. J. Völkl, G. Wollenweber, K. H. Klatt, G. Alefeld, Z. Naturforschung 26a, 922 (1971).
23. G. Sicking, H. Buchold, Z. Naturforschung 26a, 1973 (1971).
24. C. P. Flynn, A. M. Stoneham, Phys. Rev. B 1, 3966 (1970).
25. A. M. Stoneham, J. Nucl. Mat. 69, 109 (1978).
26. H. Pfeiffer, H. Peisl, Phys. Lett. A, to be published.
27. V. Lottner, A. Heim, K. W. Kehr, T. Springer, IAEA Report SM-219/27, Vienna (1978).

28. D. Emin, M. I. Baskes, W. D. Wilson, Proceedings of the First International Conference on SR, Rohrschach, Switzerland, September 1978, to be published in Hyperfine Interactions.
29. H. Teichler, to be published.
30. For an introduction to quasielastic neutron scattering, see T. Springer, Springer Tracts of Modern Physics 64 (1972).
31. L. Van Hove, Phys. Rev. 95, 249 (1954).
32. C. T. Chudley, R. J. Elliott, Proc. Phys. Soc. 77, 353 (1961).
33. J. M. Rowe, J. J. Rush, H. G. Smith, M. Mostoller, and H. Flotow, Phys. Rev. Lett. 33, 1297 (1974).
34. L. A. de Graaf, J. J. Rush, H. E. Flotow, J. M. Rowe, J. Chem. Phys. 56, 4574 (1972).
35. J. M. Rowe, J. J. Rush, H. E. Flotow, Phys. Rev. B 9, 5039 (1974).
36. N. Stump, W. Gissler, R. Rubin, phys. stat. sol. (b) 295 (1972).
37. N. Wakabayashi, G. Alefeld, K. W. Kehr, T. Springer, Solid State Commun. 15, 503 (1974).
38. W. Gissler, N. Stump, Physica 65, 109 (1973).
39. V. Lottner, J. W. Haus, A. Heim, K. W. Kehr, to be published.
40. D. Richter, K. W. Kehr, T. Springer in Proceedings of the Conference on Neutron Scattering, Gatlinburg, TN (Ed. R.M. Moon)CONF.-76 0601-P1, Vol.1, 568 (1976).
41. D. Emin, Phys. Rev. Lett. 25, 1751 (1970).
42. D. Emin, Phys. Rev. B 3, 1321 (1971).
43. T. Springer, in Ref. (1), Chapter 4.
44. A. Magerl, B. Berre, G. Alefeld, phys. stat. sol. (a) 36, 161 (1976).
45. A. Magerl, N. Stump, W. D. Teuchert, V. Wagner, G. Alefeld, J. Phys. C 10, 2783 (1977).
46. G. S. Bauer, W. Schmatz, W. Just, Proc. Second Congress International "1" Hydrogène dans les Métaux" Paris, 6-10 June (1977).
47. J. Buchholz, J. Völkl, G. Alefeld, Phys. Rev. Lett. 30, 318 (1973).
48. Y. Nakagawa, A. D. B. Woods in Lattice Dynamics, Ed. R. F. Wallis (Pergamon, Oxfd, 1965), p. 39.
49. J. M. Rowe, N. Vagelatos, J. J. Rush, H. E. Flotow, Phys. Rev. B 12, 2959 (1975).

50. In the case of heavy interstitials, such a resonant mode provides a low barrier diffusion mechanism, Thereby assuming that the particle is moving in a smooth effective potential provided by the vibrating host atoms. Under the assumption of a sinusoidal potential, the activation energy can be estimated in terms of the static Green function. For Nb and model 2 this yields the very low activation energy of 65 meV. However, such a mechanism does not explain the isotope effect observed in the H-diffusion in Nb.
51. P. Kofstadt, W. E. Wallace, L. J. Hyvönen, J. Am. Chem. Soc. 81, 5015 (1959).
52. P. Kofstadt, W. E. Wallace, J. Am. Chem. Soc. 81, 5019 (1959).
53. O. J. Kleppa, P. Dantzer, M. E. Melnichak, J. Chem. Phys. 61, 4048 (1974).
54. C. Wert, Fiz. Nizkikh Temp. 1, 626 (1975).
55. G. Pfeiffer, H. Wipf, J. Phys. F 6, 167 (1976).
56. W. Münzing, J. Völkl, H. Wipf, G. Alefeld, Scripta Metall. 8, 1327 (1975).
57. C. Baker, H. K. Birnbaum, Acta Metall. 21, 865 (1973).
58. T. Matsumoto, J. Phys. Soc. Japan 42, 1583 (1977).
59. D. Richter, A. Kollmar, Annex of the Annual Report of the ILL for 1977, Grenoble, France (1978).
60. K. W. Kehr, D. Richter, Solid State Commun. 20, 477 (1976).
61. T. R. Waite, Phys. Rev. 107, 463, 471 (1954)
62. M. Borghini, T. O. Niinikoski, J. C. Soulie, O. Hartmann, E. Karlsson, L. O. Norlin, K. Pernestal, K. W. Kehr, D. Richter, E. Walker, Phys. Rev. Lett. 40, 1723 (1978).
63. R. H. Swendsen, D. Richter, K. W. Kehr, unpublished.
64. K. W. Kehr, D. Richter, R. H. Swendsen, J. Phys. F 8, 433 (1978).
65. D. Zamir, Phys. Rev. 140 A, 271 (1965).
66. Cannelli, L. Verdini, Ricerca Sci. 36, 98 (1966).
67. R. F. Mattas, H. K. Birnbaum, Acta Metall. 23, 973 (1975).
68. Carstanjen, private communication.
69. P. Schiller and H. Nijman, phys. stat. sol. 31, K77 (1975).
70. H. Horner, H. Wagner, J. Phys. 7, 3305 (1974).
71. G. J. Sellers, A. C. Anderson, H. K. Birnbaum, Phys. Rev. B 10, 2771 (1974).
72. H. K. Birnbaum, C. P. Flynn, Phys. Rev. Lett. 37, 25 (1976).
73. T. Springer, G. Alefeld, H. Wipf, N. Stump, D. Richter, Annex of the Annual Report of the ILL for 1977, Grenoble, France (1978).

LIGHT-INTERSTITIAL DIFFUSION IN METALS

H. Teichler

Institut für Metallphysik der Universität
and Sonderforschungsbereich 126
D-3400 Göttingen, Germany

INTRODUCTION

At present the motion of positively charged light interstitials such as positive muons or the nuclei of the hydrogen isotopes in metals is analyzed and interpreted in terms of the quantum theory of diffusion (QTD) which has been proposed by Flynn and Stoneham[1] and which later on has been put forward, e.g., by Kagan and Klinger[2]. In the following lectures some of the basic ideas of the QTD shall be outlined (further aspects are discussed in the article of A.M. Stoneham in this issue) and the actual status of our understanding of the light-interstitial diffusion will be reviewed. (Besides the QTD there recently have been some attempts[3,4] to describe the hydrogen motion in metals within the so-called "Quantum Occurance Theory", a phenomenological generalization of the basic ideas of the theory of phonon-assisted defect transitions[5]. But since hitherto preliminary communications only have been published about the application of this method to the interstitial diffusion we shall focus our attention in the following on the QTD).

According to the QTD the properties of the positive muons and of the hydrogen nuclei in metals should be intimately related[6,7,8], so we shall discuss these particles on the same footing. Nevertheless, as we shall become aware later on, the big difference in mass between these particles (which yields large differences in the zero-point energies and in the eigenfrequencies of these interstitials) may perhaps give rise to different diffusion mechanisms[8].

Qualitatively the basic idea of the QTD is as follows: the light interstitials occupy localized bound states in the potential minima of the metal. They are assumed to be capable to carry out tunneling transitions to adjacent crystallographically equivalent interstitial sites in the lattice. The interstitials couple strongly the lattice and induce marked lattice distortions which in the case of a transition have to be shifted from the initial to the final site. The necessity for shifting the deformation cloud during a transition yields an appreciable reduction of the transition probability and is the origin of the strong temperature dependence of the transition process.

The above-specified model of deformation-assisted defect transitions is of importance in other areas of solid state theory, too, e.g., in the theory of hopping conductivity of small polarons[5,9,10], in the theory of reorientation of dipolar centers in alkali halides[11], and it even has been used to discuss the mobility of point defects in quantum crystals[12,13]. In these cases as well as in the QTD one is concerned with a situation where relatively light particles (e.g., localized excess-electrons or holes in small polarons, off-centre substitutional impurities like Li^+ in the dipolar centers, vacancies in quantum crystals) move amongst an ensemble of particles of greater mass, where the mobility of the lighter defects is determined by the stochastic motion of the heavier host-material particles. The light particles are assumed to be severely localized ("self-trapped"[5]) within the displacement holes induced by themselves in the host-lattice and are considered as combined quasi-particles consisting of the interstitial and a dressing deformation cloud. Historically, the investigation of the hopping mobility of small polarons was the first of these theories, so one often calls all the application of the idea of deformation-assisted defect transitions "small polaron theories".

In order to characterize the present status of these small polaron theories we may state that the earlier approaches, although differing appreciably in their basic approximations, had some simplifying assumptions in common, e.g.,

- the Born-Oppenheimer approximation (the electrons of the system follow adiabatically the motion of the defects and of the host material atoms),
- the adiabatic approximation (the defects follow adiabatically the motion of the host-material atoms),
- the linear-coupling approximation (the defect-lattice interaction is linear in the displacements of the host atoms),
- the Condon approximation (the defect transfer matrix elements are independent of the configuration of the host lattice.

Recent years now have seen many efforts to overcome at least the more severe of these restrictions: Theories beyond the adiabatic approximation have been developed, e.g., by Kagan and Klinger[14], Pavlovich and Rudko[15], and by Wagner[16]. A small-polaron theory with quadratic defect-lattice coupling has been formulated by Tonks and Dick[17]. Treatments of phonon-assisted defect transitions without making use of the Condon approximation have also been presented[1,14,15] (but in these investigations rather simplified models have been studied only), a more general theory has been discussed recently[18].

We shall start our lectures by classifying the various types of transition modes which may occur for light interstitials in metals. Then a quantitative analysis of experimental μ^+-diffusion data (in Cu) will be given in terms of direct, incoherent multi-phonon transitions between groundstate levels. Since there are some indications[1,3,4,8] that the Condon approximation (e.g.[1]) is not applicable to hydrogen diffusion in metals, the theory of incoherent direct and indirect multi-phonon transitions beyond the Condon-approximation will be outlined in some detail and results of numerical calculations will be mentioned, which indicate that the QTD beyond the Condon approximation is capable to describe the basic features of the diffusion of hydrogen in metals. In particular the differences between the diffusion mechanisms of positive muons and of the hydrogen nuclei shall be emphasized which are primarily due to the big mass difference between these particles.

CLASSIFICATION OF THE TRANSITION MODES

As disscussed, e.g., by Flynn and Stoneham[1] and by Kagan and Klinger[2] the transition of the interstitials between the individual interstitial sites in the lattice either may be coherent or incoherent. In the case of a coherent transition the final state of the quasiparticle consisting of the interstitial and its deformation cloud is identical to its initial state up to a translation operation of the lattice. In particular, initial and final state of the system are related by well-defined phase-differences and hence all the localized interstitial states may be combined to coherent Bloch-waves extending through the whole crystal. Coherent motion will be the predominant propagation mode at low temperatures. If the scattering by thermal phonons is the dominating mechanism limiting the coherent motion then the mobility via coherent transitions should decrease exponentially with increasing temperature[2]. However, there may exist another mechanism which limits the coherent motion, the "dynamical destruction of the interstitial band"[2]. This is due to fluctuations of the relative positions of the interstitial levels at neighbouring sites caused by the interaction of the interstitial with the lattice vibrations and with impurities in the crystal. As shown by Anderson[19] a coherent, delocalized Bloch-type state can exist only as long as the random shifts δE of the energy levels are smaller than the transfer integrals between adjacent sites. Otherwise localization occurs. Because of the small bandwith expected for the deformation-dressed impurities the coherent motion is very sensitive to all kinds of lattice imperfections and, in addition, will be strongly reduced at higher temperatures by thermal phonons. At higher temperatures incoherent transitions may become important. During these transitions there is a redistribution of energy among those phonon-modes which couple to the deformation cloud. By this mechanism there are so many more final states accessible for transition between adjacent interstitial sites that at higher temperatures the coherent transitions are completely overwhelmed[1]. Because of the redistribution of phonons during such a transition the coherence between the initial and final quantum state of the interstitial is destroyed. In this case the interstitial motion may be visualized as incoherent hopping between adjacent interstitial sites.

The incoherent transitions may be classified according to the number of phonons which participate in the exchange of energy during a transition. At lower temperatures "few-phonon" processes may occur where, e.g., two, three, or four... phonons are involved. At higher temperatures (typically above 1/10 or 1/5 of Debye's temperature) almost all phonon modes take part in the exchange or energy and hence "multi-phonon" processes have to be considered. The distinction between 'few-phonon' and 'multi-phonon' transitions is in particular a technical one since the evaluation of the transition rates in the two limiting cases needs different mathematical approximations[1]. The two-phonon processes at low temperatures yield a T^7 temperature dependence of the transition rate (and thus of the diffusion constant) whereas multi-phonon transitions at temperatures above Debye's temperature result in an Arrhenius law (but with temperature dependent preexponential factor).

Within the individual potential wells of the lattice interstitials may exist in different localized harmonic-oscillator-like interstitial states. In such multilevel systems one has to distinguish between direct incoherent transitions during which the quantum state of the interstitial is conserved (up to the translation to a neighbouring site) and indirect transitions during which the interstitial changes its configuration and even its energy.

QUANTITATIVE ANALYSIS OF μ^+ DIFFUSION DATA WITHIN THE FRAMEWORK OF THE QTD

By the positive muon nature has provided us with a unique tool to investigate the predictions of the QTD. The outstanding features which make μ^+ -diffusion experiments so unique in comparison with genuine light interstitial diffusion investigations in metals are:

- in μ^+SR experiments μ^+-diffusion is studied on a microscopic scale given by the lattice constant a_o rather than on a macroscopic scale. Hence μ^+-investigations can be carried out at such low temperatures where diffusion via excited states will be negligible,
- because of the small mass of the muons they have large zero-point energies and the energy levels of the excited states within the potential wells are well separated from the ground state level,
- due to the small mass of the muon the tunneling transitions will be much more important in the temperature ranges studied than 'over-barrier transitions'[2,8],
- in the temperature range used for μ^+-diffusion experiments quantum effects, i.e., deviations from the Arrhenius law should be detectable.

At present the most complete experimental μ^+-diffusion data exist for μ^+ in Cu[20,21]. These date have been analyzed within the QTD in [22,23]. In the analysis[22] the following assumptions have been used:

- the Born-Oppenheimer approximation,
- the adiabatic decoupling,
- the Condon approximation,
- the muons are localized in potential wells at the octahedral interstitial sites of the fcc lattice,
- muon-diffusion takes place via direct incoherent tunneling transitions between the ground state levels in adjacent octahedral interstitial sites (coherent transitions are neglected),
- the muon-lattice interaction is linear in the displacements of the host-metal atoms,
- a muon at on octahedral site exerts radial forces of strength g on its six nearest-neighbour host-metal atoms only,
- the host-metal lattice may be described in the harmonic approximation where nearest-neighbour host-metal atoms (with mass M_o) interact with each other via spring forces with spring constant f.

Then the theory predicts for the diffusion constant D(T) the expression[22]

$$D(T) = \frac{a_o^2}{\hbar} \sqrt{\frac{\pi}{4E_a kT_o h_1(T/T_o)}} \, |J|^2 \exp\left[- \frac{E_a}{kT_o \cdot h_2(T/T_o)} \right] \quad (3.1)$$

where

$$k\,T_o = \hbar\sqrt{2f/M_o} \approx k\Theta_D/2 \quad , \quad (3.2)$$

(Θ_D: Debye's temperature). J means the muon transfer integral in the Condon approximation.

The quantity E_a measures the energy necessary for creating a deformation which equalizes the ground state energy of the elastically relaxed initial site and the ground state energy of the non-relaxed final site of the interstitial. In the model specified above, where nearest-neighbour host-metal atoms interact with each other, E_a can be evaluated for fcc metals yielding[22]

$$E_a = 0.258\, g^2/f \quad . \quad (3.3)$$

Beyond this model E_a may be calculated by use of the lattice Green's function for Cu[24,25] provided that the forces exerted by the muons are known. Let $\underset{\sim}{g}_\alpha^i$ denote the force exerted on the host-lattice atome $\underset{\sim}{R}_\alpha$ by the interstitial in its initial site, $\underset{\sim}{g}_\alpha^f$ the force exerted in its final state, and $\underset{\approx}{G}_{\alpha\alpha'}$, the matrix elements of the lattice Green's function. Then (as may be deduced from[1])

$$E_a = \frac{1}{8} \sum_{\alpha\alpha'} (\underset{\sim}{g}_\alpha^i - \underset{\sim}{g}_\alpha^f) \cdot \underset{\approx}{G}_{\alpha\alpha'} \cdot (\underset{\sim}{g}_{\alpha'}^i - \underset{\sim}{g}_{\alpha'}^f) \quad (3.4)$$

(For Cu the E_a estimated from (3.3) with $f = a_o c_{44}$ [26] and from (3.4) agree within 5%.)

In (3.1) the $h_i(T/T_o)$ are universal functions for all fcc lattices. They have been evaluated numerically[22] and are shown in Figure 1. For temperatures above Θ_D these functions may be substituted by T/T_o. Eq. (3.1) then reduces to the well known expression for the diffusion constant of light interstitials evaluated by Flynn and Stoneham[1]. The deviations of $h_1(T/T_o)$, $h_2(T/T_o)$ from T/T_o measure the influence of the zero-point vibrations of the host-metal atoms compared to the thermally excited phonons in the lattice. From the explicit formulae for the $h_i(T/T_o)$[22] analytical expressions may be deduced by introducting an Einstein-type approximation. Then one finds

$$h_1\left(\frac{T}{T_o}\right) = \frac{\hbar\omega_E}{2kT_o} \operatorname{csch}\left(\frac{\hbar\omega_E}{2kT_o} \cdot \frac{T_o}{T}\right) \quad , \quad (3.5)$$

$$h_2\left(\frac{T}{T_o}\right) = \frac{\hbar\omega_E}{4kT_o} \coth\left(\frac{\hbar\omega_E}{4kT_o} \cdot \frac{T_o}{T}\right) \quad , \quad (3.6)$$

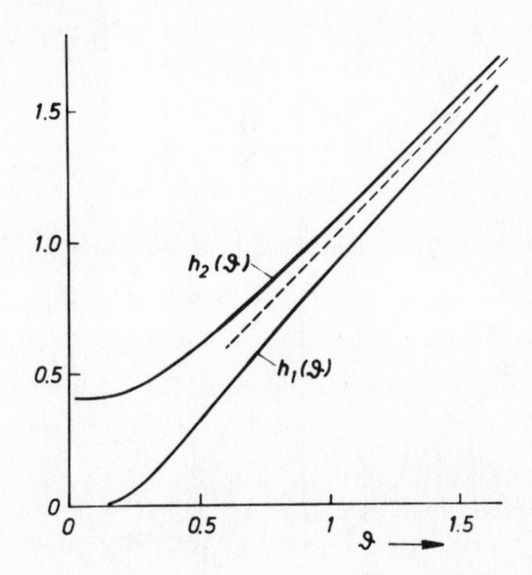

Fig. 1. Auxiliary functions $h_1(\theta)$ and $h_2(\theta)$ versus $\theta = T/T_o$.

where

$$\hbar\omega_E \approx 1.6kT_o \approx 0.8k\Theta_D \quad . \tag{3.7}$$

Eq. (3.1) predicts a linear relationship between

$$\ln \tilde{D}(T) = \ln D(T) + \frac{1}{2} \ln h_1(T/T_o) \tag{3.8}$$

and $h_2(T/T_o)^{-1}$, the slope being $-E_a/kT_o$. As proved[22] this result of the theory is an excellent agreement with the experimental data of Grebnnik et al.[21]. Figure 2 displays the data[21] analyzed according to (3.8) by use of the value $T_o = 181$ K (which follows for Cu from $f = a_o c_{44}$ [26]). The linear relationship is well satisfied for temperatures below 250 K. The deviations occurring in the monocrystal data above 250 K may be explained by assuming that at higher temperatures other transition modes than ground-state sub-barrier-tunneling become important, e.g., over-barrier hopping[21]. The values of the transfer matrix element J and of E_a estimated by a least mean square analysis of the experimental data of Grebinnik et al.[21] are shown in table I as well as values of ga_o deduced from E_a by means of (3.3).

From the parameters obtained by analyzing the μ^+ diffusion data[21] in Cu within the model of incoherent tunneling McMullen and Bergersen[23] estimated that coherent motion of positive muons in Cu should be the predominant propagation mode below 190 K provided that the coherent motion is limited by the scattering of thermal phonons only. Since the coherent motion decreases with increasing temperature whereas the incoherent motion increases the mobility of μ^+ therefore should exhibit a minimum at about 190 K. Such a minimum has not yet been found in the μ^+ spin depolarisation rate in Cu and therefrom McMullen and Bergersen[23] concluded that

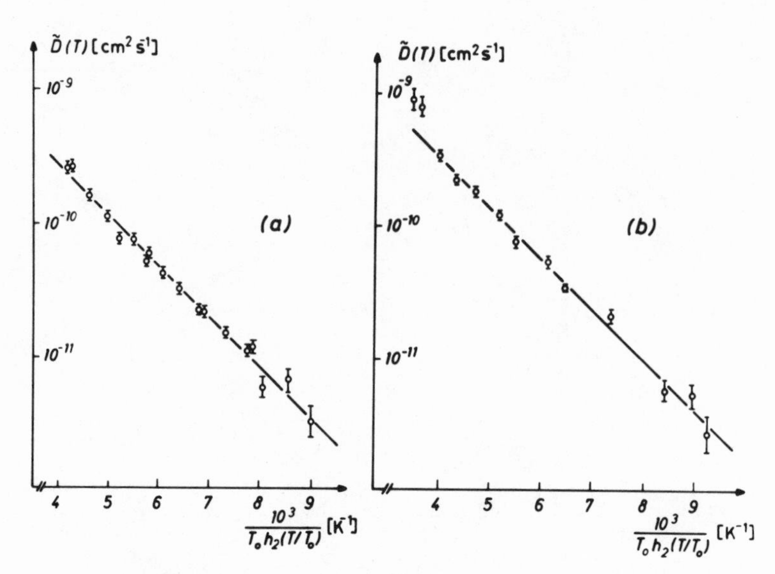

Fig. 2. Semi-logarithmic plot of $\tilde{D}(T)$ versus $h_2(T/T_o)^{-1}$ for μ^+ in polycrystalline (a) und monocrystalline (b) Cu.

the coherent motion may be limited in the Cu specimen investigated so far by the "dynamical destruction of the μ^+-band" due to defects and impurities in the lattice.

In this context one should be aware of a recent paper by Fujii and Uemura[27] where the authors point out that for ferromagnetic metals in the low-temperature regime with predominating coherent motion the diffusion constant D(T) and the muon spin depolarization rate $1/\tau_s$ are by no means proportional to each other (as is the case for incoherent motion) but show completely different temperature dependences. Qualitatively their arguments are as follows: In the high-temperature regime of incoherent motion both quantities, D(T) and $1/\tau_s$, follow approximately an Arrhenius law and are monotonically decreasing with decreasing temperature. At low temperature with predominating coherent motion the thermal-phonon-scattering of the muon out of its coherent states (build up by the muon and the surrounding electrons) decrease with decreasing temperatures. This yields a decrease of the depolarization rate

Table I. Transfer integral J, energy E_a and strength of radial μ^+-host-metal forces determined by analysing the experimental μ^+ diffusion data within the quantum theory of diffusion.

	$J[10^{-6}eV]$	$E_a[10^{-3}eV]$	$g[eV/a_o]$
polycrystalline Cu	16.6	74.8	2.64
monocrystalline Cu	18.4	75.2	2.65

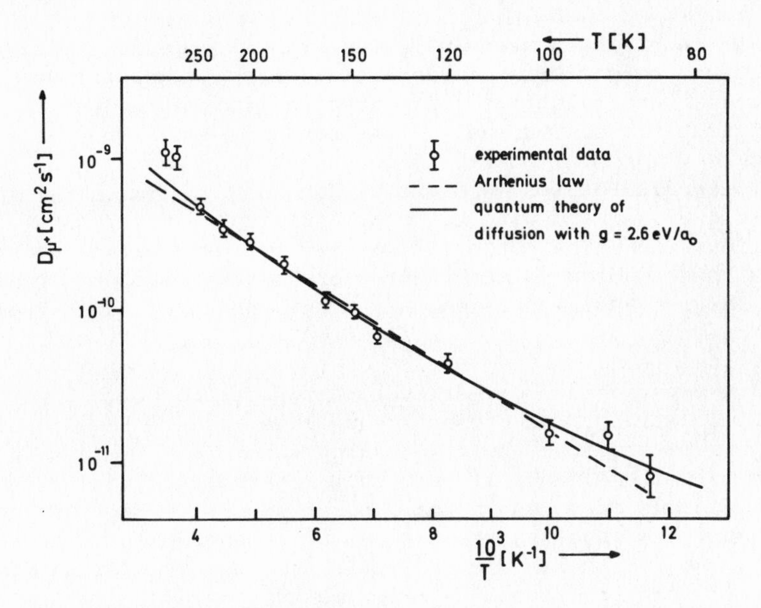

Fig. 3. Arrhenius plot of D(T) for μ^+ in monocrystalline Cu.

$1/\tau_s$ but it means an increase in the mobility, i.e., in D(T). Due to this D(T) should exhibit a minimum at that temperature where the muon motion changes from incoherent to coherent transitions whereas $1/\tau_s$ should be more or less decreasing in the whole temperature regime. As shown by Fujii[28] the theory is capable to model the temperature dependence of $1/\tau_s$ measured for μ^+ in α-Fe[29].

MULTI-PHONON TRANSITIONS BEYOND THE CONDON APPROXIMATION

Although the coupling of the interstitials to the lattice vibrations and lattice distortions is the basic idea of the QTD in almost all of the hitherto published theories the transfer matrix elements J_{if}, characterizing the transition amplitude of the interstitial from its initial site i to the final site f, are described in the Condon approximation, i.e., J_{if} is considered to be independend of the host-lattice configuration. There are some treatments of the QTD (e.g.[1,14,15]) which go beyond this approach, but these investigations include the dependence of J_{if} on the lattice configuration in a rather schematic way only. Since there are indications[1,3,4,8,18] that the Condon approximation is not well suited to describe the diffusion of hydrogen in metals in the whole temperature range we shall consider in the following the theory[18] of multiphonon-transitions beyond the Condon approximation in some detail.

The implications and weakness of the Condon approximation may be seen by the following considerations: in the adiabetic approximation (which has been used in[18]) the transition rate for transitions between neighbouring interstitial sites in general depends on the instantanous lattice configuration produced by the actual phonon fluctuations. Obviously the thermal average of the transition rate incorporates with large probability transitions at small phonon fluctuation (where the phonons may enhance the under-barrier-tunneling) and with minor probability transitions at large fluctuations where the potential barrier between initial and final site has been reduced so much that over-barrier hopping (characterized by large transition rates) may occur. Which of these two transition modes predominates will depend on a variety of properties of the system (e.g. on the mass m_i of the tunneling particle, the shape of the potential barrier, the change of the barrier with phonon fluctuations, the phonon frequencies, and on the temperature) and may not be predicted by simple arguments. Nevertheless one is inclined to believe that for extremely light particles (e.g., the excess electrons or holes in small polarons or, perhaps, the positive muons) the tunneling transitions with minor phonon-activation may yield the overwhelming contribution to the transition rate at all temperatures accessible to experiments. Therefore the Condon approximation may be applicable in these cases. On the other hand for heavier particles it may happen that tunneling transitions predominate at lower temperatures but that at an intermediate temperature the "phonon-fluctuation induced over-barrier" transitions may become more effective than tunneling. Then the Condon approximation would mean a rather poor approach to model the transfer matrix element in the whole temperature range.

By assuming the applicability of the Born-Oppenheimer approximation, of the adiabatic decoupling,and of the linear coupling model the theory[18] yields in the case of incoherent, indirect multiphonon transitions for the transition rate between adjacent interstitial sites beyond the Condon approximation at temperatures above Θ_D

$$W_{if} = \frac{1}{\hbar} \sqrt{\frac{\pi}{4E_a kT}} \; < |J_{if}|^2 > e^{-E_{if}/kT} \quad . \tag{4.1}$$

In deriving Eq. (4.1) the transitions of the interstitials between adjacent sites have been considered as rare processes and have been treated in second order perturbation theory. The formal structure of expression (4.1) is similar to that deduced by Flynn and Stoneham[1] for direct transitions in the Condon approximation. However, the meaning of the quantities partly has changed. As in [1] the quantity E_a denotes the energy necessary for preparing the lattice configuration allowing a transition of the interstitial from its initial state i to the final state f.

$$E_{if} = (E_f - E_i + 4E_a)^2/16E_a \quad , \tag{4.2}$$

E_i, E_f denote the energy values of the static (i.e. without phonons) relaxed host-metal-interstitials system for the interstitial in its initial/final state. The effective transfer integral $\langle|J_{if}|^2\rangle$ entering the transition rate (4.1) is given by a complicated thermal average of the square of $J_{if}(\underset{\sim}{u}_\alpha)$ ($\underset{\sim}{u}_\alpha$ characterizes the static displacements of the atoms plus their thermal vibrations around the displaced equlibrium positions).

For the particular case that $J_{if}(\underset{\sim}{u}_\alpha)$ depends on the $\underset{\sim}{u}_\alpha$ in terms of the height of the potential barrier separating initial and final site, $U_{sp}(\underset{\sim}{u}_\alpha)$, above the energy of the incident particle, $\varepsilon_i(\underset{\sim}{u}_\alpha)$, that means assuming

$$J_{if}(\underset{\sim}{u}_\alpha) \equiv J_o(U_{sp}(\underset{\sim}{u}_\alpha) - \varepsilon_i(\underset{\sim}{u}_\alpha)) \quad , \tag{4.3}$$

and using the linear coupling approximation for the height of the barrier and the energy of the interstitial

$$U_{sp}(\underset{\sim}{u}_\alpha) = U_{sp}^o - \sum_\alpha \underset{\sim}{g}_\alpha^{sp} \cdot \underset{\sim}{u}_\alpha \quad , \tag{4.4}$$

$$\varepsilon_i(\underset{\sim}{u}_\alpha) = \varepsilon_i^o - \sum_\alpha \underset{\sim}{g}_\alpha^i \cdot \underset{\sim}{u}_\alpha \quad , \tag{4.5}$$

a rather simple expression for $\langle|J_{if}|^2\rangle$ has been obtained[18]. At temperatures above Θ_D the average transfer integral entering W_{if} then is given by

$$\langle|J_{if}|^2\rangle = \int d\varepsilon\,|J_o(\varepsilon)|^2 \exp\left[-\frac{(\varepsilon-\Delta U_{eff})^2}{2\sigma^2}\right] / (\sqrt{2\pi}\sigma) \quad . \tag{4.6}$$

ΔU_{eff} denotes the "effective" height of the potential barrier which means the energy difference between the saddle point of the barrier and the energy of the penetrating particle within the particular lattice configuration where the energy levels of the initial and final state of the interstitial are equalized by a suitable fluctuation and which has minimum energy stored in the deformed lattice. σ^2 measures the fluctuations of the barrier height induced by the fluctuations in the phonon system and is in the high-temperature limit of the form

$$\sigma^2 \approx 8\,kT \cdot E_b q \quad . \tag{4.7}$$

The quantity

$$E_b = \frac{1}{8} \sum_{\alpha\alpha'} (\underset{\sim}{g}_\alpha^{sp} - \underset{\sim}{g}_\alpha^i) \cdot \underset{\approx}{G}_{\alpha\alpha'} \cdot (\underset{\sim}{g}_{\alpha'}^i - \underset{\sim}{g}_{\alpha'}^f) \tag{4.8}$$

measures the temperature dependence of free fluctuations of the barrier height,

$$q = 1 - E_m^2/(E_a \cdot E_b) \quad , \tag{4.9}$$

$$E_m = \frac{1}{8} \sum_{\alpha\alpha'} (\underset{\sim}{g}_\alpha^{sp} - \underset{\sim}{g}_\alpha^{i}) \cdot \underset{=}{G}_{\alpha\alpha'} \cdot (\underset{\sim}{g}_{\alpha'}^{i} - \underset{\sim}{g}_{\alpha'}^{f}) \quad , \tag{4.10}$$

characterizes corrections of the magnitude of the fluctuations due to correlations between the barrier height and the distortions around the initial and final site of the interstitial.

The transmission coefficient $|J_o(\varepsilon)|^2$ entering (4.6) can be expressed in closed forms for a one-dimensional model where the potential barrier around the saddle point is simulated by a parabola

$$U(x) = U_{sp} - \frac{1}{2} Kx^2 \quad . \tag{4.11}$$

Then[18]

$$|J_o(\varepsilon)|^2 = J_\infty^2 \cdot \left[\exp\ (\varepsilon/\varepsilon_1) + 1\right]^{-1} \tag{4.12}$$

$$\varepsilon_1 = \frac{\hbar}{2\pi} \sqrt{K/m_i} \quad , \tag{4.13}$$

(m_i mass of the diffusing particle). At high temperatures (4.12) may be simulated by a step function giving approximately[18]

$$\langle|J_{if}|^2\rangle = J_\infty^2 \frac{\sqrt{4kTE_b q/\pi}}{\Delta U_{eff}} \exp\left[- \frac{\Delta U_{eff}^2}{16E_b qkT}\right] \quad . \tag{4.13}$$

According to that $\langle|J_{if}|^2\rangle$ contributes appreciably to the activation energy of the diffusion process in the high temperature region. This result already has been deduced in the pioneering work of Flynn and Stoneham[1] for the "lattice-activated processes" in the particular case of vanishing correlation between the fluctuation of the barrier height and of the distortions around the initial and final interstitial states. Flynn and Stoneham[1] argued that the rather high activation energies found for hydrogen diffusion in fcc metals[30] may be due to lattice-activation of the transitions involved. This prediction has been confirmed by recent numerical calculations[8] of the diffusion constant for hydrogen in Cu which will be discussed in the next section.

In the intermediate temperature regime around and below Θ_D expressions (4.1) and (4.6) are modified since the host-lattice zero-point vibrations become of importance. In order to estimate the temperature dependence of $\langle|J_{if}|^2\rangle$ in this region the phonon spectrum has been modeled in[18] by an Einstein spectrum with frequency ω_E. Then $\langle|J_{if}|^2\rangle$ depends on $\Delta U_{eff}/\varepsilon_1$, on $\sqrt{\hbar\omega_E E_b}/\varepsilon_1$, and, separately, on the parameter q. Figure 4 shows a semi-logarithmic plot of $\langle|J_{if}|^2\rangle$ as function of $\hbar\omega_E/kT$ obtained[18] for a one-dimensional parabolic barrier with parameters $\Delta U_{eff}/\varepsilon_1 = 5$, $\sqrt{\hbar\omega_E E_b}/\varepsilon_1 = 1$ and q varying between 0.1 and 1. In the low-temperature regime tunneling transitions predominate where, however, the host-metal zero-point vibrations markedly increase the tunneling

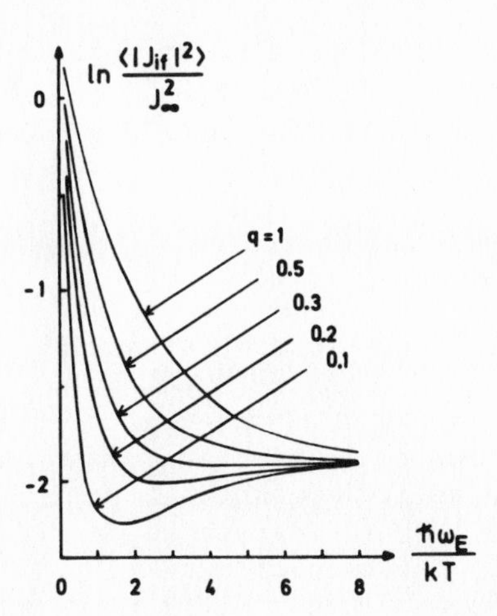

Fig. 4. Semi-logarithmic plot of $\langle|J_{if}|^2\rangle$ as function of $\hbar\omega_E/kT$.

rate above its value in the rigid lattice. For $kT >> \hbar\omega_E$ lattice-activated processes predominate. At intermediate temperatures $\langle|J_{if}|^2\rangle$ depends strongly on q showing a rather smooth transition from low-temperature to high-temperature behaviour for $q \gtrsim 0.3$ and a rather sharp transition for q = 0.1 and 0.2.

The correlation measured by q is determined by the geometry of the lattice and by the interaction potential between the hydrogen nucleus and the host-metal atoms. The quantum mechanical model for hydrogen in Cu[8,31] discussed below yields that for hydrogen at octahedral interstitial sites in fcc Cu the parameter q will be of the order of 0.5. For hydrgogen in bcc metals at tetrahedral interstitial sites there are indications that due to the geometry of the system q should be rather small. But since for bcc transition metals reliable hydrogen-host-metal interaction potentials are not available no quantitative estimates for q can be given for these cases at present. Nevertheless, as stated in[18], the finding that beyond the Condon approximation the effective transfer integral changes markedly with temperature and shows a rather sharp transition from weak to strong temperature dependence around Θ_D may explain the change in the activation energy for diffusion observed for H in Nb and for H and D in Ta at 250 K[30].

MICROSCOPIC CALCULATIONS FOR μ^+ AND HYDROGEN IN Cu

In this section we present the results of some numerical calculations[8,31] by which it has been studied, whether the QTD permits a quantitative description of the diffusion of positive muons and of the hydrogen nuclei in metals. The calculation has been carried out for interstitials in Cu, since for this material extensive experimental data exist about muon diffusion[20,21] and about hydrogen diffusion[32,33] and since light positively charged particles in metallic Cu are amenable to a rather simple quantum-mechanical treatment by means of a pseudo-potential approach which in this simplicity is not applicable to transition metals.

Calculation of the Lattice Deformation around μ^+ and Hydrogen

In order to test whether current theories are able to reproduce the vale $g = 2.65\ \mathrm{eV}/a_o$ [22] obtained for the radial force between μ^+ and neighbouring Cu-atoms from the μ^+ diffusion data the expectation value of g has been evaluated for muons and hydrogen nuclei in Cu within a quantum-mechanical model[31]. In this model the interstitials are described as positive point charges dressed by a screening electron cloud which was taken from Maysenhölders[34] self-consistent nonlinear-screening calculations of the electron density around point charges in jellium. The interaction potential $V(\underset{\sim}{R}-\underset{\sim}{R}')$ between the screened interstitial at position $\underset{\sim}{R}$ and a Cu ion at $\underset{\sim}{R}'$ then has been calculated by simulating the Cu ion by the local pseudo-potential of Nikulin[35]. From $V(\underset{\sim}{R}-\underset{\sim}{R}')$ the expectation value of the force $\underset{\sim}{g}$ exerted by the interstitials on the neighouring Cu atoms is determined according to

$$\underset{\sim}{g}(\underset{\sim}{R}-\underset{\sim}{R}') = - \nabla_{R'} \int d^3\underset{\sim}{r}\ |\ \psi\ (\underset{\sim}{r}-\underset{\sim}{R})\ |^2\ V(\underset{\sim}{r}-\underset{\sim}{R}') \quad , \tag{5.1}$$

where ψ means the muon (or proton) wave function.

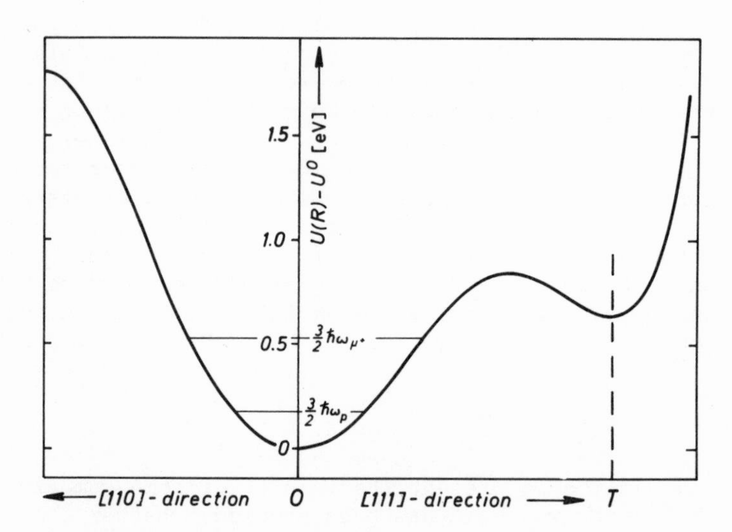

Fig. 5. Potential energy of positive point charges in Cu calculated in the rigid lattice approximation and ground-state levels for μ^+ and p at O sites.

ψ has been estimated for an unrelaxed Cu lattice by using the potential energy $U(\underset{\sim}{R})$ obtained from superposition of the interstitial — Cu-atom potential of 14 shells of atoms around an octahedral interstitial site. The potential $U(\underset{\sim}{R})$[31] is shown in figure 5. According to this calculation[31] the energy at octahedral interstitial sites (O) is lower than the energy at the tetrahedral sites (T) by about 0.635 eV. In agreement with the experimental observations of Camani et al.[36] the calculation thus yields that positive muons and hydrogen in Cu in their ground-states are localized at O sites. In addition to the ground state excited localized states may occur in the minima around O and T sites. The wave functions of these states have been estimated from $U(\underset{\sim}{R})$ by substituting the true potentials around O and T by their spherical averages. Then for the muon localized harmonic oscillator states are obtained at O sites, for the hydrogen isotopes at O and T sites[8]. The frequencies ω of these states and the resulting force constants g for the interaction between muon (hydrogen) and adjacent Cu atoms as given in[8] are shown in table II.

The force constants calculated for the different particles differ markedly. This reflects the mass-dependence of the extension of the wave functions and the variation of $\delta V(R)/\delta R$ with varying R. At O sites in particular a rather strong 'isotope' dependence is found between muon and proton whereas the effect is small within the series p, d, t. The theoretical muon-Cu-atom force constant at O sites is in fair agreement with the value $g = 2.65\ \mathrm{eV}/a_o$ estimated[8] from the experiments. From that is was concluded[8] that the pseudopotential calculations with nonlinear screening are able to describe the microscopic properties of muons (and probably of the hydrogen isotopes) in Cu and that muon diffusion proceeds in Cu via tunneling transitions of self-trapped

Table II. Harmonic-oscillator energies $\hbar\omega$, radial forces g, relative change $\delta u/(a_o/2)$ of the distance between the interstitials and nearest neighbour host-metal atoms, and lattice deformation energy E_o for μ^+, H, D, T in Cu

	$\hbar\omega$[meV]	g[eV/a_o]	$\delta u/(a_o/2)$	E_o[meV]
		O sites		
μ^+	356	2.0	0.03	90
H	120	1.34	0.02	40
D	84.5	1.27	0.02	36
T	69	1.23	0.02	34
		T sites		
H	165	4.97	0.075	371
D	117	4.78	0.072	345
T	95	4.70	0.071	334

muons between adjacent O sites.

In the harmonic approximation the static distortions around the interstitials may be estimated from the force constants g by use of the lattice response function[24,25] of Cu. Thus derived[8] radial displacements δu of the Cu atoms neighbouring the interstitials are displayed in table II as well as values of the deformation energy E_o stored in the lattice which in the linear coupling model also mean a measure for the lowering of the total energy of the system due to lattice deformations. For muons at O sites the value $\delta u/(a/2) \approx 3\%$ is in encouraging agreement with the estimate by Camani et al.[36]. Neglecting the finite extension of the muon wave function these authors derived from their measurements that the Cu atoms neighbouring the muon are displaced away by about 5%. (g = 2.65 eV/a corresponds to displacements of 4%).

Numerical Results for Hydrogen Diffusion in Cu

Under rather general conditions[2] the diffusion constant is related to the transition rates W_{if} by

$$D(T) = b^2 \sum_{i,f} w_{if}\, e^{-E_i/kT} \Big/ \sum_i e^{-E_i/kT} \quad , \tag{5.2}$$

(b: effective jump length). The evaluation of D(T) needs knowledge of the energy levels of the localized interstitial states, the forces exerted by the interstitial on the host-metal atoms, the potential barrier between adjacent potential wells, the dependence of the barrier height on the atomic displacements, and the transition matrix elements between the various interstitial states. For Cu the various parameters may be estimated from the microscopic model[8,31] described in the preceeding section.

For the hydrogen nuclei the most striking results following from this model are the large values of g and E_o at T sites. According to that the energy difference between the oscillator ground-states of hydrogen at T and O sites is reduced by lattice distortions by approximately 0.3 eV (compared to the values in an unrelaxed lattice). From this result it was concluded[8] that in Cu the diffusion of the hydrogen isotopes takes place by phonon-assisted transitions between localized states at O and T sites and vice versa where the effective activation energy for diffusion is strongly reduced by the self-trapping distortions. The O-T-O jumping corresponds to that of Sicking's model[37] for hydrogen diffusion in Pd which recently has been applied successfully to the isotope effect in the reorientation of diatomic hydrogen-impurity complexes in ferromagnetic metals by Kronmüller et al.[38]. (But note that Sicking[37] and Kronmüller et al.[38] completely neglect selftrapping effects and lattice distortions).

For the 0-T-0 path the temperature dependence of the hydrogen diffusion in Cu has been estimated[8,18] from (5.2) and (4.1) by using (4.3) and substituting $J_o(\varepsilon)$ by the transfer integral through a one-dimensional parabolic potential barrier with the curvature adapted to that of the numerically calculated potential curve along the 0-T path. In the numerical evaluation of the diffusion constant transitions of the hydrogen nuclei from the ground-state level and additional 5 excited levels at the 0 sites to the T-site levels and vice versa have been taken into account. The calculations yield that in Cu at temperatures above about 150 K for all these 0-T transitions the mechanism of phonon-activated over-barrier transitions (where phonon fluctuations reduce the effective barrier height) predominates over quantum-mechanical tunneling transitions[18]. Due to this the pre-exponential of D(T) should exhibit a much weaker dependence on the hydrogen isotope mass as the exponential isotope dependence predicted from tunneling transitions. (The experiments[32,33] roughly show a $m_i^{-1/2}$dependence). Within this model the temperature dependence of D(T) has been evaluated for H, D, T in Cu. D(T) shows slight deviations from the Arrhenius law. According to the numerical calculations[8,18] D(T) can be characterized in the temperature regime 700 K $<$ T $<$ 1200 K where the experiments[32,33] have been carried out by average activation energies E_m displayed in table III. Table III also shows the experimental activation enthalpies H_m for H, D, T in Cu of Katz et al.[33]. The numerical calculation yields the correct order of magnitude for the activation energies and is capable to describe the 'inverse' isotope dependence found in the series H, D, T in Cu and Ni[33]. The rather large E_m values reflect the fact that transitions from low-lying 0 states need strong phonon fluctuations and therefore have large activation energies whereas transitions from higher states (which need minor lattice activation) incorporate thermal excitation of the hydrogen nuclei into these levels.

CONCLUDING REMARKS

In the preceding lectures we have learned that the positive muon may be considered as an outstanding probe to test the QTD. We have in particular seen that for muons in Cu the theoretical predictions and the experimental results for the activation energy for diffusion seem to be in fair agreement, and so one may believe

Table III. Calculated activation energies E_m and measured activation enthalpies H_m[33] for H, D, T in Cu

	E_m [meV] T≈900 K	H_m [meV]
H	282	403
D	270	382
T	264	378

that the basic principles of the muon diffusion process now are well understood. Nevertheless it has to be stated that there exist some open questions, in particular concerning the diffusion of muons in the bcc transition metals Nb, V, Ta. At present it not is well established whether intrinsic muon diffusion ever has been seen in these metals. It may be that the diffusion observed so far in these materials basically is due to impurities and lattice defects and that the mobility of muons in pure and perfect crystals is of a comparable order of magnitude as in high-purity Al (for experimental details see, e.g., the contribution of E. Karlsson at this school).

With respect to hydrogen in metals we have learned that for H, D, T in Cu the theoretical and experimental values for the activation energy of diffusion also fit fairly well together. When comparing the results of the quantum mechanical model calculations for positive muons and for hydrogen in Cu we find two major differences in the corresponding diffusion mechanisms (besides the fact that the hydrogen nuclei may have additional localized excited states at the tetrahedral interstitial sites): Muon diffusion seems to take place via direct incoherent multi-phonon transitions between muon ground-state levels at adjacent interstitial sites whereas hydrogen diffusion seems to proceed via lattice-activated indirect incoherent transitions between ground-state and excited levels. This difference is a consequence of the rather big difference in the masses of these particles. Due to the larger masses the hydrogen isotopes have much smaller zero-point energies and the excitation energies for higher levels are much smaller than in the case of positive muons.

For other metals, in particular for the bcc transition metals, theoretical calculations of the hydrogen and muon diffusion constant have not yet been published. (As far as related calculations exist they concern the energy surfaces for positive point charges in the unrelaxed lattice or in the relaxed ground-state of the system but not the diffusion step itself.) The "Quantum Occurence Theory" has been shown[3,4] to be capable of modeling the temperature and isotope dependence of the diffusion constant observed for hydrogen in Nb (but this model underestimates the muon diffusion at lower temperatures). From our discussion of the QTD beyond the Condon approximation one may conclude that the change from small low-temperature to larger high-temperature activation energy found at about 250 K for H in Nb and for H and D in Ta indicates the onset of predominating lattice-activated processes at higher temperatures. (A slightly different interpretation has been given in the original analysis of the experimental data[39] where a change from tunneling to over-barrier transitions has been proposed.) According to that in Nb the low-temperature activation energy for hydrogen diffusion and the activation energy for muon diffusion should be related. However, as may be seen by constructing tentatively from the hydrogen zero-point frequencies equi-potential lines for hydrogen and positive muons in Nb, the muon diffusion in Nb is (due to the large zero-point energy of the muon) a subtile problem of self-trapping rather than a problem of static potential barriers.

REFERENCES

1. C. P. Flynn and A. M. Stoneham, Phys. Rev. B1, 3966 (1970)
2. Yu. Kagan and M. I. Klinger, J. Phys. C: Solid State Phys. 7, 2791 (1974)
3. D. Emin, M. J. Baskes and W. D. Wilson in: "Proc. First International Topical Meeting on μSR", Rorschach, Switzerland, 1978, J. Hyperfine Interactions 6, 255 (1979)
4. D. Emin, M. I. Baskes and W. D. Wilson, Phys. Rev. Lett. 42, 791 (1979)
5. T. Holstein, Ann. Phys. (N.Y.) 8, 343 (1959)
6. A. Seeger and H. Teichler, in "La diffusion dans les milieux condensés - theories et applications", Vol. I (I.N.S.T.N., Saclay, France 1976) p. 217
7. A. Seeger, in "Hydrogen in Metals I", ed. by G. Alefeld and J. Völkl (Topics in Applied Physics, Vol. 28), Springer-Verlag, Berlin-Heidelberg-New York (1978)
8. H. Teichler, in "Proc. First International Topical Meeting on μSR", Rorschach, Schwitzerland, 1978, J. Hyperfine Interactions 6, 251 (1979)
9. J. Appel, in "Solid State Physics, Vol. 21" ed. by F. Seitz, D. Turnbull, H. Ehrenreich, Academic Press, New York and London (1968)
10. M. I. Klinger, Rep. Progr. Phys. 31, 225 (1968)
11. H. B. Shore and L. M. Sander, Phys. Rev. B12, 1546 (1975)
12. A. F. Andreev and I. M. Lifshitz, Sov. Phys. - JETP 29, 1107 (1969)
13. A. Landesman, J. Low Temp. Phys. 30, 117 (1978)
14. Yu. Kagan and M. I. Klinger, Sov. Phys. - JETP 43, 132 (1976)
15. V. N. Pavlovich and V. N. Rudko, Phys. Stat. Sol. (b) 88, 407 (1978)
16. M. Wagner, Phys. Stat. Sol. (b) 88, 517 (1978)
17. D. L. Tonks and B. G. Dick, Phys. Rev. B19, 1136 (1979)
18. H. Teichler, in: "Proc. International Meeting on Hydrogen in Metals", Münster, Germany, 1979 (in press)
19. P. W. Anderson, Phys. Rev. 109, 1492 (1958)
20. I. I. Gurevich, E. A. Meléshko, I. A. Muratova, B. A. Nikolškii, V. S. Roganov, V. I. Selivanov, and B. V. Sokolov, Phys. Letters 40A, 143 (1972)
21. V. Grebinnik, I. I. Gurevich, V. A. Zhukov, A. P. Manych, E. A. Meléshko, I. A. Muratova, B. A. Nikolškii, V. I. Selivanov and V. A. Suetin, Sov. Phys. - JETP 41, 777 (1976)

22. H. Teichler, Phys. Letters 64A, 78 (1977)
23. T. McMullen and B. Bergersen, Solid State Commun. 28, 31 (1978)
24. R. Bullogh and V. K. Tewary, in "Interatomic potentials and simulation of lattice defects" ed. by P. C. Gehlen, J. R. Beeler and R. I. Jaffee, Plenum, New York (1972)
25. H. R. Schober, M. Mostoller, and P. H. Dederichs, Phys. Stat. Sol. (b) 64, 173 (1974)
26. J. De Launay, in: "Solid State Physics, Vol. 2" ed. F. Seitz and D. Turnbull, Academic Press, New York (1956)
27. S. Fujii and Y. Uemura, Solid State Commun. 26, 761 (1978)
28. S. Fujii, J. Phys. Soc. Jap. (in press)
29. R. I. Grynszpan, N. Nishida, K. Nagamine, R. S. Hayano, T. Yamazaki, J. H. Brewer, and D. G. Fleming, Solid State Commun. 29, 143 (1979)
30. G. Alefeld and J. Völkl, in: "Hydrogen in Metals I" ed. by G. Alefeld and J. Völkl (Topics in Applied Physics, Vol. 28), Springer-Verlag, Berlin-Heidelberg-New York (1978)
31. H. Teichler, Phys. Letters, 67A, 313 (1978)
32. W. Eichenauer, W. Löser, H. Witte, Z. Metallkunde 56, 287 (1965)
33. L. Katz, M. Guinan and R. J. Borg, Phys. Rev. B4, 330 (1971)
34. M. Maysenhölder, Diploma-work, University Stuttgart, 1978
35. V. K. Nikulin, Sov. Phys.-Solid State 7, 2189 (1966)
36. M. Camani, F. N. Gygax, W. Rüegg, A. Schenk, and H. Schilling, Phys. Rev. Lett. 39, 836 (1977)
37. G. Sicking, Ber. Bunsenges. 76, 790 (1972)
38. H. Kronmüller, B. Hohler, H. Schreyer, and K. Vetter, Phil. Mag. B37, 568 (1978)
39. D. Richter, B. Alefeld, A. Heidemann, and N. Wakabayashi, J. Phys. F: Metal Phys. 7, 569 (1977)

SOME ASPECTS ON POSITIVE MUONS AS IMPURITIES IN METALS

Erik Karlsson

Institute of Physics
Uppsala University
751 21 Uppsala, Sweden

INTRODUCTION

The muon spin rotation has been used to study positive muon interactions in several metals since the beginning of the 1970's. Many of the early results were in conflict with each other and it is only the last two years that one has realized the extreme importance of having high purity metallic targets, if the true muon-metal interactions are to be studied.

The interaction of the muons with impurities, vacancies or other imperfections, which can act as traps in the metals, shows up very strongly in the damping patterns of the μSR signals. These impurity-impurity interactions form interesting problems worthy of study in their own right (although as a metallurgical tool the μSR method seems a bit expensive) but the present notes will deal mainly with the efforts to understand positive muon behavior in otherwise perfect lattices.

At present, there are several unsolved problems in this area. They are of fundamental importance since they concern a particle of intermediate mass, which may be expected to behave quantum-mechanically in a way different from both its light counterpart (the positron) and its heavier analogy (the proton).

Its mass is however closer to the proton mass, $m_\mu \approx (1/9) m_p$, and it is expected that many of the results obtained with muons should be comparable to the case of protons in metals. Specifically, it is of interest to study positive muons in connection with the problems of site determinations and electronic screening in the hydrogen-in-metal case.

In the following, three basic problems concerning the interaction of positive muons in metals will be taken up for discussion:

(a) the stopping sites for positive muons in metals and the electric field gradients (EFG:s) set up by the muons at surrounding lattice points;
(b) the dependence of μ^+ diffusion on the presence of other impurities in the metals;
(c) the specific effects found in the case of muons in a semimetal (antimony), where the density of conduction electrons is low.

In the interpretation of the experimental results on positive muons in metals and semimetals it is assumed that the incoming muons slow down and get thermalized in a very short time (<1 ns) without any loss of polarization in these first processes. Furthermore, it is assumed that, in the last stage of slowing down, the muons leave the "spur" of the radiation damage behind them and come to rest at sites which are typical of the undamaged material to be studied. The validity of these assumptions is supported by comparisons with theoretically expected and experimentally observed polarizations after stopping and by estimates of the diffusion length of muons after the last knock-out of a metal atom in the stopping process. The latter has been estimated to be of the order of 10^4Å. (1)

After slowing down, the classical picture of the muon at low temperatures consists of a point charge confined to a certain, small volume of space between the metal atoms (interstitial site) where there is a minimum in the electrostatic potential. The experimental indication of stopping is that the muon polarization, detected in a plane perpendicular to the applied magnetic field, is decreasing according to the formula $P(t) = P_o\exp(-\sigma^2t^2)$ and that σ becomes independent of temperature below a certain "freezing" temperature T_f (see Fig. 1). This working definition of "stopped muons" is of course related to our way of observing the muons (we consider them immobile if the mean time of stay τ is the potential minimum is longer than $\approx 10\tau_\mu$, (which is about 20 μs).

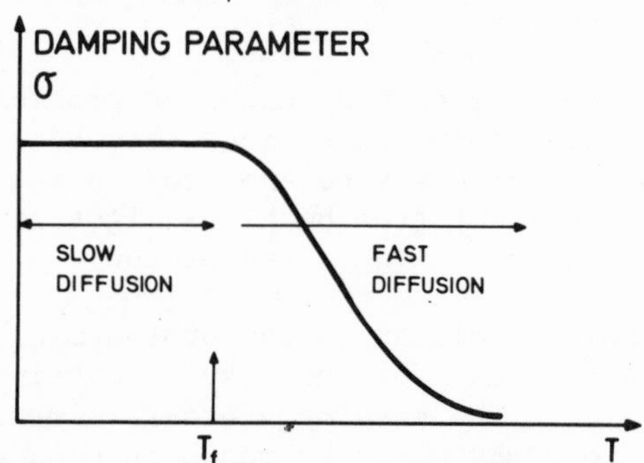

Figure 1. Temperature dependence of the μSR damping parameter as expected in a classical model.

At higher temperatures - still according to the classical picture - the muon starts to jump between interstitial sites more frequently. The jump rate, which is the inverse of the time of stay and is expected to obey a so-called Arrhenius relation

$$1/\tau = \Gamma \exp\ (-E_a/kT) \tag{1}$$

can now be obtained from the measured depolarization rate which for the case of $\sigma^2\tau^2 << 1$ has the form $P(t) = \exp\ (-2\sigma^2\tau_c t)$. The quantity τ_c is proportional to τ as will be discussed later. In this limit one uses the parameter

$$\lambda = 2\sigma^2\tau_c \tag{2}$$

to describe the damping.

In the intermediate region (around T_f) a formula of the type

$$P(t) = \exp\ \{-2\sigma^2{\tau_c}^2[\exp(-t/\tau_c)-1 + t/\tau_c]\} \tag{3}$$

is used to fit the data. It is useful to describe the damping in this case by specifying the time for decay of the polarization to 1/e of its initial value. The inverse of this time is called Λ, a parameter to be used later.

The determination of the stopping sites are performed as described in section II. The diffusion parameters, which turn out to be extremely sensitive to the purity of the metallic samples used as targets, are discussed in section IV. These phenomena can be compared to the corresponding ones for hydrogen in metals as discussed by Richter (2). The theories of diffusion are treated elsewhere in this course (see Stoneham) where quantum aspects on the diffusion mechanisms are also introduced. It is of specific interest to consider what happens to the experimental observables when the classical, point-line muon is replaced by a delocalized muon with a wave-function extending over a certain region is space. Such aspects are discussed in Section III. Finally, Section V is devoted to positive muon behavior in semi-metals.

II. STOPPING SITES IN PERFECT CRYSTALS WITH SIMPLE STRUCTURES

Most metals crystallize in one of the following simple structures: the b.c.c (body centered cubic), f.c.c. (face centered cubic, or h.c.p. (hexagonal close packed). The two cubic structures are shown in Figure 2. In both types there exist interstitial sites around which the nearest neighbor metal atoms have octahedral (O) or tetrahedral (T) symmetry. These sites are also indicated in the figures. One of the principal problems is to decide which one of the two types is the most stable site for the positive muon.

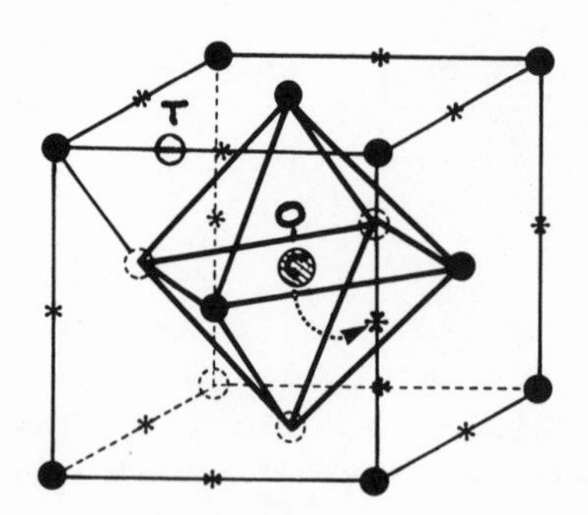

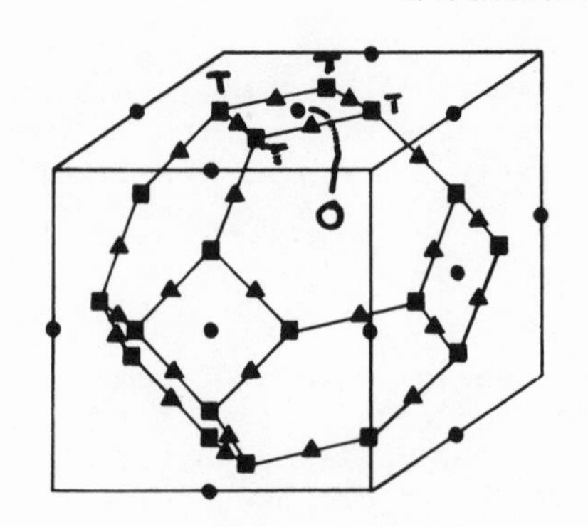

Figure 2. Interstitial sites in (a) f.c.c. crystals and (b) b.c.c. crystals.

It is expected that a muon and a hydrogen (or deuterium) atom shoul take up the same type of site, but the information on hydrogen (from channelling, neutron diffraction, etc.) is still fairly incomplete and muon data would give a valuable confirmation of the pictures of light interstitials in metals.

The magnetic interaction of the muon moment with nearby dipole moments provides a way to determine the muonic sites. Several metals are composed of isotopes with large nuclear dipole moments (Cu and Al among the f.c.c.; V, Nb, Ta among the b.c.c. metals). These moments produce fields of the order of a few Gauss (10^{-4}T) at the muonic sites. At all temperatures above a few mK these nuclear dipoles can be considered as randomly oriented and the local fields at the muons are distributed around the applied field value (B_o) wi a spread characterized by the parameter $\sigma^2 = \gamma^{\Delta B^2}{}_{dip}$ (which appeared in equation (2)). The quantity γ is the gyromagnetic factor of the muons. Calculating the dipole sums with the applied field B_o (the z-axis) in the principal directions [100], [110] and [111] one finds distinctly different values for σ for the two types of interstitial positions as illustrated by a few examples in Table 1. These are the so-called van Vleck values of σ. For the b.c.c.

Table 1. Theoretical damping parameters σ (μs^{-1}) for point-like muons in some cubic metals

Host metal	Site	Field direction [100]	[111]	[110]
Cu (f.c.c.)	octahedral	0.307	0.067	0.164
	tetrahedral	0.077	0.319	0.279
Nb (b.c.c.)	octahedral	0.740 0.380	0.197	0.205 0.459
	tetrahedral	0.188 0.449	0.308	0.124 0.393

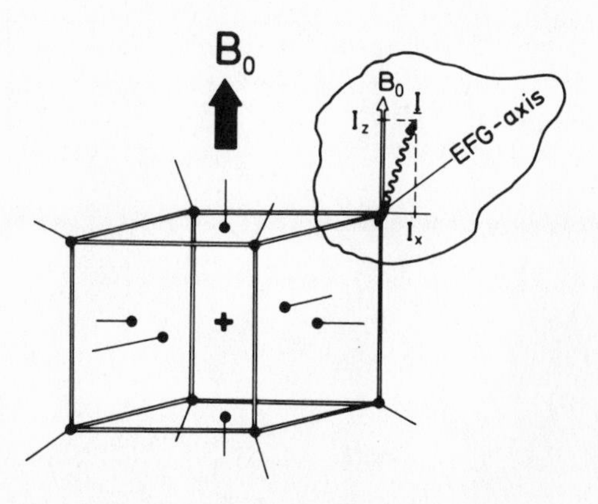

Figure 3. For calculation of the effect of the EFG (produced by a muon) on the dipolar field at the muon site.

lattice it turns out that there are also sites which are magnetically inequivalent although they are electrostatically identical. This fact shows up for certain directions of the applied field.

In principle, therefore, a measurement of σ for a single crystal sample should be enough for a site determination. By similar arguments it can be predicted that in cases where oriented electronic dipoles are present at each lattice point, as in the case of ferromagnets, the muon sites can be determined from the dipolar field $B_{dip,z}$ which is added to B_o and specific for each magnetically inequivalent site (see lecture on ferro-magnets).

The first experiments of this type (on muons in Cu single crystals) gave very small differences in σ for the three directions [100], [110] and [111]. It was later pointed out by Hartmann (3) that such experiments must be performed in sufficiently strong applied fields to give the answer expected from the simple theory of dipolar sums. The reason is that the expectation value of the dipolar field is changed by the electric quadrupole interaction experienced by the neighbor nuclei when the muonic charge is introduced in an interstice. If the muon can be considered as a point charge it produces a radially directed field gradient EFG (Fig. 3) and the nuclei are subject to a combined electric and magnetic interaction. The symmetry axis of the EFG forms, in general, a certain angle β with the direction of the applied field (z-axis) for each neighboring nucleus.

The Hamiltonian for such a combined interaction can be set up by transforming the components of the EFG-tensor to the z-system using spherical tensor algebra. The matrix elements of the Hamiltonian H in an m-representation referring to the z-axis along B_o can be written (see Matthias, Schneider and Steffen (4))

$$K_{m,m} = -y\ m + 1/2(3\ \cos^2\ \beta-1)\ [3m^2 - I(I+1)] \qquad (4a)$$

$$K_{m,m-1} = -3/2\cos\ \beta\ \sin\ \beta(1-2m)[(I-m+1)(I+m)]^{1/2} \qquad (4b)$$

$$K_{m,m-2} = 3/4\sin^2\ \beta\ [(I+m-1)(I+m)(I-m+1)(I-m+2)]^{1/2} \qquad (4c)$$

where

$$K_{mm'} = \frac{1}{\hbar\omega_E} H_{mm'}\ (\beta) \qquad (5)$$

and $y = \omega_B/\omega_E$, where ω_B is the Larmor angular frequency of the nuclei, ω_E the quadrupole interaction frequency

$$\omega_E = \frac{1}{2}\frac{eQq}{\hbar} \cdot \frac{1}{2I(2I-1)} \qquad (6)$$

and q is the electric field gradient (assumed axially symmetric).

It remains now to find the eigenstates of this Hamiltonian and evaluate the expectation values $\langle I_z \rangle_i$ and $\langle I_x \rangle_i$ (which is non-zero since the eigenstates are mixtures of m-states). Each nucleus will contribute a dipolar field component

$$\Delta B_z^{(i)} = \gamma\hbar \langle I_z \rangle_i^{stat} \frac{3\cos^2\theta-1}{r^3} + \gamma\hbar \langle I_x \rangle_i^{stat} \frac{3\sin\theta\cos\theta}{r^3} \qquad (7)$$

to the local field at the muon site. There are also oscillating components of $\langle I_z \rangle$ and $\langle I_x \rangle$ giving fluctuating fields, but these can be neglected as long as their periods do not coincide with the Larmor frequency of the muon. The lattice sums were carried out by Hartmann over about 70 nuclei around the interstitial, but the major contribution to the sums comes from the nearest neighbors. The EFG was assumed to decrease with the distance from the muon as $1/r^3$.

The resultant values of $\langle \Delta B_z^2 \rangle$ depend strongly on the parameter $y = \omega_B/\omega_E$. At large y values, where the magnetic interaction dominates, the $\langle \Delta B_z^2 \rangle$ curves approach the van Vleck values. For most nuclei the ratio of magnetic moment has such a value that this "strong field limit" is approached in a magnetic field of a few thousand gauss (0.1-0.5 T), but for strongly deformed nuclei, like Ta, it would be difficult to reach this limit.

Experiments giving the damping constant σ as function of applied magnetic fields, with the field along the main symmetry axes are shown in Figures 4a (5) and 4b (6) and compared to the theoretical

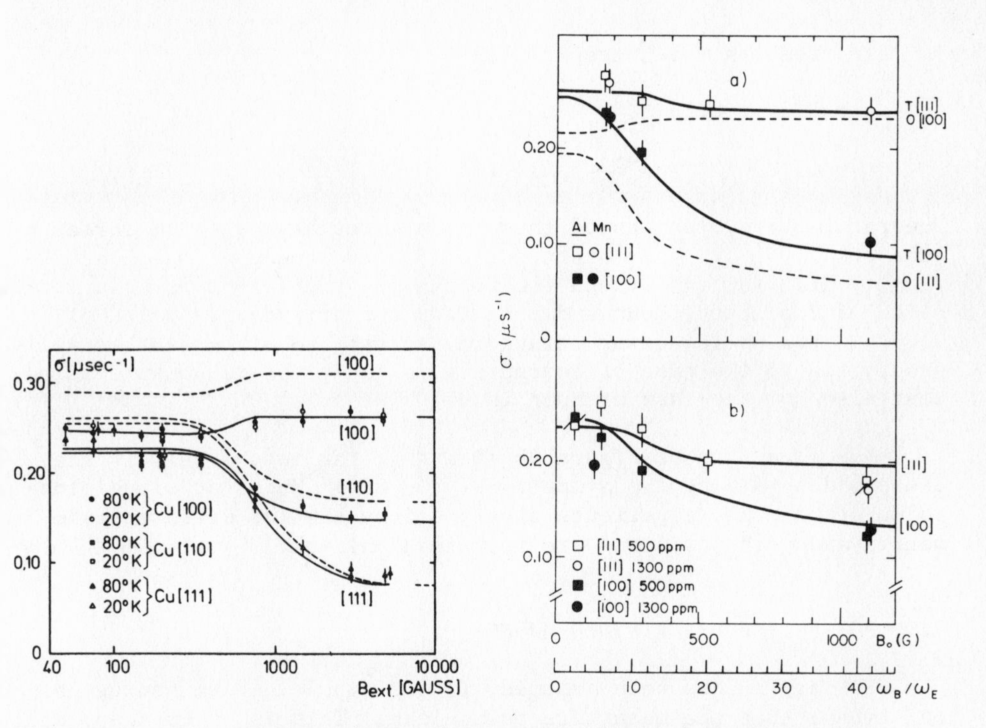

Figure 4. Fits to experimental magnetic field dependence for (a) Cu-metal (octahedral sites); from ref. (5); (b) doped Al-metal (tetrahedral sites), from ref. (6).

predictions. It is quite evident that the data can only be fitted with the assumption of octahedral sites for muons in Cu and tetrahedral sites for muons in Mn-doped Al at 15 K. In both cases it has been necessary to scale down the theoretical curves by 15-20% in order to obtain a fit. This effect is interpreted in terms of a dilation of the lattice of about 5% around the muon.

The parameter $y = \omega_B/\omega_E$ which can be determined from the fits also gives values of the quadrupole interactions of the nearest neighbor nuclei, (assuming that these are responsible for the major part of the change in σ). These values are limited in accuracy since several approximations are involved. The further deviation of q (the EFG-parameter) from equation (6) have large uncertainties because the Q's themselves are not very well known. Still, they are of great interest theoretically because they represent values of the EFG at very close distances from the impurity center and therefore provide critical tests on screening calculations.

The results were, respectively

$$q = 0.27 \pm 0.15 \ \text{Å}^{-3} \quad \text{Cu, octahedral}$$
$$q = 0.18 \ \text{Å}^{-3} \quad \text{Al, tetrahedral} \qquad (8)$$

Theoretical calculations of these quantities have been performed by Jena et al. (27). They start out from a treatment of the electron density distribution around the muon using the self-consistent formalism of Hohenberg, Kohn and Sham (28) and arrive at a value of 0.26 Å^{-3} for Al (assuming octahedral sites). Scaling Jena et al.'s prediction to the case of tetrahedral sites gives reasonable agreement also for the case of muon in Al.

The good fits of Figure 4b show that the muon itself is the center, determining the symmetry of the EFG. Muons at sites close to Mn-atoms would experience a very strong EFG (of different symmetry) and such positions are therefore unlikely in the AlMn-alloys.

III. POINT-LIKE OR EXTENDED MUONS?

So far, it has been assumed, in the dipole and screening calculations, that the muons are point-like. This is by no means self-evident since they are particles of very low mass and when placed in the potential wells of the interstitial sites, they are expected to have a relatively strong zero-point motion. Also, such a light particle should have a certain probability of "leaking through" the potential barriers between two interstitial sites i and k. Although such probabilities are quite difficult to calculate - since they are largely determined by the shifting of the center of lattice deformation from i to k - as explained by Stoneham (7) - we can introduce a phenomenological transfer or "tunneling" matrix element $J' = V_{ik}$ describing this process. From diffusion experiments and comparison with the so-called "tunneling-hopping" theory they can be estimated to be of the order of some μeV.

If no other perturbations are present the muon wave-function could in principle be distributed over several interstitial sites. In such a situation a "stopped muon" would mean a muon that is confined to a limited region of space during its lifetime and has no "drift" motion in any specific direction. Several measured properties are then expected to change, for instance:

a) the effective dipolar fields would be weighted sums over contributions to a set of nearby interstitial points. Although there are correlations between two nearby sites, since they are partially determined by the same nuclear dipoles, the observed σ-values will decrease quite rapidly when the muon wave-function is extended over a few lattice distances.

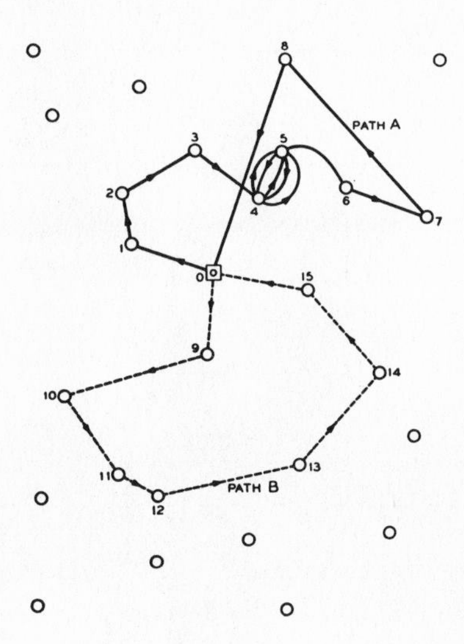

Figure 5. The Anderson localization mechanism (8) (illustrating perturbation expansion around initial site).

b) the electronic screening is expected to change character if the positive charge of the muons is smeared out, providing a more shallow attractive potential for conduction electrons. No calculations of such effects exist so far.

c) the EFG:s produced at nearby nuclei should change and their directions be less well defined.

The conditions for localization and delocalization of particles in a lattice structure has been considered in a "classical" paper by P. W. Anderson (8). He finds that if the structure is disordered such that there is an energy spread ΔE among the different possible sites for the particle, then there is a critical degree of disorder represented by $(\Delta E)_c$ for which a delocalized (bandlike) state cannot develop. The condition for localization is that $(\Delta E)_c$ is larger than the tunneling matrix element J'

$$(\Delta E)_c \gtrsim J' \quad \text{localized state} \tag{9a}$$

$$(\Delta E)_c \lesssim J' \quad \text{delocalized state} \tag{9b}$$

Numerical calculations on model systems indicate that the transition between the two regimes is not very sharp and that some delocalization can still remain at $\Delta E \approx 10\,J'$ (9). Figure 5, which is taken from the Anderson paper, illustrates the perturbation theory used to find out whether the particle is likely to remain in its original region of space (closed path).

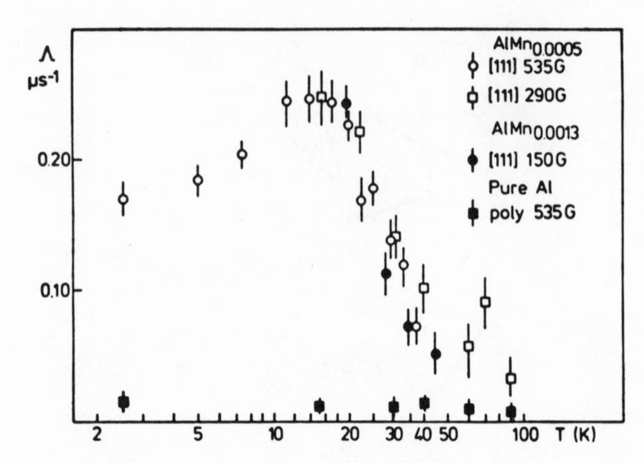

Figure 6. Temperature dependence of the damping parameter Λ for pure and Mn-doped Al samples.

Since the matrix elements J' are small, already quite low degrees of disorder will favor a localized state. An example is Al-metal where the muons are supposed to be delocalized if the metal is sufficiently pure, but localized when small amounts of certain impurities are introduced. Figure 6 illustrates the difference between the temperature dependence of the μSR damping parameter Λ for the purest obtainable Al-metal and Al doped with 500 and 1300 ppm Mn. For the latter type of target the muons are evidently stopped and fairly well localized at 15 K. The site determination illustrated in Figure 4b was made under these conditions.

The disorder introduced by the Mn-impurities (which are known not to form clusters in the sample investigated) is an elastic lattice distortion around each Mn-atom. The elastic energy decreases approximately as r^{-3} when going outward, according to a formula given by Leibfried (10)

$$E_{int} = -\frac{1}{r^3}\frac{15}{18\pi}\, d\left(\frac{\overline{C}_{11}+2\overline{C}_{12}}{3\overline{C}_{11}}\right)^2 \Delta V^{\mu}\ \Delta V^{Mn}\left|\frac{3}{5} - \Sigma_j \rho_j{}^4\right| \qquad (10)$$

where $d = C_{11} - C_{12} - C_{44}$ is the elastic anisotropy of the lattice, $\overline{C}_{12} = C_{12} + 1/5\ d$ and $\overline{C}_{11} = C_{12} + 2\ C_{44} + 2/5\ d$ are averaged elastic constants, ΔV^{μ} and ΔV^{Mn} are volume expansions induced by the defects and ρ_j are the direction cosines with respect to the cubic axes.

Introducing empirical values for the constants it is found that the energy change between adjacent sites at a distance Δr can be written

$$\Delta E_{int} = \frac{dE_{int}}{dr} \Delta r = 3 \times 0.39 \; \Delta r/r^4 \text{ eV} \tag{11}$$

For the actual impurity concentration of the $\underline{Al}$ Mn sample referred to in Figure 4a, one can estimate a value of $\overline{\Delta E} \approx 70$ μeV. The corresponding quantity for 6N-purity Al should be of the order of 0.01 μeV.

The value of J' obtained from a fit of the diffusion data above 20 K from Figure 6 is about J' = 5 μeV. It seems therefore reasonable that the Anderson localization condition is fulfilled for $\underline{Al}Mn_{0.001}$ but not for 6N Al, since

$$\Delta E(Al) < J' < \Delta E(\underline{Al}\ Mn) \tag{12}$$

Experimental studies have later been performed for $\underline{Al}$ Mn alloys with concentrations of 500 and 35 ppm Mn down to a temperature of 100 mK and for 6N Al down to 50 mK. These results are presently being analyzed (11).

Complex structures of muon depolarization curves have been reported for Al-based alloys with about 1000 ppm concentrations of other impurities (Cu, Mg, Sn, Si) as well as with cold-worked Al (12) and with vacancies induced by neutron irradiation (13). In some cases the peaks in the damping curves for certain higher temperature ranges are probably due to the fact that the muons can reach clusters of precipitated impurities or impurity-vacancy complexes. These latter experiments were performed on polycrystalline samples, from which it is difficult to obtain information on the muon sites. However, they do not contradict the picture that at low temperatures the strain-induced localization is the main mechanism in "stopping" the muons in the Al lattice. Explanations based on a very fast diffusion of Flynn-Stoneham type towards the impurities lead to unrealisitically high values of J in the low temperature range.

The widely different diffusion properties in the two f.c.c. metals Cu and Ll may possible be connected with the experimentally found (5) fact that muons localize at octahedral (O) sites in Cu but at tetrahedral (T) sites in Al. The matrix element J' for jumps between O sites is much smaller than between T sites because it decreases exponentially with distance (14, 15, 16).

Among the b.c.c. metals it was for a long time believed that positive muons did stop in the pure lattices at sufficiently low temperatures. As will be discussed in the next section this information was based on experiments with material of insufficient purity and it is now realized that the impurity content must be brought down to a level of a few ppm to observe the intrinsic muon-metal interaction. Such an experiment was performed recently and is shown in Figure 7 together with data on other Nb crystals with a

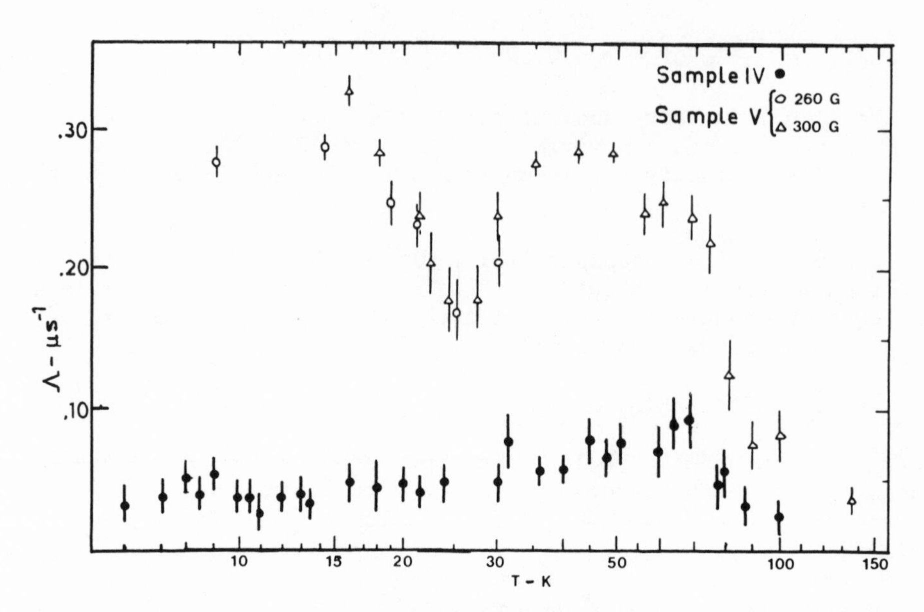

Figure 7. Comparison of damping parameters for high-purity Nb (~3 ppm) and less pure Nb (~100 ppm) samples (iv and v respectively).

content of approximately 100 ppm substitutional impurities (mainly Fe, Cr and Ta) and varying amounts of interstitial impurities, mainly N. It is presently believed that the substitutional impurities produce the same kind of strain effects as in Al below 15 K, whereas the interstitial impurities act as traps which are reached after a certain length of incoherent diffusion (see next section).

Recent preliminary experiments (17) on diffusion in the purest obtainable iron (b.c.c. Fe) seem to confirm the picture that the muons diffuse very fast, even at temperatures as low as about 10 K. As described in the lectures by Maier (18) the damping parameter is much more sensitive to changes in the jump rate in the case of magnetic materials since the dipolar field from the electronic spins enter in $\lambda = \gamma^2 B_{dip}{}^2 \tau_c$ rather than the fields from the nuclei. Furthermore the dipolar fields are ordered within each domain and the muons are supposed to jump between sites with positive and negative fields, whose effects cancel if τ_c is low enough. The role of strain fields on the jump rate at temperatures above 70 K has been investigated by introducing small amounts of carbon (50-100 ppm), since C-atoms are known to produce unusually large ΔV^{imp} in equation (10), see Table II. Fitting with Flynn-Stoneham theory produces tunneling matrix elements J of about 6 meV for the undoped Fe samples and 1-2 for the C-doped samples with activation energies for diffusion of about 50 meV in both cases.

Table II. Diffusion and tunneling parameters for muons in Fe

	E_s* (meV)	J (meV)	J'(20K) (μeV)	ΔE** (μeV)
Samples I and II	51	6	140	$\lesssim 0.1$
Sample II (Carbon-doped)	40	1.6	80	1.5

*Activation (polaron shifting) energy above 70 K
**Strain energy difference at largest distance from Carbon impurities

Because of the strong dipolar fields in the case of iron, it is also a material suitable for tests of the extension of the muonic wave-function into band-like states. As has been pointed out by Fujii and Uemura (14), the product $\gamma^2 \overline{B_{dip}^2} \tau_c$ of the hopping model should be replaced by $\gamma^2 \overline{B_{dip}^2} \tau_B$, where $\overline{B_{dip}^2}$ is an average over the volume occupied by the muon in the "band"-state and τ_B is the lifetime of the same state. The mean free path (the extension) depends on the degree of disorder but can be several times a lattice distance at low temperatures and $\overline{B_{dip}^2}$ decreases accordingly. The lifetime τ_B decreases with increasing temperature due to phonon interactions. The product is expected also to decrease when going up from T = 0 until, at a certain temperature, the band-like state collapses and the muon instead starts to diffuse incoherently. Going upwards in temperature the damping parameter is, in this model, expected first to decrease, then possibly go through a maximum and then decrease as usual in the high temperature range according to an Arrhenius law.

Estimating tunneling matrix elements at 20 K: J'(20K) = 140 μeV (undoped), 80 μeV (C-doped) from the value of J obtained above 70 K using a Debye model, we find from the inequalities (9) that a delocalization into a band-like state can in principle develop at least in parts of the sample, since ΔE is of the order of a few μeV at long distances from the impurities (Table II). However, in regions within about 15 Å from the C-impurities, localization can be expected to occur.

Our present state of understanding is therefore that in sufficiently pure b.c.c. metals, the muons should be delocalized, whereas in pure f.c.c. metals the localization depends on whether the local electronic structure favors an octahedral or tetrahedral site

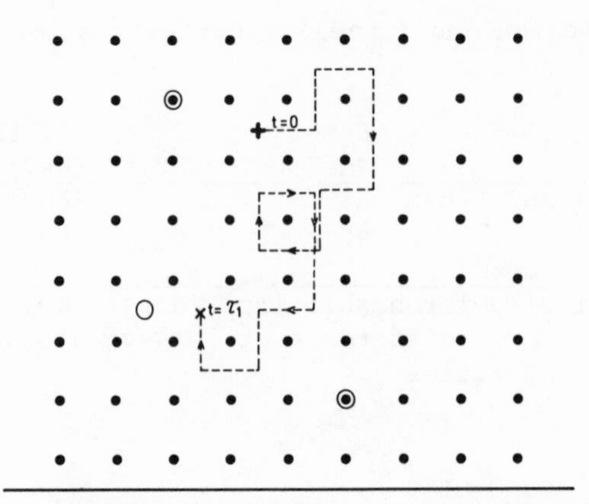

Figure 8. Schematic illustration of trap-limited diffusion. Parameters as described in text.

(and in general, of course, on the depth of the potential wells.)

The temperature conditions for localization are complicated and not well understood at present.

IV. MUON DIFFUSION IN THE PRESENCE OF TRAPS

For metals with impurities or other defects we expect, according to our previous discussion, that the diffusion processes can be described by jumps between well-defined sites in which the muons stay for average times τ_i. It there are more than one type of site we have to introduce activation energies E_i describing the depths of the potentials seen by the muon at each of these sites. Furthermore, the parameters σ are expected to be different for each site.

Intuitively, such processes can be described by a picture of the type shown in Figure 8. The muon can jump between equivalent sites in the "pure" metal (parameters E_c, τ_c). The correlation time τ_c is a measure of the time it takes to get into a region with a different dipolar field. It is proportional to the jump time τ. After an average time τ_1 it will reach one of the sites with a more attractive potential (parameters E_o, τ_o), the so-called <u>trap</u>. After leaving the first trap it can proceed to another one of the same kind or to a still deeper one, if it exists.

The muon polarization is lost at a rate

$$dP(t)/dt = dP_1(t)/dt + dP_o(t)/dt \tag{13}$$

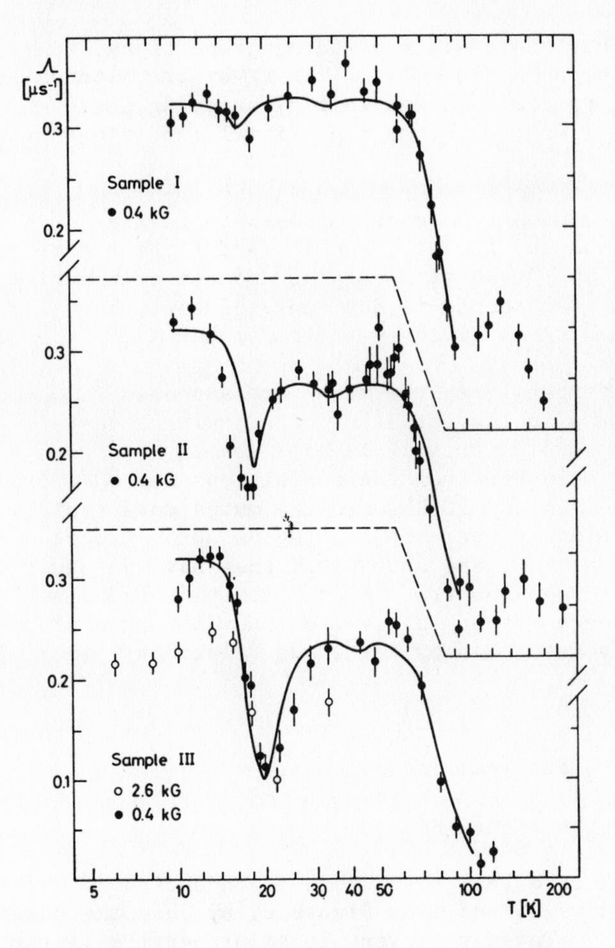

Figure 9. The dependence of the damping parameter in Nb-metal as a function of nitrogen content (18). Sample I: 3700 ppm, II: ≤ 60 ppm, III: 10-20 ppm.

where (compare equation (3))

$$dP_1(t)/dt = 2\sigma_1^2\tau_c\ [\exp\ (-t/\tau_c) - 1]$$

$$-\frac{1}{\tau_1} P_1(t) + \frac{1}{\tau_o} P_o(t) \tag{14a}$$

$$dP_o(t)/dt = 2\sigma_o^2\tau_o\ [\exp\ (-t/\tau_o)-1\ ]$$

$$\frac{1}{\tau_o} P_o(t) + \frac{1}{\tau_1} P_1(t) \tag{14b}$$

These are equations coupled by terms proportional to τ_1^{-1} and τ_o^{-1} which describe the changes of P_1 or P_o by transitions between the tw states. For randomly chosen initial stopping sites and small defect concentrations we can choose the initial conditions $P_1(0) = 1$ and $P_o(0) = 0$.

We will discuss these processes with reference to Figure 9, which gives the experimental results and fitted theoretical damping curves for muons in Nb with varying amounts of impurities. The detailed time dependence is not analyzed here, but only the gross structure of P(t) as given by the parameter Λ defined earlier (29).

At low temperature, the muons are supposed to occupy certain fixed positions in the lattice ($\tau_c >> \tau_\mu$) with a characteristic value of $\sigma_1 = 0.33\ \mu s^{-1}$ (a polycrystalline sample was used). Above 14 K they can be activated from these positions and a "motional narrowing effect is seen around 20 K where the muons move rapidly before they are caught in the traps. In the region 30-50 K most of the muons have reached a trap, and above 60 K they have enough thermal energy to diffuse between the traps ("trap limited diffusion"). The temperature dependence of each activation and diffusion process can be approximately described by Arrhenius relations

$$1/\tau_c = \Gamma_c \exp(-E_c/kT) \tag{15a}$$

$$1/\tau_1 = \Gamma_1 \exp(-E_1/kT) \tag{15b}$$

$$1/\tau_o = \Gamma_o \exp(-E_o/kT) \tag{15c}$$

In the actual case the temperature dependencies $\Lambda(T)$ for all three samples could be fitted (see Figure 9) by the same parameters E_c, Γ_c, E_o and Γ_o. These represent local properties of the jump processes and should not depend on the impurity concentration. Likewise, E_1 was taken equal to E_c since the trapping process in stage 1 should be diffusion controlled.

The time τ_1 decreases of course with increasing concentration of nitrogen atoms. The disappearance of the dip at about 18 K at larger impurity concentrations is directly related to the shorter lifetimes of the state of free diffusion. The ratio τ_1/τ_c at 18 K obtained from the fit is however larger than expected for random jumps and indicates a drift where the muons perform a more or less directed motion towards the traps (the trap changes the interstitial potentials over a relatively long range). The prefactor Γ_o which describes the equilibrium of escape and recapture by the traps, is much smaller than expected, which can also be explained by the same kind of drift motion.

The values obtained for $E_c = E_1$ are 0.020 eV and 0.048 eV, respectively. The corresponding values for hydrogen diffusion are

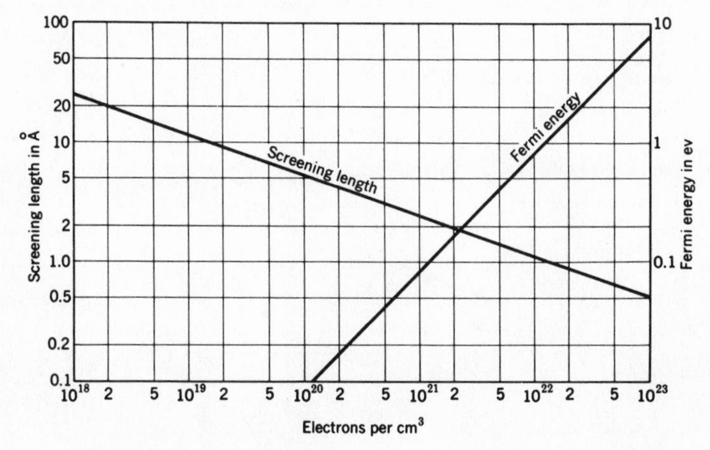

Figure 10. Electronic screening radium r_s as a function of electron density at Fermi surface, n_o. From ref. (22).

0.068 eV and 0.167 eV (for hydrogen bound to a nitrogen impurity) in niobium (19). The differences in binding energy can most probably be attributed to the difference in the zero-point energies for the two particles.

The exact nature of the trapping sites (position relative to the impurity atom, lattice dilations, etc) remains to be investigated. The polycrystal data of Figure 9 only indicates that the "gross" damping factor Λ in the trapping region is not very much different from that at low temperatures. Improved P(t) data for single crystals of doped Nb-samples with fields applied along certain selected axes are expected (20) to give new information, both on the sites in the low-T region for the impure samples (which may be associated with strain trapping) and the nature of the traps. The release and trapping processes could in principle be separated in the time spectra. Muons in the traps are also expected to have other electronic screening conditions than those in an otherwise pure lattice, which might be visible in the Knight shifts. No such measurements have been reported so far.

V. MUONS IN SEMI-METALS

It is known that muons have a high probability to form muonium when placed in insulators, and muonium or muonium-like states are also found in semi-conductors. In the latter case some to them seem to form very weakly bound states with electrons (so-called shallow bound states) where the electron wave function is extended over many lattice sites. These states have a very low hyperfine field compared to free muonium (21). In metals, on the other

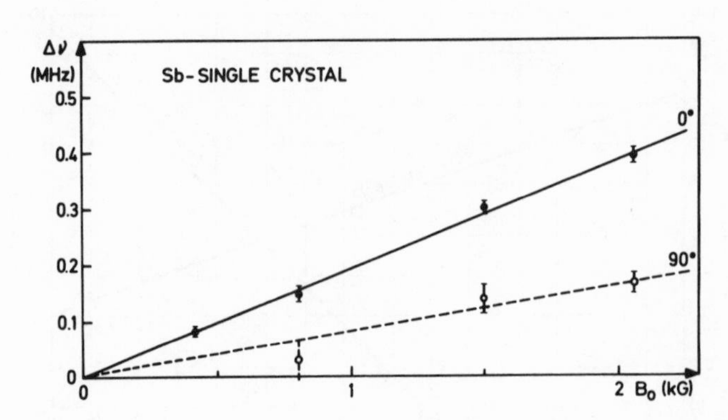

Figure 11. Frequency shift of the μSR signal in Sb-metal (24).

hand, bound states are not found and the muonic charge is instead screened by the conduction electrons. This screening charge density is set up by electrons in scattering states.

Calculations based on the free electron gas model predict a simple relation between the screening radius r_s and the density $n_o(E_F)$ of conduction electrons at the Fermi surface in the metal (Figure 10). The densities for normal metals lie in the range 10^{22}–10^{23} cm^{-3} corresponding to screening radii less than 1 Å. The semi-metals, on the other hand, have much lower densities (As: $2 \cdot 10^{20}$; Sb: $5 \cdot 10^{19}$; Bi: $3 \cdot 10^{16} cm^{-3}$) which would lead to screening radii of 4, 7, and 20 Å, respectively, if the model applies.

Calculations of the probability for forming bound states when the electron density is increased have been performed by Jena, Singwi and Nieminen (23); also for a free electron gas. Already for $r_s = 5$, they predict a bound state with a binding energy of about 0.2 eV, which should be responsible for about 80% of the s screening and be extended over a large region (out to about 20 Å around the impurity). The magnetic splitting of such states was also considered. If they exist they are expected to produce additional magnetic fields at the muon sites.

Experiments with muons in all three semimetals have been performed. The largest magnetic effect was seen in Sb whereas only a small effect was seen in As and no effect at all (within limits of error) in Bi (24).

The experimental data for Sb are shown in Figures 11-14. In the range of fields applied (up to 0.2 T) there is a field shift at the muons which increases linearly with the applied field, like a Knight shift. This shift is however of the order of 1% whereas

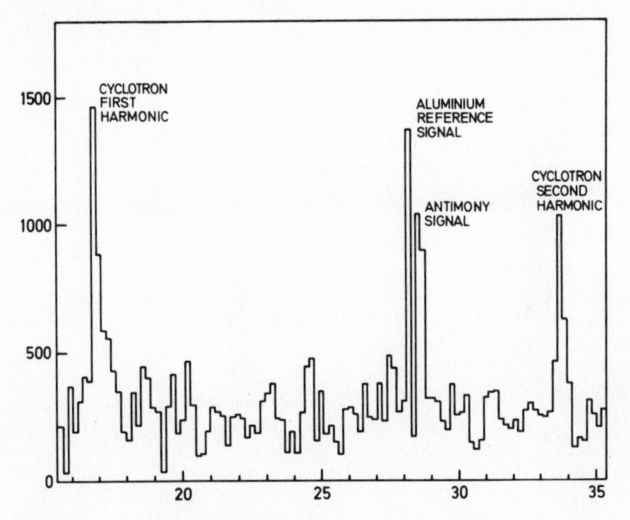

Figure 12. Fourier spectrum of Sb-metal at B_o = 2.0 KG and T = 2 K. The reference signal comes from an Al-layer added to the target (24).

muon Knight shifts in the normal metals are about 100 times smaller. At low temperatures and strong fields the Sb-signal can be easily distinguished from the reference signal (muons stopped in Al) (Figure 12).

Using a single crystal target of Sb it can also be observed that the shift is strongly orientational dependent, but with symmetry around the hexagonal c-axis (Figure 13). It indicates that the screening electrons - whether they are bound or not - are not in spherically symmetric states. If they are shallow bound states this anisotropy can be interpreted in terms of different effective masses for different orbits ($m^*_\perp$ and $m^*_{||}$). The hyperfine field from a bound electron gives an orbital contribution

$$B_{hf} \propto \langle 1/r^3 \rangle \propto (m^*/m)^3 \tag{16}$$

The Knight shifts for Sb-nuclei have also such an anisotropy, but these phenomena are not necessarily connected (25).

The most remarkable aspect of the experimental data is their temperature dependence. The shifts increase as 1/T when the temperature is lowered, down to about 20 K, below which the effect is saturated. Normal Knight shifts are only very weakly temperature dependent. Figure 14 is an attempt to fit the temperature dependence with a Brillouin function, which is the behavior expected if the field at the muon (B_{hf}) followed the orientation of the magnetic

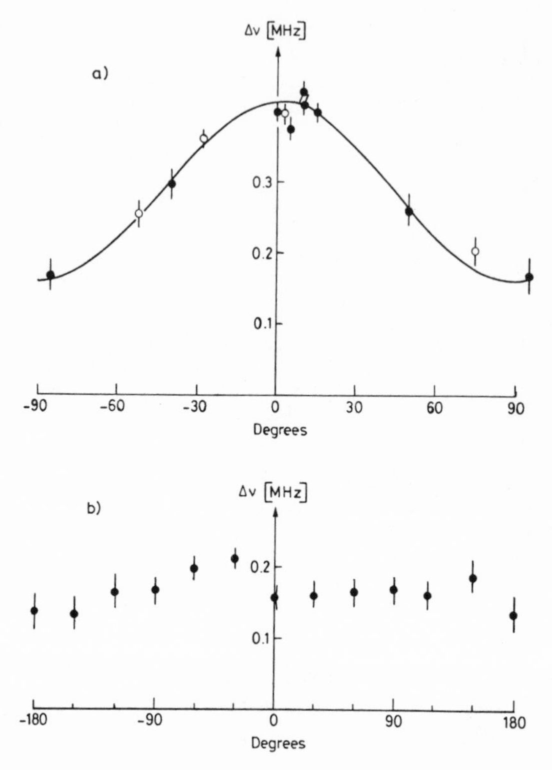

Figure 13. Orientation dependence of the frequency shift in a single crystal of Sb-metal (24). a) Field applied in a plane containing the c-axis; b) Field applied in the plane perpendicular to the c-axis.

moment of a bound electron. There is indeed a reasonable fit to the data, but the fit requires that fields of the order of 30 T act on the electron instead of the applied 0.2 T. There is also a conflict between the field dependence which is characteristic for unbound electrons (Pauli paramagnetism) and the temperature dependence which indicates a local moment (following Boltzmann statistics). There are, possibly, three ways out of this dilemma, some aspects of which are under present investigation. These are:

1. The field acting on the bound electrons is not the external field directly, but instead an exchange field existing when the unbound electrons (which are also present) are polarized by the external field ahd have a strong overlap with the bound electrons.

2. The electrons producing the field shift are bound electrons,

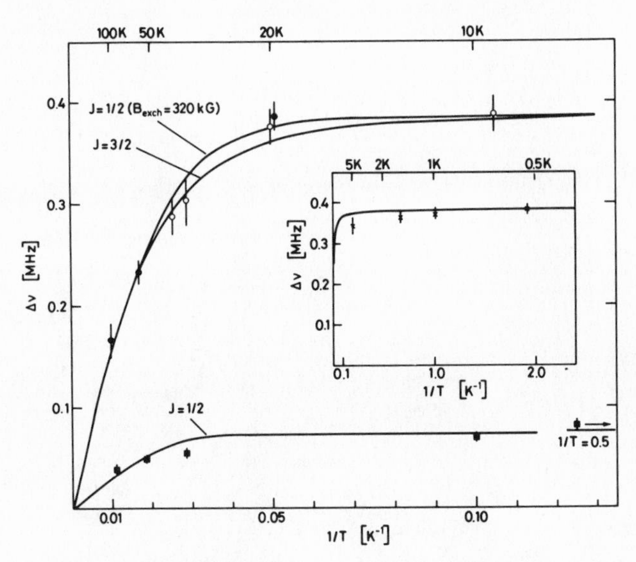

Figure 14. (1/T)-dependence of the frequency shift in Sb-metal, fitted to a Brillouin function (24).

but their binding energy (and therefore the hyperfine field) increases with the applied field. This is a phenomenon known from semi-conductor physics (26). The linearity of such a field dependence is not expected to be strictly followed when the field increases about a certain value.

3. There is simply an onset of diffusional motion above 20 K which disrupts the bound state or the screening charge characteristic of muons fixed in the lattice, and happens to give rise to a (1/T)-like dependence of the shift.

Much remains to be done about muons in the semi-metals. Since all three are isomorphic (see Figure 15) they can be easily alloyed with each other, and alloying produces large changes in the band structure (especially the so-called "semimetallic gap") of these metals.

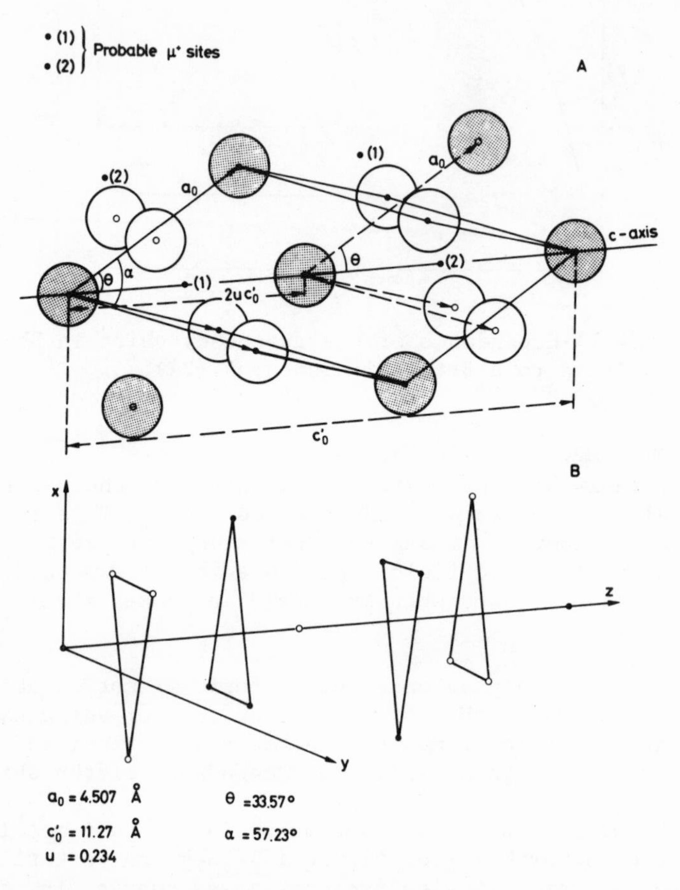

Figure 15. Crystal structure of Sb-metal with possible interstitial sites indicated (24).

ACKNOWLEDGEMENTS

Most of the work treated here has been performed within the CERN μSR Collaboration. Many discussions with Drs. Norlin, Hartmann, Pernestål, Niinikoski, Richter, Kehr, Chappert, Yaouanc, Dufresne and others have contributed to the picture of muon interactions in metals presented in these talks.

REFERENCES

1. D. K. Brice, Physics Letters 66A, 53 (1978).
2. D. Richter, Lecture Notes, this school.
3. O. Hartmann, Phys. Rev. Letters 39, 832 (1977).
4. E. Matthias, W. Schneider, and R. M. Steffen, Phys. Rev. 125, 261 (1962).
5. M. Camani, F. N. Gygax, W. Ruegg, A. Schenk, and H. Schilling, Phys. Rev. Letters 39, 836 (1977).
6. O. Hartmann, E. Karlsson, L.-O. Norlin, D. Richter, and T. Niinikoski, Phys. Rev. Letters 41, 1055 (1978).
7. A. M. Stoneham, Lecture Notes, this school.
8. P. W. Anderson, Phys. Rev. 109, 1492 (1958).
9. S. Yashino and M. Okazaki, J. Phys. Soc. Japan 43, 415 (1977).
10. G. Leibfreid, Z. Phys. 135, 23 (1953).
11. E. Karlsson et al., to be published.
12. W. J. Kossler, A. T. Fiory, W. F. Lankford, K. G. Lynn, R. P. Minnich, and C. E. Stronach, Hyperfine Interactions 6, 295 (1979).
13. D. Herlach, W. Decker, M. Gladisch, W. Mansel, H. Metz, H. Orth, G. zu Putlitz, A. Seeger, W. Wahl, and M. Wigand, Hyperfine Interactions 6, 295 (1979).
14. S. Fujii and Y. Uemura, Solid State Cpmm. 26, 761 (1978).
15. S. Fujii, to be published in J. of Phys. Soc. of Japan (1979).
16. H. Teichler, Phys. Letters A64, 78 (1977).
17. A. Yaouanc, J. F. Dufresne, R. Longobardi, J. P. Pezzetti, J. Chappert, O. Hartmann, E. Karlsson, and L.-O. Norlin, (to be published).
18. P. Meier, Lecture Notes, this school.
19. M. Borghini, T. O. Niinikoski, J. C. Soulie, O. Hartmann, E. Karlsson, L.-O. Norlin, K. Pernestal, K. W. Kehr, D. Richter, and E. Walker, Phys. Rev. Letters 40, 1723 (1978).
20. L.-O. Norlin et al., preliminary results.
21. J. S.-Y. Wang and C. Kittel, Phys. Rev. B7, 713 (1973).
22. C. Kittel, Introduction to Solid State Physics, 4th Edition (John Wiley, New York 1971), p. 281.
23. P. Jena, K. S. Singwi, and R. M. Nieminen, Phys. Rev. B17, 301 (1978).
24. M. Borghini, O. Hartmann, E. Karlsson, T. O. Niinikoski, L.-O. Norlin, and K. Pernestal, to be published in Physica Scripta (1979).

25. A. D. Goldsmith and R. R. Hewitt, Phys. Rev. B12, 4650 (1975).
26. W. Jones and N. H. March, Theoretical Solid State Physics (Wiley Interscience Publ., New York 1973) Vol. 2. Chapter 10.5.
27. P. Jena, S. G. Das, and K. S. Singwi, Phys. Rev. Letters 40, 264 (1978).
28. P. Hohenberg and W. Kohn, Phys. Rev. 136B, 864 (1964). W. Kohn and L. J. Sham, Phys. Rev. 140A, 1133 (1965).
29. K. W. Kehr, D. Richter and G. Honig, Hyperfine Interactions 6, 219 (1979). See also Z. Physik B (in press).

LAMPF RESULTS ON μ^+ DIFFUSION IN METALS*

M. Leon

University of California, Los Alamos Scientific Laboratory
Los Alamos, New Mexico 87545 USA

Some recent results from LAMPF on the diffusion of positive muons in metals are discussed. The depolarization rate in very high purity vanadium falls below the rates found earlier in lower purity targets at LAMPF and SREL, showing that the claim of Fiory et al. that the intrinsic μ^+ hopping rate was being measured is not justified. The addition of 500 ppm of oxygen to this very high purity V sample gives rise to a broad peak in the depolarization rate, with its maximum at 15 K but extending to room temperature (Fig. 1).

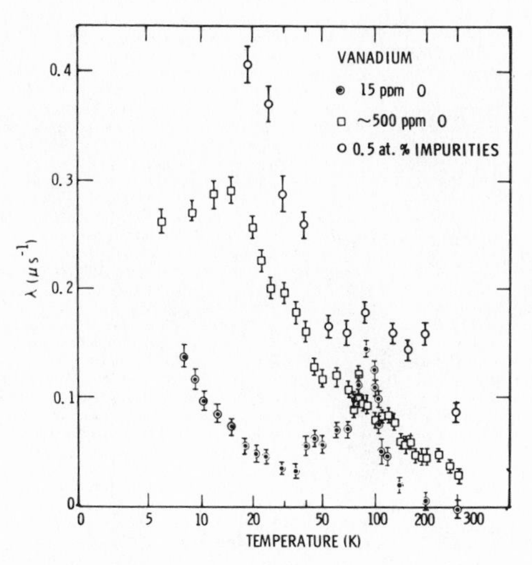

Figure 1. Exponential depolarization rates for high purity, lower purity, and oxygen-doped vanadium.

Of the Nb samples investigated, the two with significant substitutional impurities (Figs. 2a and 2b) show the familiar "double-humped" behavior; in the third (Fig. 2c), which has much of these substitutional impurities removed but which contains about 300 ppm of C, this structure is gone and the depolarization rate is much lower for temperatures less than 60 K. We believe that the remaining structure is due to trapping and escape from impurities and defects.

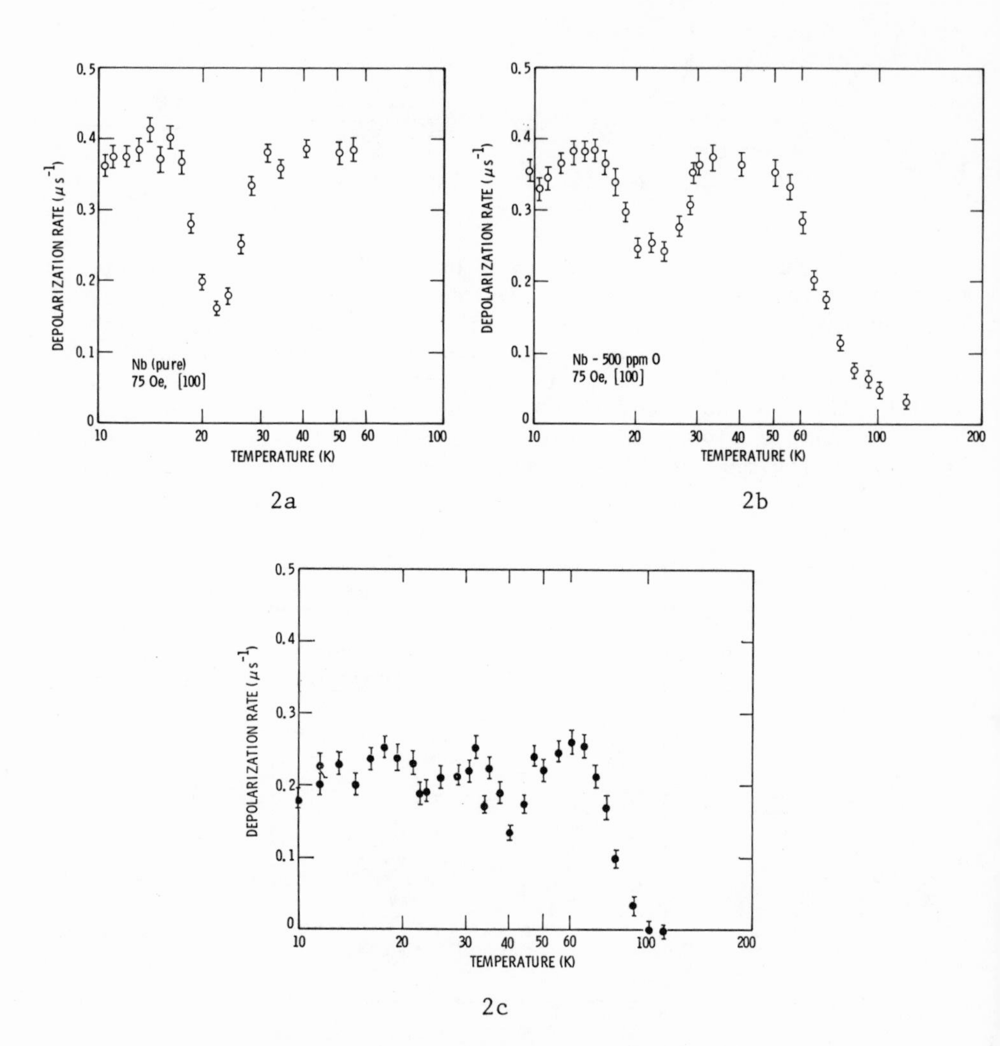

Fig. 2 Exponential depolarization rates for niobium targets.

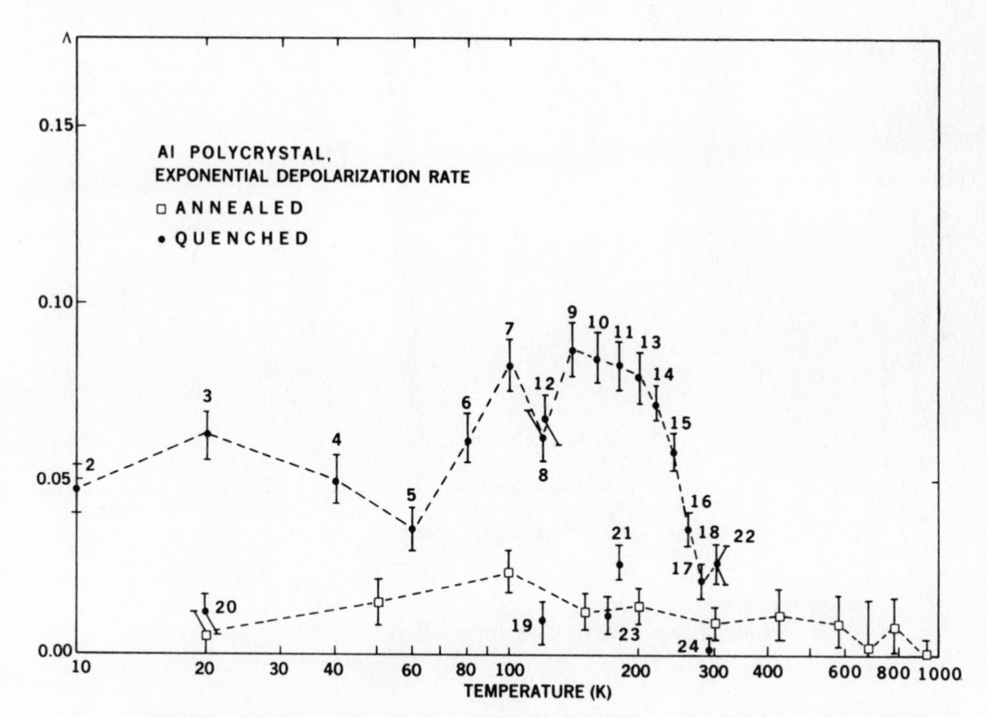

Fig. 3 Exponential depolarization rates from quenched and annealed aluminum. The numbers indicate the order of the temperature points; the effect of annealing for T ≥ 200 K is apparent.

We have compared the depolarization from annealed and quenched aluminum (Fig. 3). We suspect that the rather low depolarization structure ($\Lambda \leq .025\ \mu s^{-1}$) in the annealed Al is due to "weak traps" and indeed that the μ^+ mobility in Al is dominated by such traps. The quenched Al shows a strong depolarization structure ($\Lambda \leqslant .09\ \mu s^{-1}$) which is presumably due to vacancy trapping. The structure disappears as the sample temperature is raised to 200 - 300 K. It is clear that the μ^+ can be profitably employed to study the mechanics of vacancy annealing in Al.

We have done a series of experiments involving paramagnetic impurities in Al and Ag. The impurity ions have electronic magnetic moments which are $\sim 10^3$ times larger than the nuclear magnetic moments, so that a μ^+ trapped in the vicinity of such an impurity will suffer rapid depolarization. Because the impurity moments have a very short relaxation time, the depolarization is exponential rather than Gaussian and there is no motional narrowing as μ^+ moves from one impurity to another. Strong room temperature depolarization is seen in Al-Gd (Fig. 4), Al-Er, and Ag-Er; the

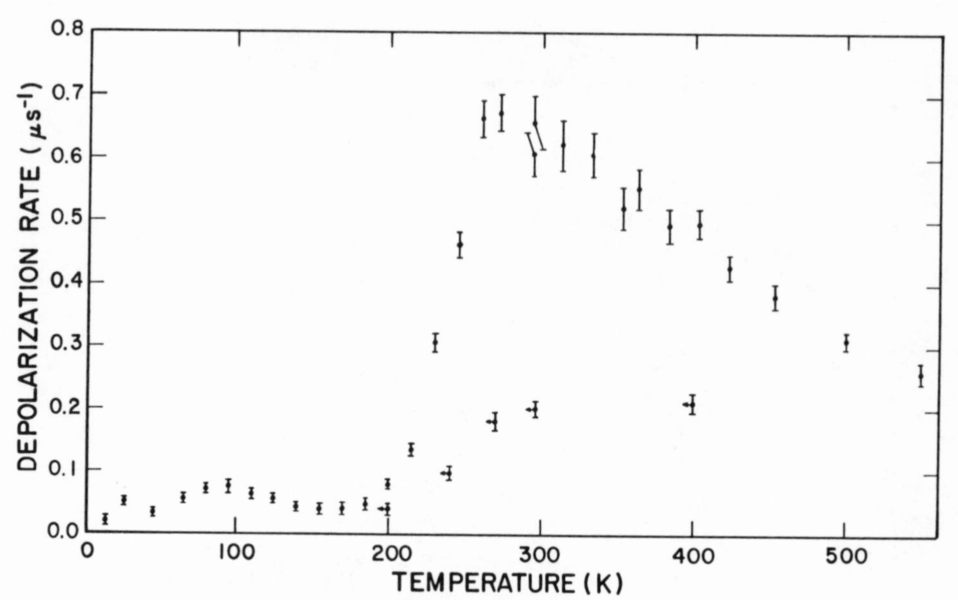

Fig. 4 Exponential depolarization rates for Al-Gd. The decrease above room temperature and lower depolarization values after heating (points marked ⫞) are thought to be due to the precipitation of Gd at elevated temperatures.

depolarization rises from much lower values in the range 200 - 300 K. We have also examined Al-La and Al-Lu targets; La and Lu should trap muons similarly to Gd and Er, but do not have electronic moments. Since not even a depolarization rate corresponding to the nuclear moments is observed in these targets, we conclude that muon is moving from trap to trap rather than remaining in a single trap. The fact that the μ^+ does not find the paramagnetic impurity until fairly high temperatures we believe is due to the mobility of the muon being significantly reduced from the intrinsic (perfect lattice) value by all the other weak traps present.

*This work was supported by the U.S. Department of Energy.

μSR IN SEMICONDUCTORS

Peter F. Meier

Physik-Institut der Universität Zürich

CH-8001 Zürich

I. INTRODUCTION

The purpose of this lecture is to give an introduction to the behaviour of muonium (μ^+e^-) in solids and to acquaint you with the present status of the μSR research in semiconductors. Most of the first applications of positive muons in solid state physics dealt with muonium and only a few experiments were made on muons in metals. Today this ratio has changed since the major part of μSR studies is concerned with metals, as you can see from the program of this school. Owing to the Fermi-Dirac distribution of the conduction electrons the muon magnetic moment reacts in metals quite differently on internal and external magnetic fields than in the paramagnetic muonium state where the interaction is dominated by the large magnetic moment of the unpaired electron.

The electronic properties of muonium are very similar to those of a hydrogen atom since the reduced masses are almost equal (m^*_{Mu} = 0.995 m^*_H). For the ground state in vacuum the binding energy is 13.539 eV and the Bohr radius is a_o = 0.532 Å. In a solid, the electron wave-function may be distorted due to the interaction with the electronic structure of the environment. This distortion is particularly strong for muonium in semiconductors. The change in the electron wave-function leads to a change in the hyperfine interaction which is directly measured by the μSR technique.

Although hydrogen is known to be present in various chemical forms in most semiconductors, only indirect information on the nature of the hydrogen centers in Si and Ge has been obtained by infrared absorption techniques. From μSR measuremements one therefore hopes to get more information as to the nature of the hydrogen impurities. Just as the muon or proton in metals can be regarded as the simplest impurity problem in conductors, the muonium, as a substitute for hydrogen, is a prototype of an impurity center in a non-metal.

Since measurements in Si [1] have revealed the existence of two different muonium states the μSR activity has concentrated on the investigations of these states in Si and Ge. I am going to report on these studies in a way which seems most appropriate for a pedagogical presentation. Thereby, the historical developments and the individual achievements of the various groups are not properly accounted for. These aspects can be found in the review articles about this topic [2,3,4].

II. SPIN-HAMILTONIAN FOR MUONIUM

The Fermi contact interaction between the magnetic moments of a muon and an electron is given by

$$H_{cont} = -\frac{8\pi}{3}|\psi(0)|^2\, \vec{\mu}_\mu \cdot \vec{\mu}_e \tag{1}$$

where $|\psi(0)|^2$ is the electron density at the muon and where the operators for the magnetic moments are

$$\vec{\mu}_\mu = -\, g_\mu\, \mu_\mu\, \vec{S}_\mu \tag{2}$$

$$\vec{\mu}_e = -\, g_e\, \mu_B\, \vec{S}_e \tag{3}$$

$g_e \cong +2$ and $g_\mu \cong -2$ are the g-factors of the electron and the muon in the muonium state. The contact term is a special case of the general hyperfine interaction between two particles with magnetic moments which results if the electron wave-function around the muon is spherically symmetric. The general interaction can be written as

$$H_{hf} = -\, \vec{\mu}_\mu \cdot \vec{B}_{hf}\,(\vec{r}, \vec{S}_e) \tag{4}$$

where $\vec{B}_{hf}$ is an operator and depends on the electron coordinate $\vec{r}$ and its orbital momentum $\vec{\ell}$:

$$\vec{B}_{hf}(\vec{r},\vec{S}_e) = -\, 2\mu_B \left\{\frac{8\pi}{3}\, \vec{S}_e \delta(\vec{r}) + \frac{\vec{\ell}}{r^3} + \frac{3\vec{r}(\vec{S}_e.\vec{r})}{r^5} - \frac{\vec{S}_e}{r^3}\right\}. \tag{5}$$

In this section we consider a muonium atom in its 1S-ground state in an external field $\vec{B}$. The spin-Hamiltonian is

$$H_{Mu} = h\nu_o\ \vec{S}_\mu \cdot \vec{S}_e + g_e\mu_B\ \vec{S}_e \cdot \vec{B} + g_\mu\mu_\mu\ \vec{S}_\mu \cdot \vec{B} \tag{6}$$

where

$$h\nu_o = \hbar\omega_o = -\frac{8\pi}{3}\ |\psi(0)|^2\ g_e\ \mu_B\ g_\mu\ \mu_\mu \tag{7}$$

is the hyperfine energy splitting. The most recent [5] precise value for ν_o in vacuum is

$$\nu_o = 4463.30235\ (52)\ \text{MHz} . \tag{8}$$

The nice thing about the Hamiltonian (6) is that it can be solved analytically and that the calculation of the time dependence of the muon spin polarization is not too trivial. Therefore, the problem is a nice exercise in quantum mechanics, and for the benefit of those among you who are not already familiar with these calculations and for those who have to teach quantum mechanic courses, I have developed the theory in the Appendix.

Using the notation

$$\gamma_\mu \equiv -\frac{g_\mu\mu_\mu}{h} = 13.55\ \text{MHz /kG} \tag{9}$$

and

$$\Gamma \equiv (g_e\mu_B - g_\mu\mu_\mu)/h = 2.82\ \text{MHz/G} \tag{10}$$

the eigenvalues of (6) are given by

$$\begin{aligned} E_1/h &= \nu_1 = \nu_o/4 + \Gamma B/2 - \gamma_\mu B \\ E_3/h &= \nu_3 = \nu_o/4 - \Gamma B/2 + \gamma_\mu B \\ E_{2,4}/h &= \nu_{2,4} = -\nu_o/4 \pm \sqrt{\nu_o^2 + \Gamma^2 B^2}/2 . \end{aligned} \tag{11}$$

These energy levels are plotted against the field strength in Fig. 1 with ν_o = 2361 MHz which is the value obtained from recent μSR experiments in Ge [6]. The M = ±1 levels diverge linearly with the field because they are pure states with $m_\mu = m_e = \pm 1/2$ respectively. The two M = 0 levels are a field dependent linear combination of $m_\mu = -m_e = \pm 1/2$.

In a μSR experiment in a longitudinal field, i.e. where the initial μ^+ polarization is parallel to the applied field, only transitions obeying the selection rule $\Delta M = 0$ occur. Since the present experimental resolution is lower than 500 MHz, all terms

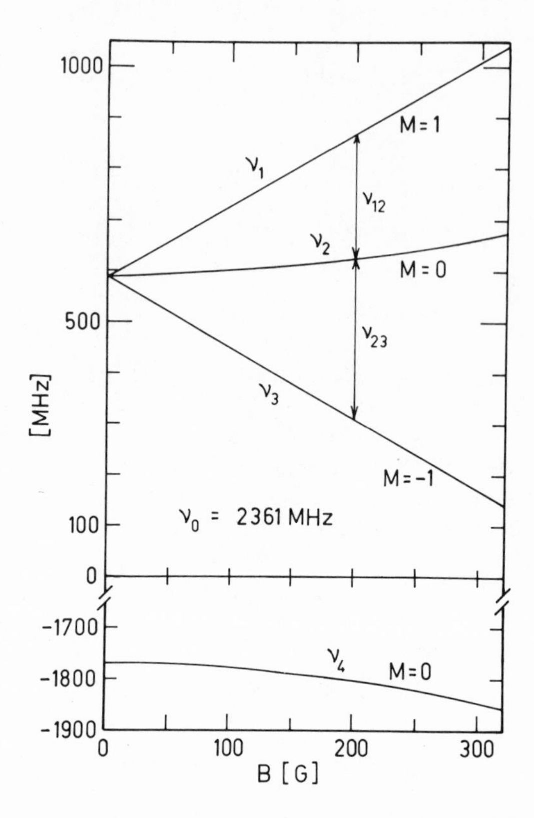

Fig. 1. Breit-Rabi diagram for muonium in Ge.

oscillating with the frequency ν_{24} are averaged to zero. Thus in a longitudinal field experiment, no μSR frequency is observed but since the ratio between the constant part and the oscillating part depends on the field, a measurement of the polarization as a function of the field yields information about ν_o. This is further discussed in Sect. 5.

III. TRANSVERSE FIELD MEASUREMENTS ON MUONIUM

In transverse fields, there occur transitions with $|\Delta M| = 1$. Again, the high frequencies ν_{14} and ν_{34} between the singlet and triplet states are averaged out. The transitions $1 \leftrightarrow 2$ and $2 \leftrightarrow 3$, however, are (for Ge) in the experimentally accessible range for fields below 300 G. The evaluation of the time dependence of the muon polarization gives (see (A.17) with $\overline{\cos\omega_{14}t} = \overline{\cos\omega_{34}t} = 0$ and $\delta \equiv 0$)

$$p_\mu^x(t) = \frac{1}{2}\cos^2\beta \cos\omega_{12}t + \frac{1}{2}\sin^2\beta \cos\omega_{23}t \tag{12}$$

where

$$\omega_{12} = 2\pi(\nu_1 - \nu_2) \; ; \quad \omega_{23} = 2\pi(\nu_2 - \nu_3) \tag{13}$$

and [cf(A.7)]

$$\cos^2\beta = \frac{1}{2}\left(1 + \frac{x}{\sqrt{1+x^2}}\right) \; ; \quad x = \frac{\Gamma B}{\nu_o} \; . \tag{14}$$

Eq. (12) can be rewritten in the form

$$p_\mu^x(t) = \frac{1}{2}\cos\Omega^+ t \cos\Omega^- t + \frac{1}{2}\frac{x}{\sqrt{1+x^2}}\sin\Omega^+ t \sin\Omega^- t \tag{15}$$

with

$$\Omega^\pm = \frac{1}{2}(\omega_{23} \pm \omega_{12}) \; . \tag{16}$$

In a field of 100 G and with ν = 2361 MHz the values are $\Omega^+ = 2\pi \times 140$ MHz, $\Omega^- = 2\pi \times 17$ MHz, and $x = 0.12$. Thus, the recorded polarization has beats. This "two frequency precession" of muonium was first observed by Gurevich et al. [7] in fused quartz, ice, and Ge.

Note that

$$\nu_{23} + \nu_{12} = (\Gamma - 2\gamma_\mu)\, B \tag{17}$$

and

$$\nu_{23} - \nu_{12} = \sqrt{\nu_o^2 + \Gamma^2 B^2} - \nu_o = \frac{\Gamma^2 B^2}{2\nu_o}\left(1 + O\left(\frac{\Gamma B}{\nu_o}\right)\right) \; . \tag{18}$$

A Fourier transform of the μSR spectrum should exhibit two lines with a center depending linearly and a splitting depending quadratically on the field strength. In Fig. 2 measured spectra in quartz and in Si are shown. From the splitting of the two lines the hyperfine frequency ν_o can be extracted. Whereas the value thus obtained for quartz agreed with the vacuum value (8), the splitting of the muonium lines in Si is considerably greater than in quartz. According to (18) this corresponds to a smaller value of ν_o (Si) and hence to a reduced electron density at the muon.

Whereas the two-frequency precession in quartz is easily detectable at room temperature, this is not the case for Si and Ge where the two transitions were first seen only at nitrogen temperatures. At SIN measurements have been made on various Si samples with different doping concentrations and at different temperatures. The data indicate that muonium may be seen in a broad range of

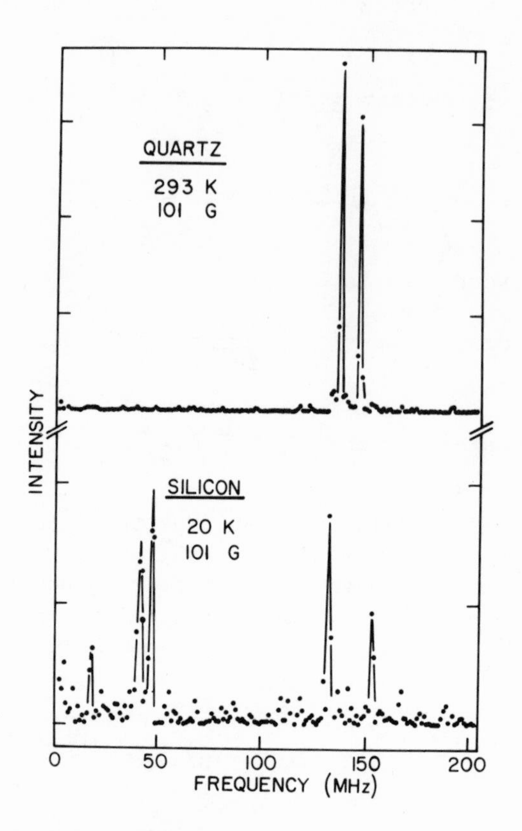

Fig. 2. Fourier transforms of μSR spectra in quartz and in slightly p-doped Si.

doping concentration and up to temperatures of around 250 K. To investigate whether the hyperfine frequency would change with temperature, an experiment on an undoped Ge sample with $\rho > 50\ \Omega$ cm was made by the Konstanz group. In Fig. 3 the results of the fitted frequencies ν_{12} and ν_{23} are shown as a function of temperature. Above 50 K the relaxation rate of both transitions suddenly increases and the signals are no longer detectable around 100 K. The result for the hyperfine frequency is [6]

$$\nu_o(\mathrm{Ge})/\nu_o(\mathrm{vac}) = 0.5290 \pm 0.0007 \tag{19}$$

which is considerably lower than the earlier value [7]. The preliminary result [8] for Si is

$$\nu_o(\mathrm{Si})/\nu_o(\mathrm{vac}) = 0.446 \pm 0.008 \ . \tag{20}$$

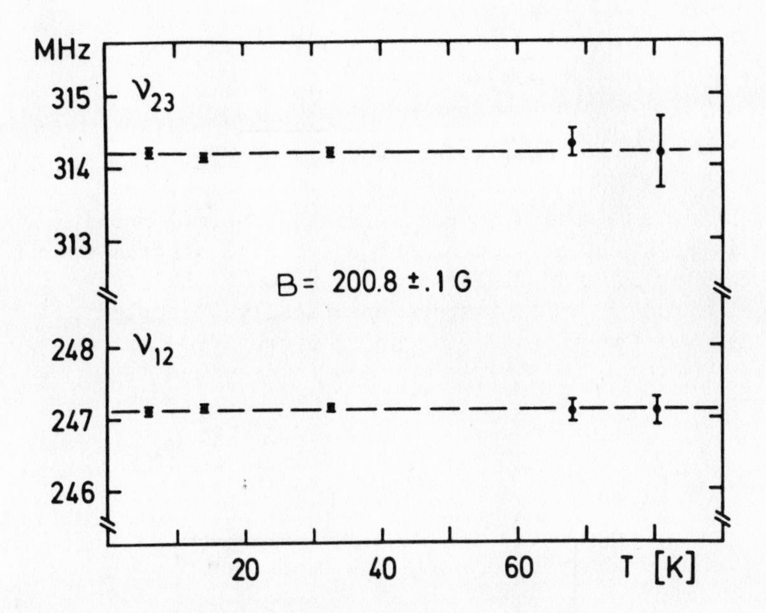

Fig. 3. Fitted transition frequencies of the muonium state in Ge versus temperature.

IV. ANOMALOUS MUONIUM

In Fig. 2 there are two additional prominent peaks in the Si spectrum around 43 MHz which do not appear in quartz. These frequencies were first observed in Berkeley [1] and later studied further in Zürich [9]. They have been assigned to an "anomalous muonium" state Mu*.

The Mu* frequencies have been observed from zero applied fields up to 5 kG and their dependence on the direction of the field has been studied. The data is well-reproduced by an anisotropic spin Hamiltonian of the form

$$H^* = h\nu\ \vec{S}_\mu\cdot\vec{S}_e + h\delta\ S_\mu{}^z S_e{}^z + g_e\mu_B\ \vec{S}_e\cdot\vec{B} + g_\mu\mu_\mu\ \vec{S}_\mu\cdot\vec{B} \qquad (21)$$

which differs from (6) only in the addition of the axial term. The z-axis in this expression corresponds to any one of the four [111] axes of the crystal.

For the external field directed along the z-axis the eigenvalues of H* can be found analytically (see Appendix) and the four frequencies are

$$\nu_1 = (\nu + \delta)/4 + (\Gamma/2 - \gamma_\mu)B$$
$$\nu_3 = (\nu + \delta)/4 - (\Gamma/2 - \gamma_\mu)B \tag{22}$$
$$\nu_{2,4} = -(\nu + \delta)/4 \pm \sqrt{\nu^2 + \Gamma^2 B^2}/2 .$$

To illustrate the difference from the normal muonium I have plotted in Fig. 4 a portion of the Breit-Rabi diagram appropriate to the data in Si. For fields exceeding 100 G the states with different electron polarization are practically decoupled and all amplitudes for transitions with $\Delta m_e \neq 0$ are zero ($\sin^2\beta \approx 0$).

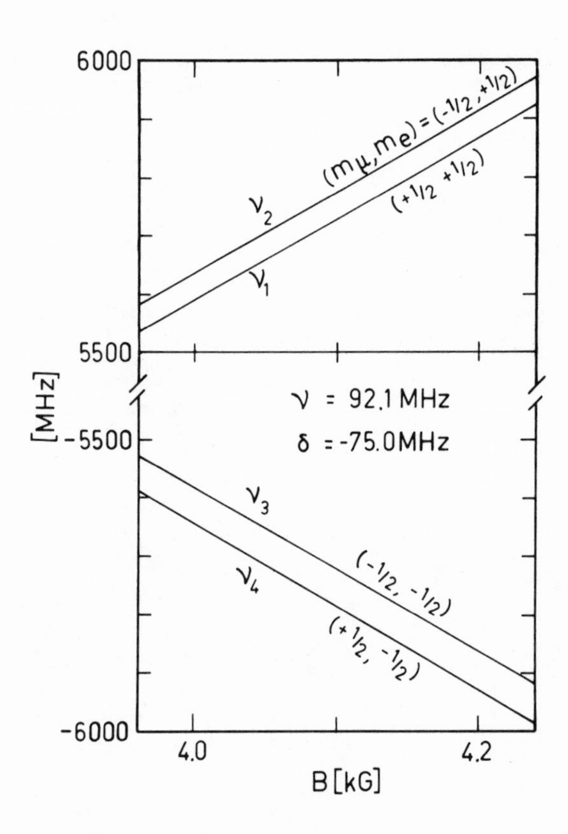

Fig. 4. Portion of the Breit-Rabi diagram for the anomalous muonium state in Si with the external field along z.

Thus one can observe the transitions ν_{21} and ν_{34}. Their values are (for an arbitrary direction of the external field with $(\hat{B}\hat{z}) = \cos\Theta$)

$$\nu_{21} = \left[(\gamma_\mu B + \frac{1}{2}(\nu + \delta\cos^2\Theta))^2 + \frac{\delta^2}{4}\sin^2\Theta\cos^2\Theta\right]^{1/2}$$
$$\nu_{34} = \left[(\gamma_\mu B - \frac{1}{2}(\nu + \delta\cos^2\Theta))^2 + \frac{\delta^2}{4}\sin^2\Theta\cos^2\Theta\right]^{1/2} . \qquad (23)$$

These values are obtained [10] from an approximate solution to the Hamiltonian (21) which, however, is for B > 100 G in very good agreement with the numerical solution.

The results of the measurements of the field dependence in Si[9] are shown in Fig. 5. The interpretation of the data is the following. Part of the muons stopped in Si form an axially symmetric (z) muonium state Mu* with a hyperfine interaction corresponding to (21). If the applied field is along a [111] crystal axis, one fourth of the ensemble of μ^+ forming Mu* has z along the field direction

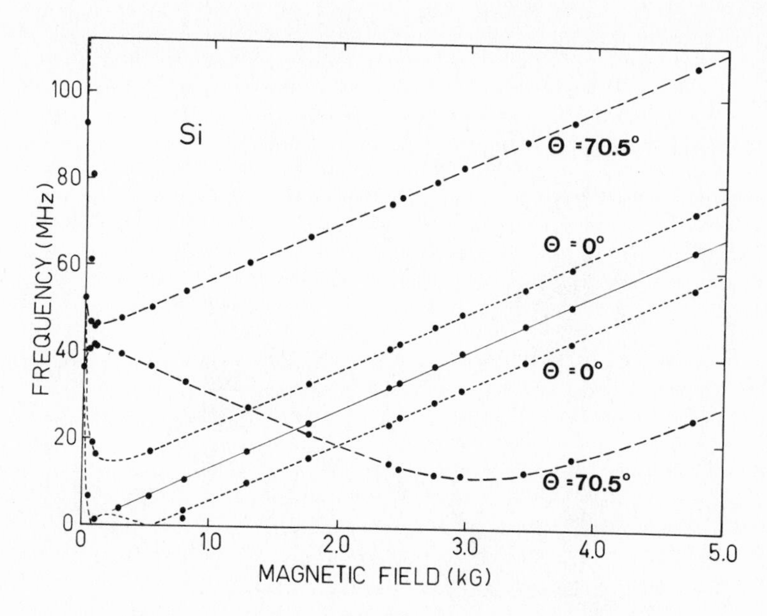

Fig. 5. Experimentally observed precession frequencies in Si as a function of the external field. Theoretical curves are included for the μ^+ components (solid line) and for the Mu* components.

($\Theta = 0^o$) leading to one pair of transition frequencies. Three quarters of the ensemble have z along $[11\bar{1}]$, $[1\bar{1}1]$ and $[\bar{1}11]$. These axes lie at an angle of $\Theta = 70.5^o$ with respect to the field axis and give rise to the other pair of frequencies. This model has further been checked by measuring the change of the frequencies as a function of the sample orientation in a fixed field of 3.83 kG.

For fields lower than about 100 G all four states are coupled. For an arbitrary direction of $\vec{B}$ the frequencies and transition amplitudes have to be calculated numerically [11]. Low field measurements [12] are shown in Fig. 6. Owing to the simultaneous occurrence of up to six frequencies,in addition to the normal muonium frequencies, and to the possible splitting of the lines due to a slight misalignment of the crystal, the experimental problems in the low field region are great. They are, however, of importance for the

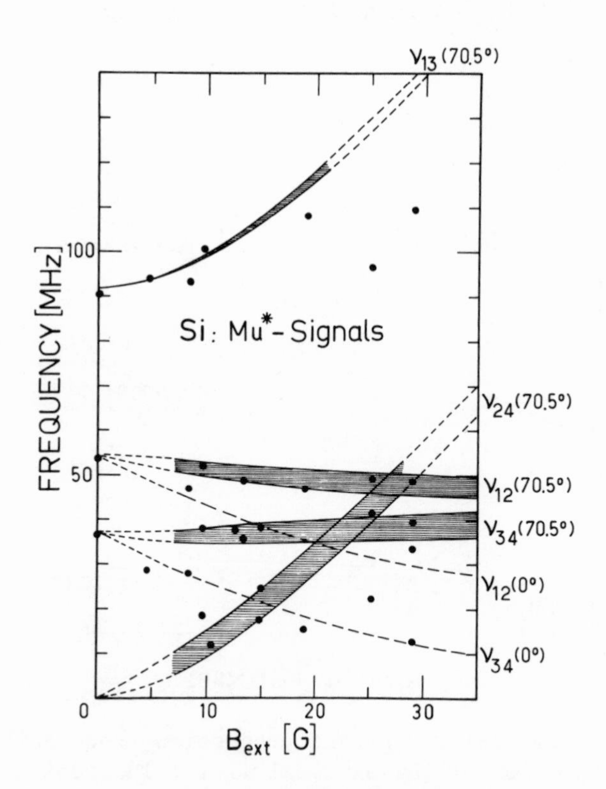

Fig. 6. Observed Mu* frequencies at low fields [12]. The bands correspond to the most intense transitions for $\Theta = 0^o$ and $\Theta = 70.5^o$. The band width corresponds to an allignment error of 10°.

search of a superhyperfine interaction with a ^{29}Si nucleus. Such an interaction could show up in low fields [13] and a clear identification of such frequencies would allow a site assignment of the Mu* state.

It seems that the data in Si can be described at all fields by the Hamiltonian H* (21) with the values

$$\nu = 92.1 \pm 0.3 \text{ MHz} \qquad \delta = -75.0 \pm 0.4 \text{ MHz}$$
$$g_\mu = -(2.01 \pm 0.01) \qquad g_e = 2.2 \pm 0.2 \quad . \tag{24}$$

In Ge, the Mu* state has been detected by the Konstanz group[14], see Fig. 7. The data are again in full agreement with the predictions from H* (21) with the exception that no signals have been observed for fields below 1 kG. The fit of the data gave for the hyperfine parameters the values

$$\nu = 130.7 \pm 1.0 \text{ MHz} \qquad \delta = -103.9 \pm 1.0 \text{ MHz}. \tag{25}$$

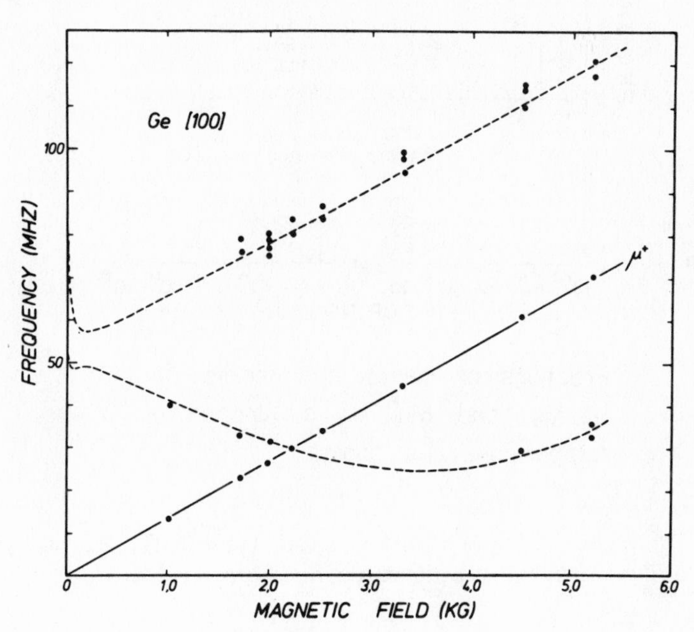

Fig. 7. Anomalous muonium frequencies in Ge with a [100] axis along the external field [14].

Concerning the occurrence of this anomalous muonium state as a function of temperature and doping concentration in Si we refer to Fig. 8 [15]. The absence of Mu and Mu* states at high temperatures and in both doping extremes is believed to be related to the absence of muonium in most metals where free charge carriers screen the electron-muon Coulomb interaction and/or relax the spin of the muonium electrons. More detailed experiments are under way at SIN [8]. They show in particular the occurrence of Mu also in some n-doped samples.

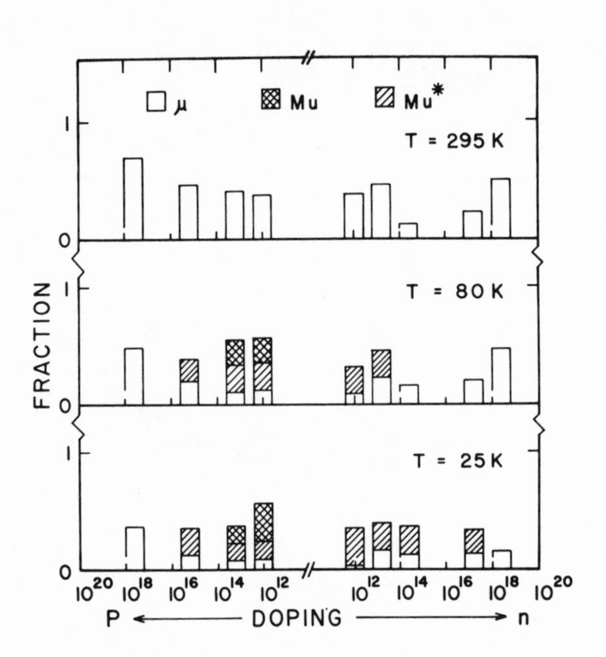

Fig. 8. Fractions of the three components of μ^+, Mu, and Mu* as a function of temperature and doping [15].

V. LONGITUDINAL FIELD MEASUREMENTS AND DEPOLARIZATION

The muon polarization $P_\mu^z(t)$ in a longitudinal field is given by (see (A.25) and (A.7))

$$P_\mu^z(t) = \frac{1}{2(1 + x^2)} (1 + 2x^2 + \cos\omega_{24}t) \qquad (26)$$

with $x = \Gamma B/\nu_o$. If the oscillations are averaged out, the observable constant polarization is

$$\overline{P}_\mu^z = \frac{1 + 2x^2}{2 + 2x^2} \, . \qquad (27)$$

In zero field ($x = 0$) the polarization is 1/2 and increases to 1 for high fields ($x >> 1$) where the electron and muon spins are decoupled (Paschen-Back effect). For a critical field of $B_c = \nu_o/\Gamma$ the polarization is 3/4.

In Fig. 9 the results [16] of a longitudinal field quenching experiment in Si at room temperature are shown. The data are consistent with Eq. (27) for a critical field of (643 ± 42)G which gives for the hyperfine frequency a value of $\nu_o/\nu_o^{vac} = 0.41 \pm 0.03$.

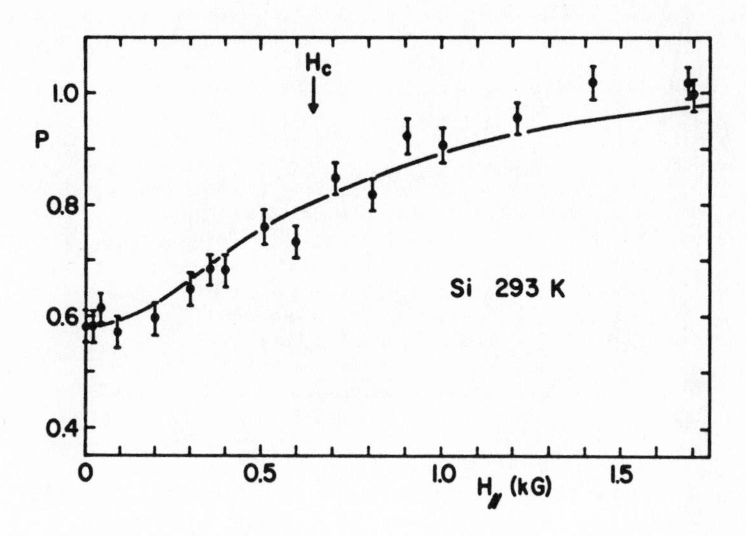

Fig. 9. Residual muon polarization versus the longitudinal magnetic field strength [16].

One may ask how does this indirect observation of a muonium state correlate with the fact that in transverse fields no precession frequencies have been found at room temperature.

One explanation is given by a reduced lifetime of the muonium state. It may happen that the muonium reacts chemically and forms a state in which the μ^+ is in a diamagnetic environment or that the state is interrupted by a ionization process. If the lifetime τ of the muonium state is shorter than a few precession periods ν_{12}^{-1}, ν_{23}^{-1}, but longer than the hyperfine period ν_o^{-1}:

$$\nu_{12}^{-1} > \tau > \nu_o^{-1} \tag{28}$$

no transition can be observed in the transverse field but, in a longitudinal field the formula (27) still holds. In a perpendicular field the free muon Larmor precession is seen, however, with an apparent change of the initial phase which is due to the precession in the muonium state from time zero to time τ. This effect has been observed by Kudinov et al. [17] who investigated Ge at intermediate temperatures (Fig. 10).

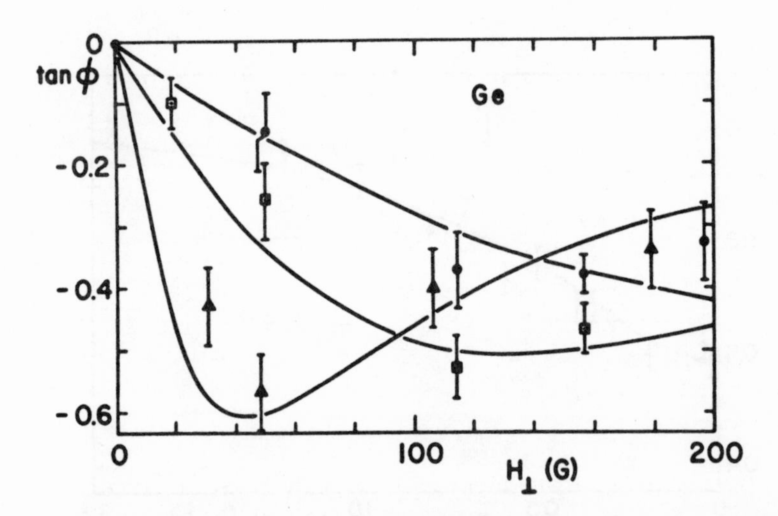

Fig. 10. Phase shifts ϕ of the free muon signal as a function of the applied transverse field [17] (●: T = 197 K, ⊡ : T = 187 K, ▲: T = 177 K).

Since the hyperfine energy is strongly reduced in the anomalous muonium state, a longitudinal field measurement would give a steep rise of P_{μ}^{z} at low fields, the critical field for Si being around 30 G. Such a rise has indeed been found [18].

Although it seems that longitudinal field measurements allow an indirect determination of the hyperfine frequency, the analysis of the data is complicated by another possible effect which is the interaction of the bound electron with charge-carriers leading to spin flips of the electron. To account for this effect one has to describe the behaviour of the muon spin polarization in a muonium state for situations where the bound electron spin interacts with the surroundings. The Hamiltonian H is decomposed into

$$H = H_{Mu} + V + H_R \tag{29}$$

where V describes the interaction of the electron spin with the reservoir whose dynamics is governed by H_R. Since we are only interested in the dynamics of the muonium subsystem it is sufficient to have the information contained in the reduced density matrix, which is obtained by summing over the degrees of freedom of the reservoir. It is possible to obtain a master equation for this open system [19] which contains in a limiting case the equations of Ivanter and Smilga [20] who accounted for the interaction V by introducing phenomenologically a spin flip frequency ν_{ex}. It is not possible here to give the details of these theories but the essential points can easily be understood.

If the exchange frequency ν_{ex} is much higher than the hyperfine frequency ν_o, the effective hyperfine field is zero and the muon behaves essentially as a free μ^+ with a relaxation time proportional to ν_{ex}/ν_o^2. This behaviour corresponds exactly to the exchange narrowing effect which is well-known in magnetic resonance experiments [21]. For $\nu_{ex} \ll \nu_{12}$, ν_{23} the reservoir acts as a weak perturbation on the muonium and causes a relaxation with a rate proportional to ν_{ex}. For intermediate values of ν_{ex} no precession signals can be seen.

The situation becomes more complicated if the effects of ν_{ex} are combined with the possibility of a chemical reaction after a mean lifetime τ. The resulting formulae then contain both ν_{ex} and τ as parameters and it is evident that the rough information obtained in a longitudinal field experiment allows freedom for several possible parameter values.

Much data has been accumulated by longitudinal field experiments in Si and Ge (see Refs. in [2] and [3]) which all give some piece of information. What is still lacking is a comprehensive experiment covering the whole temperature range from 4 K to room temperatures and looking for the behaviour of both Mu and Mu* states in transverse as well as in longitudinal fields.

VI. MODELS AND DISCUSSIONS

On the whole we may say that the μSR data on the muonium and anomalous muonium states in Si and Ge are well described by the discussed phenomenological spin Hamiltonians. The question then arises about the electronic structures of these centers. The structures are still unclear and several models exist which account for the observed hyperfine interaction in a qualitative way.

The work by Wang and Kittel [22] explains the normal muonium states as deep donors in the framework of a semi-phenomenological model. They assume a spherically symmetric potential around the μ^+ with an unscreened Coulomb potential inside a sphere of radius R. Outside this cavity the influence of the host-crystal is simulated by screening the potential with a constant dielectric function ε_0 (ε_0 = 12 for Si and ε_0 = 15.8 for Ge). To remove the discontinuity at R, a constant term is added to the interior potential:

$$\tilde{V}(r) = \begin{cases} -e^2/r + e^2(1 - 1/\varepsilon_0)/R & r < R \\ -e^2/\varepsilon_0 r & r > R \end{cases} \tag{30}$$

Inside the cavity the mass of the electron is given by the free electron mass m and outside by the effective mass m* of the conduction band electrons (m* = 0.31 m for Si and m* = 0.17 m for Ge). Solving the Schrödinger equation for this potential, Wang and Kittel [22] calculated the binding energy as a function of the cavity radius R (Fig. 11). The low binding energies for R < 1 A.U. indicate a shallow donor state whereas for higher R values a steep increase occurs leading to a deep donor state.

Si and Ge have the diamond type crystal structure where two natural interstitial sites are available [23], one with tetrahedral local symmetry and one with hexagonal local symmetry. Assuming the cavity radius of the model to be given by fitting the cavity inside the packed hard spheres of the lattice one obtains radii on the order of 2 A.U. which are also indicated in Fig. 11 for both the tetrahedral and the hexagonal site occupancy. Thus the model predicts a deep muonium state with a binding energy around 0.2 Ry. The corresponding electron densities at the muon can then be calculated. The results for $f \equiv |\psi(0)|^2 / |\psi(0)|^2_{vac}$ are hexag. site: $f_{Si} = 0.58$, $f_{Ge} = 0.51$

tetrah. site: $f_{Si} = 0.64$, $f_{Ge} = 0.57$

which have to be compared to the experimental values:

$f_{Si} = 0.45$, $f_{Ge} = 0.53$.

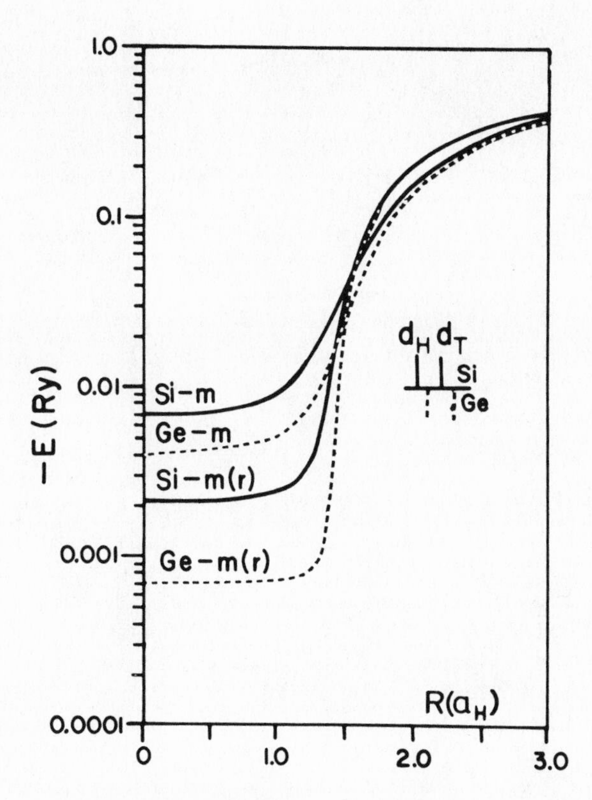

Fig. 11. Ionization energy -E as a function of the cavity radius R [22].

There is a reasonable agreement which prompted various improving modifications of the model. The above results are not essentially changed as long as the free electron mass is used inside the cavity. This assumption, however, is difficult to justify [24] in the frame of existing theories for other impurity states in semiconductors and the quantitative success of the simple model remains a puzzle.

The crudeness of the model compares well with the amount of information obtained from the μSR data. At present the sites of the muonium are not known. As elucidated in another lecture at this school the site determination requires the presence of a dipolar interaction. Si has an isotope Si^{29} with spin 1/2 and natural abundance of 4.7 %. The same applies to Ge^{73} with I = 9/2 and 7.6 %. Therefore, for some muonium states there will be an interaction of the bound electron with a neighbouring magnetic moment of a nucleus. This gives rise to the so-called super-hyperfine interaction which, at low fields, leads to additional muon precession frequencies [13,25]. Experiments on Si are presently done at SIN.

For the anomalous muonium state there are no calculations of microscopic models. A variety of models which lead to an anisotropic term have been proposed. If a paramagnetic muonium state is formed in the diamond lattice, the crystal field will perturb the electron wave function and the hyperfine interaction has to be calculated using the general Eq. (5). For a tetrahedral interstitial site, the contributions from the anisotropic terms cancel and one is again left with the scalar contact term. This is different if the μ^+ has locally hexagonal symmetry where the crystal field indeed yields an axially symmetric contribution along a [1,1,1] axis. A possible explanation is thus given by assigning the normal muonium state to muons trapped at tetrahedral interstitial sites and the anomalous state to muons at hexagonal sites [26]. Detailed calculations, however, are lacking.

Extended Hückel theory calculations for interstitial hydrogen atoms in silicon model crystals [27] showed that the tetrahedral site is a stable position, the hexagonal site is not. Since vacancies and divacancies form deep traps for light interstitials [27] one has also to consider the possibility that Mu and Mu* are formed at these sites. Recently, a channeling experiment on deuterium in silicon was reported [28]. The results indicate that at room temperature the D is located predominantly in a single interstitial site 1.6 Å along a <111> direction from a Si atom in the antibonding direction. The specific nature of the silicon-deuterium bonding, however, is not known.

The anomalous muonium state was first attributed to a shallow donor state. This is a state with a localized level in the gap near the conduction-band edge such that the electrons are weakly bound and can, by thermal activation, easily be ionized into the conduction bands. These shallow donor states in semiconductors are thoroughly investigated [29] and are well described by the effective mass theory which is especially successful for group V donors in Si and Ge, but also for interstitial Li. The electron wave-function is spread over many lattice sites. It is therefore possible that, in a center with a deep level ground state but an excited shallow state, the lifetime of the excited state is long since the overlap with the ground-state wave-function would be small. The effective mass theory, however, does not seem to be directly applicable to hydrogen or μ^+ in Si [24].

It was also suggested [1] that the anomalous muonium state can be connected with an impurity-exciton bound state involving holes which are created in the slowing-down process of the implanted muon.

In conclusion we may say that the present μSR data do not allow one to discriminate among the various possible models. It would help greatly if measurements allowed the precise determination of the electron g-factor and its asymmetry in the muonium states. As can be seen from Eq. (5) this could provide information about the electron wave-function. Compared to the ESR technique such an experiment is difficult since the most easily observable transition frequencies ν_{12} and ν_{34} do not critically depend on the electron g-factor, in contrast to ν_{14} and ν_{23}. There exists the possibility to attack this problem by inducing the transitions ν_{14} and ν_{34} by microwaves [30]. This resonance technique has been used in the famous muonium high-precision experiments of Hughes [5]. I think that this technique will sooner or later also be applied to the study of muonium states in solids. Further information as to the nature of the observed states can be obtained by careful measurements of the asymmetries and relaxation rates as a function of temperature and doping.

What do specialists working on semiconductors think about the exotic μSR results? The knowledge of impurity states in semiconductors is of primary technical importance and the behaviour of hydrogen in particular is of growing interest in connection with amorphous Si. A variety of experimental techniques exists to study these impurities. They range from infrared spectroscopy, where some absorption bands can be related to stretching vibrations of hydrogen and deuterium, to ENDOR experiments [31], where the impurity state is observed by identifying the interaction with many neighbouring nuclei and where detailed information about the wave-function can be obtained. The μSR results are to date in between these extreme cases. It is hoped, however, that the μSR technique will be developed further and I believe that as soon as precise values about electron g-factors are obtained and the muon site can be determined, the μSR method will be accepted as a unique tool to investigate the hydrogen-like impurities in semiconductors.

APPENDIX

This appendix is intended to be an instructive exercise in quantum mechanics with relevant results for μSR in semiconductors.

The aim is to calculate the time dependence of the muon spin polarization if the muonium system is described by the spin operator

$$H = \frac{h\nu}{4} \vec{\sigma}_\mu \cdot \vec{\sigma}_e + \frac{h\delta}{4} \sigma_\mu{}^z \sigma_e{}^z + \frac{g_e\mu_B}{2} B\sigma_e{}^z + \frac{g_\mu\mu_\mu}{2} B\sigma_\mu{}^z \ . \qquad \text{(A 1)}$$

This Hamiltonian describes the anomalous muonium state if the external field is parallel to the z-direction defined by the anisotropic term. At the same time we obtain the results appropriate to the normal muonium state (6) simply by putting $\delta = 0$.

We take as a basis set the four vectors

$$\chi^{\alpha\beta} = \chi_\mu{}^\alpha \chi_e{}^\beta \qquad \alpha,\beta = \pm 1 \qquad \text{(A 2)}$$

where

$$\sigma_\mu{}^z \chi_\mu{}^\alpha = \alpha \chi_\mu{}^\alpha \quad , \quad \sigma_e{}^z \chi_e^\beta = \beta \chi_e{}^\beta \qquad \text{(A 3)}$$

with $\vec{\sigma} = (\sigma^x, \sigma^y, \sigma^z)$ denoting the three Pauli spin matrices.

Since

$$\begin{aligned} \vec{\sigma}_\mu \cdot \vec{\sigma}_e \chi^{\alpha\beta} &= (\sigma_\mu{}^z \sigma_e{}^z + \tfrac{1}{2}(\sigma_\mu{}^+ \sigma_e{}^- + \sigma_\mu{}^- \sigma_e{}^+)) \chi^{\alpha\beta} \\ &= \alpha\beta \chi^{\alpha\beta} + \tfrac{1}{2}(\alpha - \beta)^2 \chi^{\beta\alpha} \end{aligned} \qquad \text{(A 4)}$$

it is easy to see that χ^{++} and χ^{--} are eigenstates of (A1) with quantum numbers ($F = 1$, $M = \pm 1$) and eigenvalues

$$E_{1,3}/h = \nu_{1,3} = \frac{1}{4}(\nu+\delta) \pm \frac{1}{2h}(g_e\mu_B + g_\mu\mu_\mu) B \ . \qquad \text{(A 5)}$$

The other two eigenvalues correspond to eigenvectors ϕ_2 and ϕ_4 which are obtained by rotating χ^{-+} and χ^{+-} according to

$$\begin{pmatrix} \phi_2 \\ \phi_4 \end{pmatrix} = \begin{pmatrix} \cos\beta & \sin\beta \\ -\sin\beta & \cos\beta \end{pmatrix} \begin{pmatrix} \chi^{-+} \\ \chi^{+-} \end{pmatrix} \qquad \text{(A 6)}$$

with

$$\operatorname{ctg} 2\beta = x \ . \qquad \text{(A 7)}$$

Here we have used the abbreviation

$$x = \Gamma B / \nu \tag{A 8}$$

with

$$\Gamma = \frac{1}{h}(g_e \mu_B - g_\mu \mu_\mu) . \tag{A 9}$$

The eigenvalues E_2 and E_4 are given by

$$E_{2,4}/h = \nu_{2,4} = -\frac{1}{4}(\nu + \delta) \pm \frac{1}{2}(\nu^2 + \Gamma^2 B^2)^{1/2} . \tag{A 10}$$

We now calculate the time dependence of the muon polarization for the transverse field case in an elementary way. If initially the μ^+ is polarized in the x direction the initial muon state $|\chi^o\rangle$ has to fulfill

$$\sigma_\mu^x \, | \chi_\mu^o\rangle = |\chi_\mu^o\rangle \tag{A 11}$$

and in terms of our basis vectors:

$$|\chi_\mu^o\rangle = \frac{1}{\sqrt{2}}(|\chi_\mu^+\rangle + |\chi_\mu^-\rangle) . \tag{A 12}$$

Since the electrons in the sample are unpolarized we have to calculate the time evolution of a mixed state with components $\chi_\mu^o \chi_e^+$ and $\chi_\mu^o \chi_e^-$ each with weight 1/2. We develop the component $|\chi_1\rangle = |\chi_\mu^o \chi_e^+\rangle$ in terms of the eigenvectors of H:

$$|\chi_1\rangle = \frac{1}{\sqrt{2}}\{|\chi^{++}\rangle + \cos\beta|\phi_2\rangle - \sin\beta|\phi_4\rangle\} . \tag{A 13}$$

Hence

$$e^{-iHt/\hbar}|\chi_1\rangle = \frac{1}{\sqrt{2}}\{e^{-i\omega_1 t}|\chi^{++}\rangle + (\cos^2\beta e^{-i\omega_2 t} + \sin^2\beta e^{-i\omega_4 t})|\chi^{-+}\rangle + \sin\beta\cos\beta(e^{-i\omega_2 t} - e^{-i\omega_4 t})|\chi^{+-}\rangle\} \tag{A 14}$$

and

$$\sigma_\mu^x e^{-iHt/\hbar}|\chi_1\rangle = \frac{1}{\sqrt{2}}\{e^{-i\omega_1 t}|\chi^{-+}\rangle + K|\chi^{++}\rangle + L|\chi^{--}\rangle\} \tag{A 15}$$

where the abbreviations K and L can be read off from (A 14). The matrix element of interest is thus given by

$$\langle\chi_1|e^{iHt/\hbar}\sigma_\mu^x e^{-iHt/\hbar}|\chi_1\rangle = \frac{1}{2}\{\cos^2\beta e^{i(\omega_2-\omega_1)t} + \sin^2\beta e^{i(\omega_4-\omega_1)t}\} + \text{c.c.} = \cos^2\beta \cos\omega_{21}t + \sin^2\beta \cos\omega_{41}t. \tag{A 16}$$

Similarly the time dependence of σ_μ^x can be calculated for the other component and the result for the muon spin polarization is

$$P_\mu^x(t) = \frac{1}{2}\{\cos^2\beta\,[\cos\omega_{21}t + \cos\omega_{43}t] + \sin^2\beta\,[\cos\omega_{41}t + \cos\omega_{23}t]\}\ . \qquad (A\ 17)$$

Although this calculation is elementary, it is more convenient to use the appropriate quantum mechanical tool for dealing with non-pure states, i.e. the density matrix. If the incoming muon has a polarization vector $\vec{p}$, then its density matrix is given by

$$\rho_\mu = \frac{1}{2}(\mathbb{1} + \vec{p}.\vec{\sigma}_\mu) \qquad (A\ 18)$$

since

$$<\vec{\sigma}_\mu> = \mathrm{Tr}\,(\rho_\mu\vec{\sigma}_\mu) = \vec{p} \qquad (A\ 19)$$

where the trace is over the two-dimensional spin space of the muon. The electrons which are captured in the muonium formation are unpolarized and are to be described by the density matrix

$$\rho_e = \frac{1}{2}\mathbb{1}\ . \qquad (A\ 20)$$

The density matrix for muonium is then simply the tensor product of ρ_μ and ρ_e:

$$\rho_{Mu} = \frac{1}{4}(\mathbb{1} + \vec{p}.\vec{\sigma}_\mu) \qquad (A\ 21)$$

and the time dependence of the muon polarization is determined by

$$\vec{p}_\mu(t) = \mathrm{Tr}\,(\rho_{Mu}\, e^{iHt/\hbar}\vec{\sigma}_\mu\, e^{-iHt/\hbar})\ . \qquad (A\ 22)$$

As an example we calculate $p_\mu^z(t)$ for the longitudinal-field case where $\rho_{Mu} = \frac{1}{4}(\mathbb{1} + \sigma_\mu^z)$. Using as basis for the evaluation of the trace in (A 22) the states $\phi_1 \equiv \chi_1$, ϕ_2, $\phi_3 \equiv \chi_3$ and ϕ_4, we get

$$p_\mu^z(t) = \frac{1}{4}\sum_{j,k=1}^{4} <\phi_j|\mathbb{1} + \sigma_\mu^z|\phi_k><\phi_k|\sigma_\mu^z|\phi_j>e^{i(\omega_k-\omega_j)t} \qquad (A\ 23)$$

Since the only non-diagonal elements different from zero are $<\phi_2|\sigma_\mu^z|\phi_4> = <\phi_4|\sigma_\mu^z|\phi_2> = \sin(2\beta)$, the oscillating part of p_μ^z can be read off from (A 23) to be

$$p_\mu^{z(t)}\big|_{osc.} = \frac{1}{4}\sin^2(2\beta) \times 2\cos\omega_{24}t\ . \qquad (A\ 24)$$

Determining the constant part from $p_\mu{}^z(t = 0) = 1$, we readily obtain

$$p_\mu{}^z(t) = \frac{1}{2}\,(1 + \cos^2(2\beta) + \sin^2(2\beta)\,\cos\omega_{24}t)\ . \qquad \text{(A 25)}$$

For an arbitrary direction of the external field, the Hamiltonian describing the anomalous muonium state has to be diagonalized numerically. Denoting the eigenvalues by $|\phi\rangle$:

$$H\,|\,\phi_m\rangle = \hbar\,\omega_m\,|\,\phi_m\rangle \qquad \text{(A 26)}$$

and the unitary matrix connecting $|\phi\rangle$ with the basis $|\chi\rangle$ by U:

$$|\phi_m\rangle = \sum_n U_{mn}\,|\chi_n\rangle \qquad \text{(A 27)}$$

one obtains for arbitrary initial polarization $\vec{p}$:

$$\vec{p}_\mu(t) = \frac{1}{4} \sum_{\substack{m,n,m',n',\\ k,\ell}} U_{km}^{+}\, U_{nn'}\, U_{\ell n}^{+}\, U_{mm'}\ \cdot$$

$$\cdot\ e^{i(\omega_n-\omega_m)t} \langle\chi_k|\,\mathbb{1} + \vec{p}.\vec{\sigma}_\mu|\chi_{n'}\rangle\ \langle\chi_\ell|\vec{\sigma}_\mu|\chi_{m'}\rangle\ . \qquad \text{(A 28)}$$

REFERENCES

1. J.H. Brewer, K.M. Crowe, F.N. Gygax, R.F. Johnson, B.D. Patterson, D.G. Fleming, and A. Schenck, Phys. Rev. Lett. 31, 143 (1973).
2. J.H. Brewer, K.M. Crowe, F.N. Gygax, and A. Schenck, in Muon Physics, eds. V.W. Hughes and C.S. Wu, (Academic Press, New York, 1975).
3. B.D. Patterson, Hyperfine Interactions 6, 155 (1979).
4. J.H. Brewer and K.M. Crowe, Ann. Rev. Nucl. Part. Sci. 28, 239 (1978).
5. D.E. Casperon, T.W. Crane, A.B. Denison, P.O. Egan, V.W. Hughes, F.G. Mariam, H. Orth, H.W. Reist, P.A. Souder, R.D. Stambaugh, P.A. Thompson, and G. zu Putlitz, Phys. Rev. Lett. 38, 956 (1977).
6. E. Holzschuh et al., to be published.
7. I.I. Gurevich, I.G. Ivanter, E.A. Meleshko, B.A. Nikolskii, V.S. Roganov, V.I. Selivanov, V.P. Smilga, B.V. Sokolov, and V.D. Shestakov, Zh. Eksp. Teor. Fiz. 60, 471 (1971) [Sov. Phys.-JETP 33, 253 (1971)].
8. C. Boekema et al., to be published.
9. B.D. Patterson, A. Hintermann, W. Kündig, P.F. Meier, F. Waldner, H. Graf, E. Recknagel, A. Weidinger, and Th. Wichert Phys. Rev. Lett. 40, 1347 (1978).

10. P.F. Meier, A. Hintermann, and B.D. Patterson, to be published.
11. A. Hintermann, P.F. Meier, and B.D. Patterson, Hyperfine Interactions 6, 163 (1979).
12. C. Boekema, W. Kündig, P.F. Meier, W. Reichart, K. Rüegg, H. Graf, A. Hintermann, and B.D. Patterson, Hyperfine Interactions 6, 167 (1979).
13. P.F. Meier and A. Schenck, Phys. Lett. 50 A, 107 (1974).
14. H. Graf, E. Holzschuh, E. Recknagel, A. Weidinger, and Th. Wichert, Hyperfine Interactions 6, 177 (1979).
15. B.D. Patterson, W. Kündig, P.F. Meier, F. Waldner, H. Graf, E. Recknagel, A. Weidinger, and T. Wichert, Helv. Phys. Acta 51, 442 (1978).
16. D.G. Andrianov, E.V. Minaichev, G.G. Myasishcheva, Yu.V. Obukhov, V.S. Roganov, G.I. Savel'ev, V.G. Firsov, and V.I. Fistul, Zh. Eksp. Teor. Fiz. 58, 1896 (1970), [Sov. Phys.-JETP 31, 1019 (1970)].
17. V.I. Kudinov, E.V. Minaichev, G.G. Myasishcheva, Yu.V. Obukhov, V.S. Roganov, G.I. Savel'ev, V.M. Samoilov, and V.G. Firsov, Zh. Eksp. Teor. Fiz. 70, 2041 (1976). [Sov. Phys.-JETP 43, 1065 (1976)].
18. B. Eisenstein, R. Prepost and A.M. Sachs, Phys. Rev. Lett. 5, 515 (1960).
19. P.F. Meier, to be published.
20. I.G. Ivanter and V.P. Smilga, Zh Eksp. Teor. Fiz. 54, 559 (1968) [Sov. Phys.-JETP 27, 301 (1968)].
21. P.W. Anderson and P.R. Weiss, Rev. Mod. Phys. 25, 269 (1953).
22. J.S. Wang and C. Kittel, Phys. Rev. B 7, 713 (1973).
23. K. Weiser, Phys. Rev. 126, 1427 (1962).
24. S.T. Pantelides, Hyperfine Interactions 6, 145 (1979).
25. R. Beck, P.F. Meier, and A. Schenck, Z. Physik B 22, 109 (1975).
26. Yu.M. Belousov, V.N. Gorelkin, and V.P. Smilga, Zh. Eksp. Teor. Fiz. 74, 629 (1978) [Sov. Phys.-JETP 47, 331 (1978)].
27. V.A. Singh, C. Weigel, J.W. Corbett, and L.M. Roth, phys. stat. sol. (b) 81, 637 (1977).
28. S.T. Picraux and F.L. Vook, Phys. Rev. B 18, 2066 (1978).
29. for a recent review, see S.T. Pantelides, Rev. Mod. Phys. 50, 797 (1978).
30. A.B. Denison, private communication.
31. G. Fehér, Phys. Rev. 114, 1219 (1959).

µSR IN FERROMAGNETIC METALS

Peter F. Meier

Physik-Institut der Universität Zürich

CH-8001 Zürich

I. INTRODUCTION

Since the first observation of muon precession in Ni and Fe[1] much µSR work has been devoted to the study of muons in ferromagnetic materials. This work has increased our understanding of the behaviour of muons in solids by a great extent. It has revealed the advantages as well as the limitations of using positive muons as probes in solids.

In this lecture I shall discuss some fundamental aspects of µSR in ferromagnetic metals. There is no claim of a comprehensive treatment. The examples were selected in order to acquaint you with the basic problems and results. Most of the µSR experiments in ferro- and antiferromagnetic substances are summarized in Ref.[2] where appropriate references can be found. For the recent work at SIN I refer to a forthcoming review article[3].

I shall restrict myself mainly to a discussion of the classic ferromagnetic transition metals Fe, Co, and Ni and to the rare-earth metals Gd and Dy. In the following list the lattice structure, the Curie temperature T_C, and the saturation magnetization M_S at T = 0 K are summarized:

		T_C(K)	M_S(G)
Fe	bcc	1043	1740
Co	hcp	1388	1446
Ni	fcc	627	510
Gd	hcp	292	2010
Dy	orthorh.	85	2920

In Fig. 1 a μSR histogram of an experiment in Co at zero applied field is shown[4]. The oscillations correspond to a precession

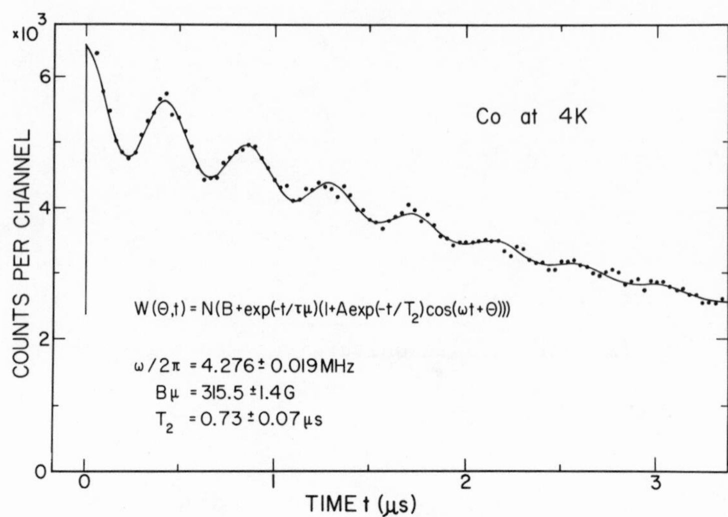

Fig. 1. Time spectrum of a μSR experiment in Co at zero applied field[4].

of the muons in a field of 316 G. The macroscopic magnetization of this Co sample was zero and one has to ask which field then is observed in that μSR experiment. Therefore, using the example of cobalt, I will give you first a short review of some aspects of ferromagnetism along with their implications on μSR.

On a microscopic scale all magnetic moments of a ferromagnet are lined up along certain crystallographic axes, the directions of easy magnetization. This is due to the effect of the anisotropy energy. In Co, the hexagonal axis is the direction of easy magnetization at room temperature and the anisotropy energy is given up to higher order terms by

$$E_{an} = K_1 \sin^2 \Theta + K_2 \sin^4 \Theta \qquad (1)$$

where Θ is the angle between the magnetization and the hexagonal axis. Depending on the sizes and signs of K_1 and K_2, the anisotropy energy is minimized for a different angle Θ. For Co at room temperature $K_1 > 0$ and $K_2 = 0.24\ K_1$ which implies that $\Theta = 0^o$.

On a macroscopic scale, however, the net magnetization of a Co specimen is zero because the sample is composed of a great number of domains within each of which the local magnetization is saturated. The directions of the magnetization of the different domains are distributed randomly in a polycrystalline sample. For a single crystal almost all domains are magnetized along the hexagonal axis, either parallel or antiparallel. Similarly, for a single crystal of iron, where the (1,0,0) axis is easy, there is an equal volume of domains oriented in either of the six (1,0,0) directions.

Polarized muons stopped in an unsaturated sample precess about different axes according to the direction of magnetization of the domain in which they come to rest. Their precession frequency, however, is the same. It is only the amplitude of the precession which is influenced by the direction of the external field. For a random distribution of domain directions, the asymmetry is reduced to 2/3 as compared to precession in a transverse field.

If an external field B_{ext} is applied along the c-axis of a single crystal Co sphere the volume occupied by domains with a magnetization direction parallel to B_{ext} increases at the expense of the unfavourably orientated domains until at a certain field, the saturation field B_s, the whole crystal is one single ferromagnetic domain. A further increase of B_{ext} then simply adds to the internal fields. If the direction of the external field is changed with respect to an easy axis of magnetization, the direction of magnetization changes at an amount which is given by the minimum of the anisotropy energy and the Zeeman and demagnetization energies.

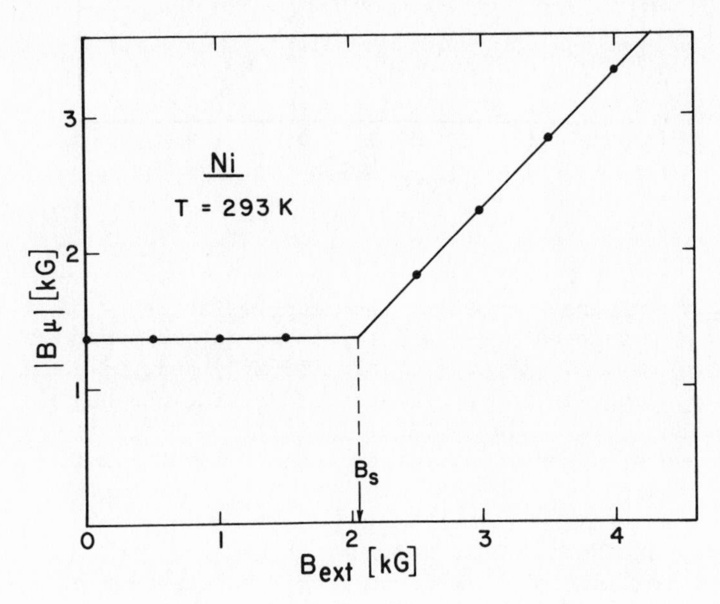

Fig. 2. Observed local field in Ni as a function of the applied field.

In Fig. 2, the observed local field B_μ at a muon in Ni is shown as a function of the applied field. For $0 \leq B_{ext} \leq B_s$, the reorientation of domains takes place without influencing the local field. Above B_s the difference $B_{ext} - B_s$ adds to the local field. This experiment shows that the local field is positive, i.e. that B_μ is parallel to the domain magnetization M_s or at least that the angle between $\vec{M}_s$ and $\vec{B}_\mu$ is less than 90°. This is in contrast to the case of Fe, see Fig. 3, where B_μ below saturation is negative. Above B_s the, per definition, positive contribution of B_{ext} reduces first the absolute value of the local field which becomes zero at about 4 kG.

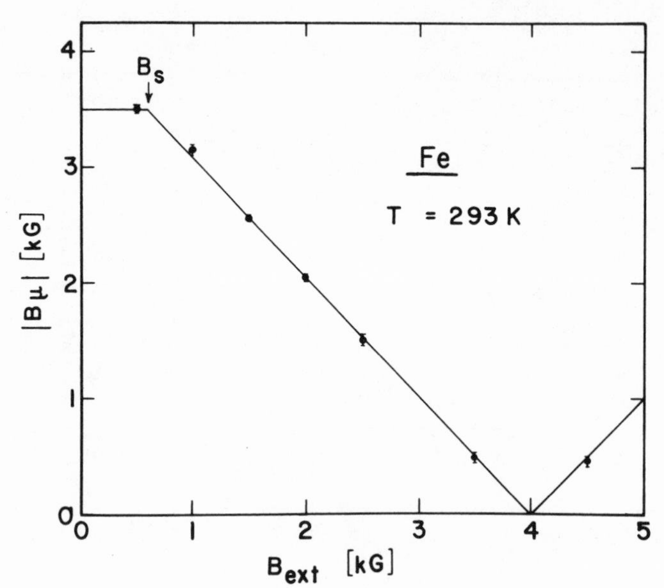

Fig. 3. Local field in Fe as a function of B_{ext}.

The temperature dependence of the saturation magnetization $M_s(T)$ is roughly reproduced by the mean-field theory of magnetism except at low temperatures, where spin-waves or Stoner excitations produce a $T^{3/2}$ or T^2 dependence of $M(T)-M(0)$ and, of course, in the critical region near the Curie temperature T_c. µSR has so far not been applied to investigate these features of $M_s(T)$ in detail, but only the overall temperature dependence of the local field $B_\mu(T)$ has been studied.

For Ni and Fe the observed µSR frequencies varied with temperatures approximately as expected from the behaviour of $M_s(T)$. This was not the case for Co and Gd, as is shown in Fig. 4, where the observed frequencies in Co are plotted as a function of temperature.

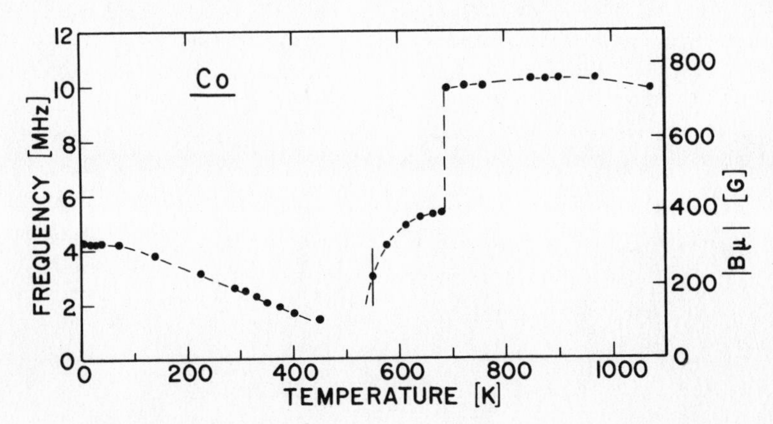

Fig. 4. Observed precession frequencies in Co as a function of temperature.

The magnetization of Co shows a smooth decrease as T is varied from 0 K up to 690 K where, in conjunction with a structural phase change into a fcc structure, the magnetization jumps by about 2%. This behaviour is shown by the curve $B_L = 4\pi/3\ M_s$ in Fig. 5.

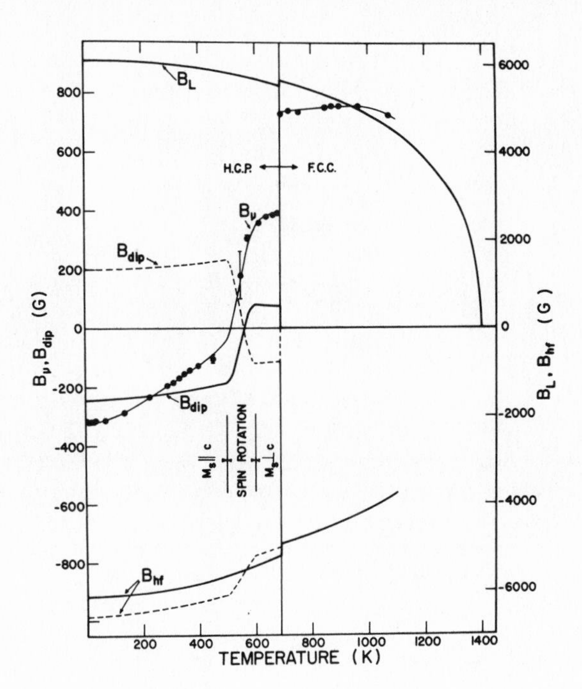

Fig. 5. Temperature dependence of B_μ, B_L, and B_{hf} in Co [4], see text.

In independent experiments the signs of B_μ have been determined to be negative at room temperature and positive at 670 K and 875 K. Therefore $B_\mu(T)$ has the temperature dependence shown by the data points (circles) indicated in Fig. 5 [4]. Thus the local field $B_\mu(T)$ in Co behaves quite differently from $M_s(T)$. To understand this we have to disentangle the various contributions to the local field at a muon.

II. CONTRIBUTIONS TO THE LOCAL FIELD

In zero applied field the local field $\vec{B}_\mu$ at the muon site $\vec{R}_\mu$ is given by

$$\vec{B}_\mu\ (\vec{R}_\mu) = \vec{B}_{hf}\ (\vec{R}_\mu) + \vec{B}_{dip}(\vec{R}_\mu). \tag{2}$$

The hyperfine field $\vec{B}_{hf}(\vec{R}_\mu)$ results from the interaction with the polarized conduction electron cloud around the muon and is given by

$$B_{hf,i}(\vec{R}_\mu) = \int d^3r\ D_{ij}(\vec{r}-\vec{R}_\mu)\ m_j(\vec{r}). \tag{3}$$

Here $\vec{m}(\vec{r})$ is the conduction electron magnetization and

$$D_{ij}(\vec{x}) = (\nabla_i\nabla_j - \frac{1}{3}\ \delta_{ij}\Delta)\frac{1}{x} - \frac{2}{3}\ \delta_{ij}\Delta\frac{1}{x}\ . \tag{4}$$

The first term in (4) transforms as a spherical harmonic of order 2. Therefore, for a spherically symmetric screening cloud, only the second term in (4) contributes to the hyperfine field. This yields the contact field

$$\vec{B}_{hf}(\vec{R}_\mu) = \frac{8\pi}{3}\ \vec{m}\ (\vec{R}_\mu)\ . \tag{5}$$

Other contributions may arise from the non-spherical parts of the screening cloud. They would give rise to a pseudodipolar field contribution to the hyperfine field since they have the same symmetry properties as the dipolar field to be discussed below but are of short range only. The contribution to the local field which arises from the conduction electron magnetization far away from the muon is calculated together with that from the localized moments.

The dipolar interaction with all magnetic moments $\vec{\mu}(\vec{n})$ localized at the lattice sites $\vec{n}$ gives rise to the dipolar field $\vec{B}_{dip}(\vec{R}_\mu)$:

$$B_{dip,i}(\vec{R}_\mu) = \sum_{\vec{n}} D_{ij}(\vec{n} - \vec{R}_\mu)\ \mu_j(\vec{n}) \tag{6}$$

where D_{ij} is given by eq. (4) except that we may neglect here the contact terms since the μ^+ is assumed to avoid the occupied lattice sites $\vec{n}$. A justification for replacing the magnetization of the ions by the dipole moment $\vec{\mu}$ localized at the lattice sites will be given later.

For a ferromagnetic alignment of the localized moments, the dipolar sum (6) over an infinite lattice is only conditionally convergent. For a finite sample, this means that the sum is shape dependent. The usual procedure is first to sum over all dipoles within a volume Ω around $\vec{R}_\mu$ and to replace the sum over the remaining volume V-Ω by an integration:

$$B_{dip,i}(\vec{R}_\mu) = \sum_{\vec{n}\in\Omega} D_{ij}(\vec{n}-\vec{R}_\mu)\mu_j(\vec{n}) + \frac{N}{V}\int_{V-\Omega} d^3r\ D_{ij}(\vec{r})\mu_j(\vec{r}). \quad (7)$$

The first term results in a contribution $\vec{B}'_{dip}(\vec{R}_\mu;\Omega)$, the residual dipolar field, which depends on the lattice structure, the muon site, and the volume Ω. If Ω is chosen to be a sphere around $\vec{R}_\mu$ one can also account for the conduction electron magnetization $\vec{m}(\vec{r})$ by replacing in the second term $\mu_j(\vec{r})$ by $\mu_j^{tot}(\vec{r}) = \mu_j(\vec{r}) + m_j(\vec{r})$, the average value of μ^{tot} being the saturation magnetization M_s. By partial integration, the second term is transformed into two surface integrals. The value from the inner surface is dependent on the volume Ω and is called the Lorentz field $\vec{B}_L(\Omega)$. For a sphere, one gets

$$\vec{B}_L = \frac{4\pi}{3}\vec{M}_s \quad (8)$$

where M_s is the magnetization of the single domain in which $\vec{R}_\mu$ is located. By symmetry, for a spherical volume Ω, the residual dipolar field sum $\vec{B}_{dip}(\vec{R}_\mu,\Omega)$ is independent of Ω if the radius of the sphere is chosen large enough. The integral from the outer surface depends on the state of magnetization of the sample. For zero net magnetization, the contributions from the differently aligned domains cancel each other. If an external field is present, we obtain the shape-dependent demagnetization field $\vec{B}_{dem}(V,\vec{B}_{ext})$. For ferromagnetic materials, the dipolar field interaction of the muon with the localized moments of the sample can thus be split into the contributions:

$$\vec{B}_{dip}(\vec{R}_\mu) = \vec{B}'_{dip}(\vec{R}_\mu) + \vec{B}_L + \vec{B}_{dem} \ . \quad (9)$$

The values of the residual dipolar field, $\vec{B}'_{dip}(\vec{R}_\mu)$, have been discussed in [5] for various lattice structures and muon sites $\vec{R}_\mu$. Here we just note that $\vec{B}'_{dip}(\vec{R}_\mu)$ can be expressed as

$$B'_{dip,i}(\vec{R}_\mu) = E'_{ij}(\vec{R}_\mu)\ \mu_j \quad (10)$$

where the dipolar tensor $E'_{ij}(\vec{R}_\mu)$ is symmetric and traceless. This implies that the residual dipolar field is zero if the magnetic moments have an arrangement which is of cubic symmetry with respect to $\vec{R}_\mu$. This applies to both octahedral and tetrahedral interstitial sites (as well as to substitutional sites) in fcc lattices.

In Fig. 6 we show the dependence of the tensor components E'_{ii} for a hcp lattice as a function of the c/a ratio. Since the numerical sums are very slowly convergent, an Ewald procedure has been used to obtain these values [5].

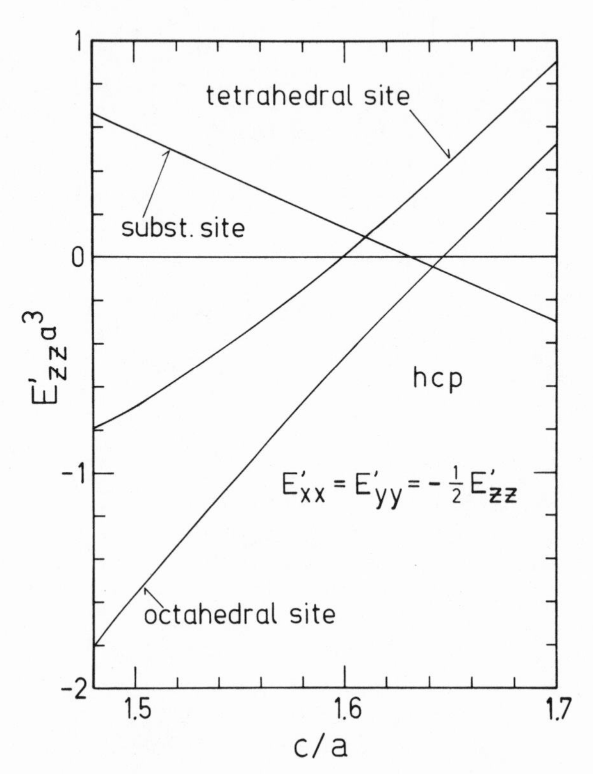

Fig. 6. Dipolar field tensor component E'_{zz} as a function of the c/a ratio for the sites in the hcp lattice.

III. SITE DETERMINATION

For a light impurity there are two natural interstitial sites in an hcp lattice, the octahedral and the tetrahedral site (Fig. 7). For both, as well as for the substitutional site, the symmetry is such that the dipolar tensor E'_{ij} has the structure

$$E' = \begin{bmatrix} E'_{xx} & 0 & 0 \\ 0 & E'_{xx} & 0 \\ 0 & 0 & -2E'_{xx} \end{bmatrix} . \tag{11}$$

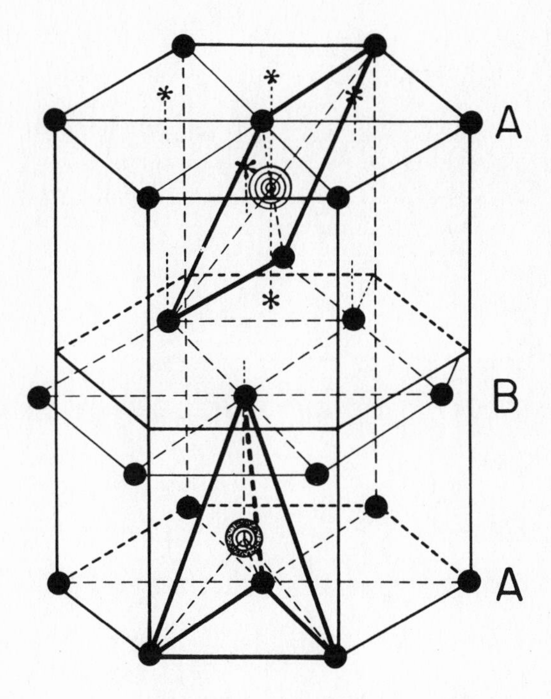

Fig. 7. Octahedral and tetrahedral interstitial sites in hcp lattices.

According to the discussion in the previous sections the local field at the muon in zero external field is given by

$$\vec{B}_\mu(\vec{R}_\mu) = \vec{B}_{hf}(\vec{R}_\mu) + \frac{4\pi}{3}\vec{M}_s + \vec{B}'_{dip}(\vec{R}_\mu) . \tag{12}$$

The site determination of the muon rests on the fact that upon a change of the direction of magnetization with respect to the crystal axes the residual dipolar field $\vec{B}'_{dip}$ transforms differently from $\vec{M}_s$ and $\vec{B}_{hf}$. Therefore, we combine the latter two fields to a vector

$$\vec{X} = \vec{B}_{hf} + \frac{4\pi}{3} \vec{M}_s . \tag{13}$$

Denoting, as in Section 2, the angle between the hexagonal axis and the direction of $\vec{M}_s$ by Θ, we write

$$\vec{X} = X\ (\sin\Theta,\ 0,\ \cos\Theta)\ . \tag{14}$$

The contribution $\vec{B}'_{dip}$, on the other hand, has according to (10) and (11) quite a different dependence on Θ:

$$\vec{B}'_{dip} = \mu E'_{xx}\ (\sin\Theta,\ 0,\ -2\cos\Theta)\ . \tag{15}$$

In general, the anisotropy energy is temperature dependent due to spin-wave contributions to the spin-orbit interactions and the dependence of the crystal field on lattice vibrations and expansion. In Co it is known that K_1 changes sign at 520 K and $K_1 < -K_2$ for $T > 600$ K. Eq. (1) then implies that

$$\begin{aligned} \Theta &= 0^\circ && \text{for} \quad T<520\ \text{K} \\ \Theta &= \arc\sin(-K_1/2K_2)^{1/2} && \text{for}\ 520\ \text{K}<T<600\ \text{K} \\ \Theta &= 90^\circ && \text{for} \quad T>600\ \text{K}\ . \end{aligned} \tag{16}$$

and from (14) and (15) we have

$$\begin{aligned} B_\mu &= X - 2\mu\, E'_{xx} && \text{for} \quad T<520\ \text{K} \\ B_\mu &= X + \mu\, E'_{xx} && \text{for} \quad T>600\ \text{K}\ . \end{aligned} \tag{17}$$

The steep rise of B_μ through the spin rotation region, as shown in Fig. 5, is then understandable if E'_{xx} is positive. Now, for the substitutional and tetrahedral interstitial sites one finds $E'_{xx} < 0$ for the appropriate c/a ratio, but $E'_{xx} > 0$ for the octahedral site. This leads to the conclusion that the μ^+ is at the octahedral interstitial site. The argument can of course be made more quantitative [3,4]. One has to consider the temperature dependence of the lattice parameters and of the magnetization in the calculation of the Co-moments $\mu(T) = 1.61\ \mu_B\ M_s(T)/M_s(0)$.

Knowing the site and B'_{dip} one obtains from the measured B_μ the vector $\vec{X}$ (13). For metals with simple ferromagnetic structures it may be assumed that $\vec{B}_{hf}$ is always parallel or antiparallel to $\vec{M}_s$. Therefore, knowing $\vec{X}$, one can easily determine $\vec{B}_{hf}$. The result is shown in Fig. 5. $B_{hf}(T)$ has, except for the jump at the structural phase transition, a smooth temperature variation for the octahedral site assignment (solid line) whereas for the tetrahedral site (dashed line) a large discontinuity occurs.

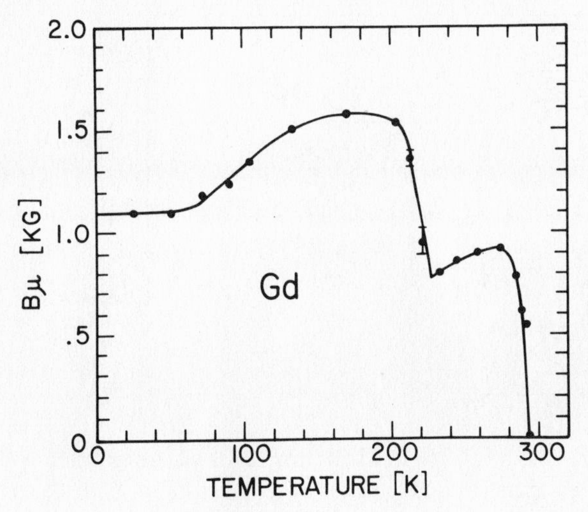

Fig. 8. Local field B_μ in Gd versus temperature [6].

The influence of the residual dipolar field on the field at a muon is also apparent in the experiments on Gd. In Fig. 8 the observed frequencies are shown as a function of temperature [6]. As in the case of Co, the anisotropy constants K_1 and K_2 of Gd are known to be strongly temperature dependent. This leads to a variation of the angle Θ, which has been determined by neutron scattering [7] and torque measurements [8,9]. The resulting values for Θ scatter slightly, around 160 K gross differences occur, but for 250 K<T<T_C=292 K, the angle $\Theta = 0^o$ is definitely known. The measured values for B_μ have been analyzed for arbitrary values of Θ, by calculating the residual dipolar contribution B'_{dip} for the tetrahedral, octahedral, and substitutional site assignment and using $\mu = 7.55\ \mu_B\ M_s(T)/M_s(0)$. Combining Eqs. (12) to (15) the following equation for X results:

$$X^2 + 2X\mu E'_{xx}\ (1 - 3\cos^2\Theta) + \mu^2 E'^{\,2}_{xx}\ (1 + 3\cos^2\Theta) = B_\mu^2. \quad (18)$$

This equation is solved for X with the angle Θ as a parameter and the results are shown in Fig. 9. Eq. (18) has two solutions, which depend on the sign of B_μ. Since the latter could not be determined with the polycrystalline sample used there are altogether six possibilities which are displayed in Fig. 9. The crosses denote points where the angle $\Theta = 0^o$ is assumed. It is seen that the assumption of a tetrahedral or substitutional site would for both signs of B_μ lead to a quite wild variation of X(T). Only for the octahedral site assignment and $B_\mu > 0$ is a reasonable X(T) obtained. The solid curve in Fig. 9 is obtained by assuming that X(T) follows the function $M_s(T)$ when the high temperature points denoted by the crosses

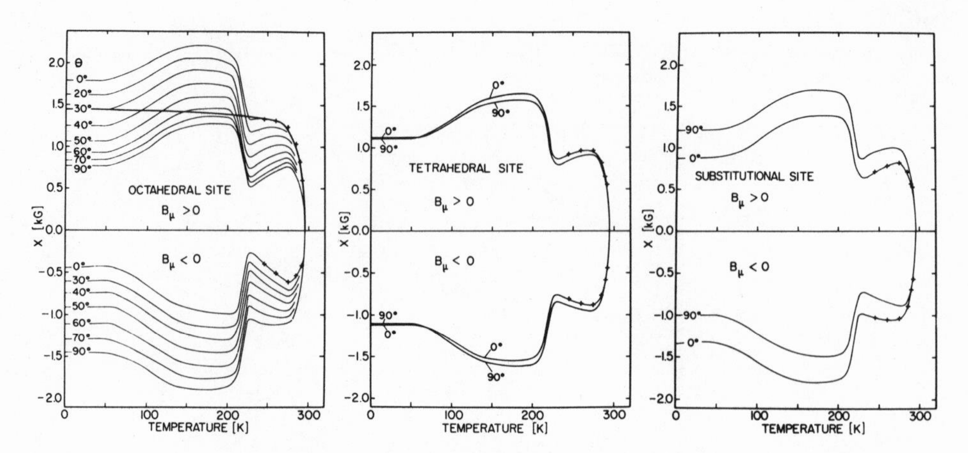

Fig. 9. Curves X versus T with angle Θ as a parameter, for the three sites in Gd [3].

are fixed. It is thus possible to extract the temperature dependence of the angle Θ from the μSR data [10]. The results compare well with the neutron scattering data and some of the torque measurements (see Fig. 10). They definitely show that the maximal angle Θ is lower than 65° in contrast to the claims of Ref. [8].

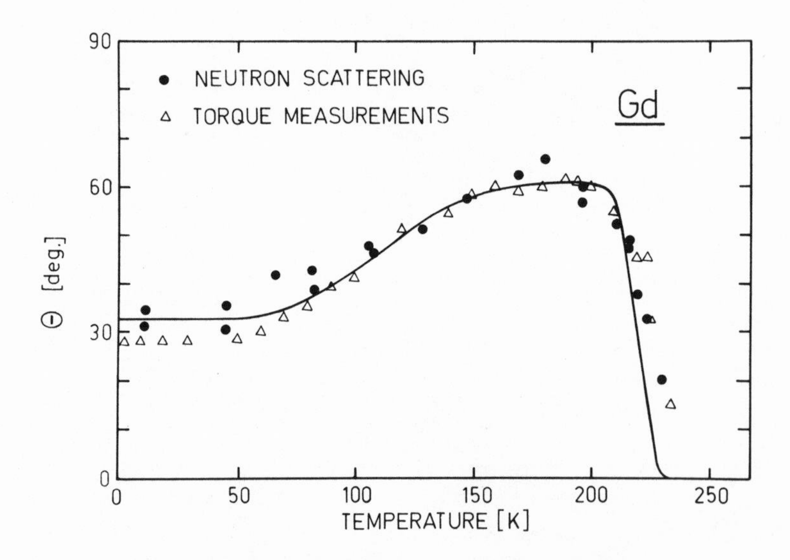

Fig. 10. Angle Θ versus T in Gd. The solid line is the angle obtained from the analysis of the μSR data [10].

The site assignment of the μ^+ in Co and Gd was made possible by the change of the angle Θ as a function of temperature. In principle it is possible to force a change of Θ by applying an external field which has to be strong enough to overcome the anisotropy field. Such a measurement which of course requires a single crystal sample would also give information about the anisotropy energy.

Up to this point I have considered the dipolar field values at points $\vec{R}_\mu$ which are geometrical points in the unit cell of the lattice and the above site assignments have to be specified in two senses. Firstly, the experiments in Co and Ni do not determine whether the muon is sitting at one and the same octahedral interstitial site or whether it is diffusing rapidly among octahedral sites. This question will be taken up again in the next section. Secondly, the muon has to be described by a wave-function ψ which may have an appreciable extension. For a light particle in particular, the formation of a local mode with large zero-point motion is expected. It was noted [5], however, that as long as the wave-function ψ is spherically symmetric, the dipolar field values at $\vec{R}_\mu$ are not changed by the finite spatial extension. The argument is again based on the symmetry properties of the dipolar interaction given by (6) and (4). The dipolar field arising from a ion at $\vec{R}$ but averaged over the finite extension of ψ is given by the convolution

$$\left\langle B'_{dip,i}(\vec{R}_\mu)\right\rangle_\psi = \int d^3r \, |\psi(\vec{r})|^2 \, B'_{dip,i}(\vec{R}_\mu - \vec{r}) \tag{19}$$
$$= (\nabla_i \nabla_j - \frac{1}{3}\delta_{ij}\Delta) \int d^3r \, |\psi(\vec{r})|^2 \frac{1}{|\vec{R} - \vec{R}_\mu + \vec{r}|} \mu_j$$

where the derivatives act on $\vec{R}$. If the extension of the wave-function is limited to the interstitial region, i.e. if $\psi \equiv 0$ for $|\vec{r}| > |\vec{R} - \vec{R}_\mu|$ one has

$$\frac{1}{|\vec{R} - \vec{R}_\mu + \vec{r}|} = \sum_{\ell=0}^{\infty} (2\ell+1) \, P_\ell \, (\cos\Theta) \frac{r^\ell}{|\vec{R} - \vec{R}_\mu|^{\ell+1}} \, . \tag{20}$$

Inserting this expression into (19) one sees that for a spherically symmetric wave-function $\psi(r)$, only the term with $\ell = 0$ is non-vanishing and one is left with

$$\left\langle B'_{dip,i}(\vec{R}_\mu)\right\rangle_\psi = B'_{dip,i}(\vec{R}_\mu) \, . \tag{21}$$

The same argument applies of course also to spherically harmonic oscillations of the host nuclei around the lattice sites $\vec{R}$. Furthermore, it shows that for calculating the residual dipolar field it is justified to replace the spherically symmetric magnetization around an ion by a dipole moment $\vec{\mu}$ at the nucleus.

A complication arises from the fact that due to the presence of an interstitial muon a lattice relaxation takes place. A radial change of the distances of the first neighbours does not alter significantlythe calculated values for E'_{ij} for the appropriate c/a ratios.Since, however, also the ions further away will relax one has to be aware of the possibility that the residual dipolar field might be slightly changed. Since the electric field gradient produced by an interstitial muon in a metal exhibits an oscillatory behaviour as a function of the distance it is to be expected that most of the changes in the residual dipolar field due to displacements of host ions are compensated. The measurements in Co and Gd done so far could well be interpreted by assuming an unperturbed host lattice.

IV. DIFFUSION ?

You have learned in other lectures at this school about the μSR studies of diffusion in paramagnetic metals and about the importance of impurities and vacancies. In ferromagnets these perturbations play an even greater role in that not only are the relaxation times affected, but also the precession frequency itself. There are magnetic structures where chemically equivalent but magnetically non-equivalent sites for the μ^+ exist. This is the case for ferromagnets with bcc lattice structure (Fe) and for all antiferromagnets.

In the bcc lattice there are again two interstitial sites, the octahedral and tetrahedral site (Fig. 11), however, the surrounding ions are arranged in a tetragonal symmetry. This means that in general there are three inequivalent octahedral sites corresponding to the three distinct tetragonal axes ([1,0,0], [0,1,0], [0,0,1]). The non-vanishing components of the dipolar field tensor E' are

$$E'_{11} = E'_{22} = -\frac{1}{2} E'_{33} \tag{22}$$

where the preferred tetragonal axis (index 3) may be any one of the cubic axes. Thus, for the easy axis of magnetization being the z-axis, the residual dipolar fields for Fe at T = 0 K are [5]:

$$\begin{aligned} B'_{dip} &= +\,18.52\ (-5.25)\ \text{kG for sites with tetragonal axis along } [0,0,1] \\ &= -\ 9.26\ (+2.63)\ \text{kG} \quad " \quad " \quad [0,1,0] \\ &= -\ 9.26\ (+2.63)\ \text{kG} \quad " \quad " \quad [1,0,0]. \end{aligned} \tag{23}$$

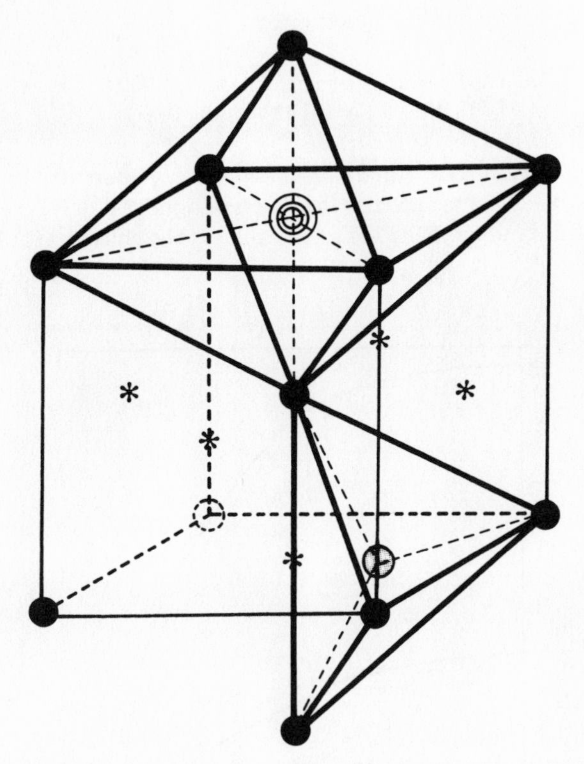

Fig. 11. Octahedral and tetrahedral sites in the bcc lattice.

The values in brackets refer to the tetrahedral interstitial sites where, by analogy to the behaviour of light interstitials in other bcc metals, the potential energy is probably lower than at octahedral sites. Since the electrostatic potential for the muon is the same at all three sites one expects, if the muon is frozen in, two different local fields. Measurements at room temperature showed a single muon frequency [1]. It was therefore concluded that the μ^+ diffuses rapidly among the interstitial sites and averages the three dipolar values of (23). It follows from a symmetry argument that the sum of the three fields B'_{dip} is zero for an arbitrary direction of the magnetization.

The hope then was to observe the localization of the μ^+ at low temperatures but the measurements showed only that the signal disappeared around a certain temperature which seems to depend strongly on the impurity content of the sample used. The lowest temperature to date for which a precession signal has been seen is 23 K [11]. The observed relaxation times, however, depend strongly on the sample quality which indicates the influence of the trapping of the muons by chemical or structural defects. Further measurements are necessary to clarify the situation.

The μSR data in Fe and the relaxation times observed in Ni indicated that the μ^+ diffuses rapidly, at least for $T \gtrsim 40$ K. It was a great surprise therefore, when the measurements in Dy [12] showed a μSR signal also in the antiferromagnetic phase above $T_C = 85$ K (Fig. 12). Below T_C dysprosium is ferromagnetic with the magnetization lying in the basal plane of the orthorhombic lattice. Between

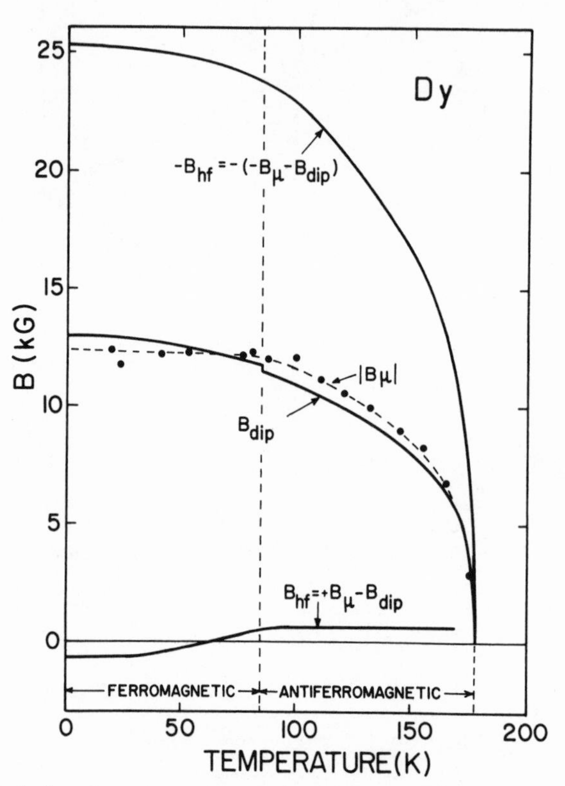

Fig. 12. μSR data for Dy. The two curves for B_{hf} correspond to $B_\mu < 0$ and $B_\mu > 0$ respectively [12].

T_C and $T_N = 178$ K, Dy is antiferromagnetic and has hcp lattice structure. The spins are confined to the basal plane and have a helical arrangement with the helix axis parallel to the hexagonal axis. The turn angle per atomic plane is temperature dependent, being 43° just below T_N. In this antiferromagnetic phase the magnetic structure is certainly very different from the lattice structure. The observation of a μSR-signal then definitely shows that most muons are frozen in. The relaxation times were rather short (150 ns) but independent of temperature. It is also remarkable that the frequencies did not show an observable change as the temperature was varied from $T < T_C$ to $T > T_C$.

Below T_C, the dipolar field is given by

$$B_{dip} \; (T < T_C) = \frac{4}{3}\pi M + B'_{dip} \; . \tag{24}$$

In the antiferromagnetic helical state the distribution into Lorentz field and residual field does not make sense since the dipolar sum (6) is convergent. $\vec{B}_{dip}(T > T_C)$ has a varying direction from one layer to the next, the absolute value depends on the turn angle and for the octahedral site one finds [13] $B_{dip}(\alpha = 26.5^o) = 0.97 \times B_{dip}(T < T_C)$. Due to the limited accuracy of the data neither the sign of B_μ nor the site could be determined in this polycrystalline sample. Further measurements would be of great help.

It should also be mentioned that μSR data [14] on α-Fe_2O_3 which is an insulating antiferromagnet demonstrate that the muon is frozen in and does not average the contributions from the two sublattice magnetizations. With increasing temperature, the relaxation rate increases and the signal disappears around 500 K. This is interpreted as the onset of thermally activated diffusion.

The observed relaxation times in Co and Gd did not exhibit a strong temperature dependence and no conclusions about the muon motion can be drawn. Since the impurity content of the samples used was not specified, it might well be that the muons are trapped by structural or chemical defects. However, the observed variation of the residual dipolar field upon a change of the easy axis, demonstrates that the magnetic environment is undisturbed because the values of B'_{dip} are very sensitive to irregularities in the magnetic structure around the muon.

V. HYPERFINE FIELDS

In the previous sections I have tried to show how it is possible to understand the origin of the measured local field B_μ. Knowing the site and $B'_{dip}(R_\mu)$ one obtains X and hence also the hyperfine field B_{hf}. We will now discuss the values obtained for $B_{hf}(T)$. Provided that the interpretation of the data is correct, i.e. octahedral site for Co and Gd, $B'_{dip} = 0$ for Ni and Fe, the following values result for the hyperfine field extrapolated to T = 0 K :

Fe	-11.1 kG
Co	-6.1 kG
Ni	-0.7 kG
Gd	-7.3 kG
Dy	-0.7 or -25.0 kG

where the two values for Dy correspond to the two possible signs for B_{μ}.

The obvious common feature of these values is their negative sign which means that the magnetization densities at the muon are in opposite direction to the averaged magnetization densities in the unit cell. Such a negative value is in complete agreement with other experiments at dilute non-magnetic impurities in ferromagnetic metals which, generally, show negative hyperfine fields for small valency Z of the impurity and positive ones for high Z. This change is explained by the Daniel-Friedel model [15] which assumes a homogeneous free electron gas with a positive spin polarization. Simulating the electrostatic potential of the impurity by a square-well potential whose depth depends on Z, the spin densities at the impurity are then calculated. Although most applications of this model have dealt with substitutional impurities it is equally well applicable to interstitial impurities. Another model for hyperfine fields was proposed by Blandin and Campbell [16] who, instead of assuming an a-priori spin polarized electron gas, considered explicitly the interaction between the localized moments of the host and the conduction electrons giving rise to an inhomogeneous spin polarization. Recently, a generalized model has been developed [17] which comprises both models of Daniel and Friedel and of Blandin and Campbell as limiting cases. It has been applied to a calculation of the hyperfine fields at muons in Gd and Dy [13] since for these rare-earth metals the basic assumption common to all models, namely the free electron gas, is not as questionable as it is for the transition metals. Before outlining this model I shall first discuss the information on the magnetization density as obtained from neutron scattering data.

Via the magnetic interaction polarized neutrons probe the magnetization density in ferromagnetic crystals locally in reciprocal space since the observed quantity, the magnetic scattering amplitude depends on the momentum transfer. The spatial distribution of the magnetization density is then obtained from a Fourier transform and it is evident that the results for the interstitial region where many Fourier components contribute are of limited accuracy. The magnetic moment distributions obtained are shown in Fig. 13 for the

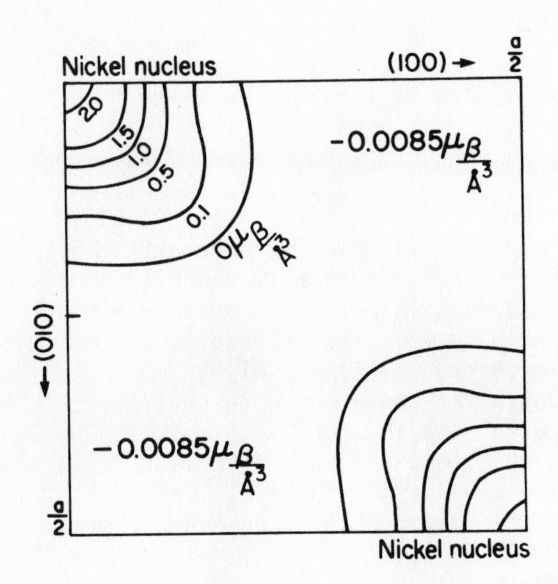

Fig. 13. Distribution of magnetization density in Ni as obtained from neutron scattering data [18].

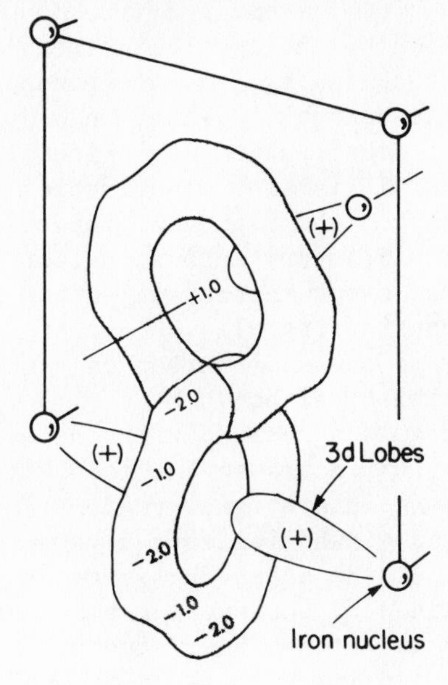

Fig. 14. Magnetization density in Fe [19].

[100] plane in Ni [18] and in Fig. 14 for Fe [19]. The small negative value in the interstitial region for Ni is obtained from a fit to the form factors and does not mean that the spin densities are constant there as is obvious from the size of the positive density near

the nuclei. This is also evident from an inspection of Fig. 14 where only the gross features (positive values at octahedral, negative at tetrahedral sites) can be regarded as guaranteed results of neutron scattering data.

The neutron scattering data from Co, Ni, and Gd show a nearly spherical distribution of positive magnetization around each atom, $m_\ell(|\vec{r} - \vec{R}_n|)$, and a small negative constant contribution m_o. Adopting this picture we can now clarify the origin of the local fields at the muon. As discussed before, the contribution to the residual dipolar field arising from the spherically symmetric magnetization m_ℓ can be replaced by that from a dipole moment μ. If the muon charge would not alter the electronic environment, the constant term m_o would give rise to a contact field $\frac{8\pi}{3}m_o$ which is sometimes referred to as the field due to the ambient spin polarization. The Lorentz field is given by $\frac{4\pi}{3}(m_o + m_\ell)$ which is of course proportional to the macroscopic saturation magnetization.

Visualizing now a magnetization density distribution as shown in figures 13 and 14 for the unperturbed lattice, we ask now what the implantation of a positive muon at, say the octahedral interstitial site may do. The screening of the muon charge leads to an appreciable distortion of the local electronic structure. Whereas, in simple metals the screening can be expected to arise from a rearrangement of s-like conduction electron states, in transition metals an additional contribution may arise from a distortion of the d-wave functions. It is also conceivable that the presence of the μ^+ then even changes the magnetic properties of the neighbouring ions. A microscopic theory accounting for these effects is very difficult. Some first steps in that direction have been made [20,21] without, however, considering all the aspects. It has to be mentioned that already in the undisturbed case a microscopic theory meets great difficulties. Although band structure calculations give results wich are in good agreement with the measurements for the magnetization distribution around the ions, they are very uncertain in predictions for the interstitial region. As an example we refer to Harmon and Freeman [22] who indicate a value of $-0.002\ \mu_B/\text{Å}^3$ for the spin density $\Delta\rho$ at the octahedral site in Gd, where the small value of $\Delta\rho$ is obtained by subtracting two large numbers. This value is about 20 times smaller than the value obtained from neutron scattering data. Accounting for the additional presence of an impurity makes a first-principle calculation a formidable task. The best thing that theory can do at the moment is to develop simplified models and to study the influence of different physical mechanisms on spin densities. In particular, such studies should be directed to answering questions as (i) What is the sign and order of magnitude of B_{hf} at the muon? (ii) Is $B_{hf}(T)$ proportional to $M(T)$?

(iii) What is the change of B_{hf} at the phase transition in Dy?

Previous models [23] concentrated on the spin density enhancement at a μ^+ in a homogeneously polarized electron gas with a polarization determined from the background density m_o measured with neutrons. It has been pointed out [13] that such a treatment is inadequate since the background m_o is not to be identified with the homogeneous conduction electron polarization ρ^o. This is evident e.g. for Gd where m_o is negative but ρ^o positive (0.55 μ_B/atom). Rather, one should consider first the exchange interaction between the localized moments and the conduction electrons. This leads already in the unperturbed case to a spatially inhomogeneous spin density $\rho^o(\vec{r})$ of the electron gas which may be negative in some regions due to oscillations as is well known from the RKKY interaction. The presence of the muon then leads to a disturbed density $\rho(\vec{r})$ and a comparison between neutron scattering results and μSR hyperfine fields must then be made through $\rho^o(\vec{R}_\mu)$ and $\rho(\vec{R}_\mu)$. Such a treatment has been put forward in [17] using rather crude wavefunctions, namely plane waves. In the framework of scattering theory, the screened potential of the muon and the exchange potentials at the ions are considered, the latter being treated in distorted-wave Born approximation. The calculated values for B_{hf} at the octahedral sites are -2.7 kG for Gd and -1.5 kG for Dy. More interesting than these numbers, which give the right sign and the correct order of magnitude, is the fact that the model predicts a nonlinear dependence of B_{hf} on the host magnetization M which implies that the function $B_{hf}(T)$ is not proportional to M(T), see Fig.15 [17]. Deviations of this kind have indeed been observed in the μSR measurements but they are also found rather frequently with other dilute non-magnetic impurities (see e.g. [24]). In addition, the model predicts a small change in B_{hf} at the ferromagnetic to helical phase transition in Dy. It is not possible here to give further details but let me just remark that the μSR results have renewed the interests of theorists in calculations of the induced hyperfine fields. Although evidently difficult the problems are at least easier than in the case of other impurities possessing ionic cores where a slight polarization of bound-state wave functions has a great influence on the hyperfine fields.

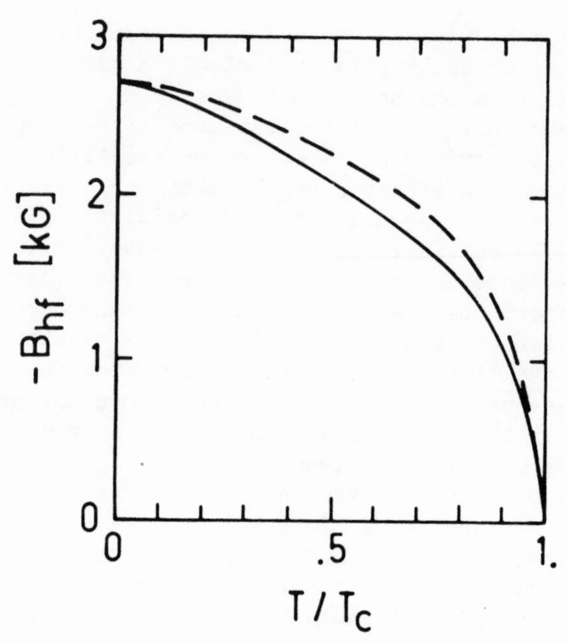

Fig. 15. Calculated hyperfine field for Gd. The dashed line is the host magnetization normalized to $B_{hf}(T = 0K) = -2.75$ kG [17].

VI. OUTLOOK

My discussion of μSR in ferromagnetic metals was restricted to the observed frequencies and the site determination. The analysis of the other μSR data, asymmetry and relaxation times, was only marginally mentioned although a great deal of insight can be gained by a careful analysis of this additional information. You will, however, hear at this school quite a lot about μSR relaxation times in connection with diffusion and trapping in other materials where the impurity content of the samples is much more under control than in the case of the ferromagnetic samples measured so far. I think that future μSR research in ferromagnets will be made with the use of pure single crystal samples. Particularly important would be an answer to the questions raised by the μSR results in Fe.

The relevance of the results obtained so far has to be considered in the frame of the development of μSR as a tool in solid state physics. A lot has been learned from these experiments which is not so much of importance for the understanding of magnetism but of extreme value for the understanding of positive muons as probes.

In particular the identification of the stopping site and the non-diffusion in Dy as well as the still unsolved Fe problem bear on the μSR research in other solids.

The phenomenon of ferromagnetism in transition and rare-earth metals is far from being understood and the research will continue for quite some time. Positive muons can be considered to be probes similar to other impurities with the advantage of a relatively simple electronic structure but with the disadvantage that the results may be masked by non-identification of the site or by diffusion properties. For the investigation of the conduction electron polarization, the μ^+ results fit well into those from other experiments [25] and μSR can contribute valuable pieces of information. As has been exemplified with Gd, future μSR work can also be used to determine the directions of easy magnetizations.

The knowledge that we have gathered from μSR in these classic ferromagnetic metals may be used to attack problems in ferromagnetism which are at the frontier of current research. Experiments are starting now, e.g. on rare-earth orthoferrites which have interesting magnetic properties and are studied intensively with various methods.

In this respect one has to mention a μSR experiment on spin glasses [26] the results of which have attracted great interest among people working in that field. By alloying e.g. a small amount of Mn into Cu one observes at low temperatures a freezing of the Mn spins which are distributed randomly. There is a transition into a magnetic phase where there exists a local order of spins in a cluster but no long-range order as in purely ferro- or antiferromagnetic cases. When the temperature is lowered below the spin-glass freezing temperature, the relaxation time of muons stopped in <u>Cu</u>Mn and in <u>Au</u>Fe shows a sudden decrease which originates from the dipolar fields and the polarization of conduction electron spin densities from the Mn or Fe local moments. Combining this information from μSR, i.e. the size and temperature dependence of the mean random field, with other experiments helps greatly in understanding the mechanisms leading to this amorphous magnetic structure.

To conclude I believe that with the know-how obtained in μSR so far the way is paved for μSR contributions to the understanding of current problems in magnetism.

I would like to thank A. Denison and W. Kündig for many helpful discussions and critical reading of the manuscript.

REFERENCES

1. M.L.G. Foy, N. Heimann, W.J. Kossler, and C.E. Stronach, Phys. Rev. Lett. 30, 1064 (1973).
2. A.T. Fiory, Hyperfine Interactions 6, 63 (1979); see also J.H. Brewer and K.M. Crowe, Ann. Rev. Nucl. Part. Sci. 28, 239 (1978).
3. A. Denison et al., to be published.
4. H. Graf, W. Kündig, B.D. Patterson, W. Reichart, P. Roggwiller, M. Camani, F.N. Gygax, W. Rüegg, A. Schenck, H. Schilling, P.F. Meier, Phys. Rev. Lett. 37, 1644 (1976).
5. P.F. Meier, W. Kündig, B.D. Patterson, and K. Rüegg, Hyperfine Interactions 5, 311 (1978).
6. H. Graf, W. Hofmann, W. Kündig, P.F. Meier, B.D. Patterson, and W. Reichart, Solid State Commun. 23, 653 (1977).
7. J.W. Cable and E. Wollan, Phys. Rev. 165, 733 (1968).
8. C.D. Graham, J. Phys. Soc. Japan 17, 1310 (1962).
9. W.D. Corner and B.K. Tanner, J. Phys. C 9, 627 (1976).
10. H. Graf, W. Kündig, P.F. Meier, and B.D. Patterson, J. Appl. Phys. 49, 1549 (1978).
11. N. Nishida, R.S. Hayano, K. Nagamine, T. Yamazaki, J.H. Brewer, D.M. Garner, D.G. Fleming, T. Takenchi, and Y. Ishikawa, Solid State Commun. 22, 235 (1977).
12. W. Hofmann, W. Kündig, P.F. Meier, B.D. Patterson, K. Rüegg, O. Echt, H. Graf, E. Recknagel, A. Weidinger, and T. Wichert, Phys. Lett. 65 A, 343 (1978).
13. P.F. Meier, W. Kündig, and K. Rüegg, Hyperfine Interactions 6, 77 (1979).
14. H. Graf, W. Hofmann, W. Kündig, P.F. Meier, B.D. Patterson, and A. Rodriguez, Solid State Commun. 25, 1079 (1978).
15. E. Daniel and J. Friedel, J. Phys. Chem. Solids 24, 1601 (1963).
16. A. Blandin and I.A. Campbell, Phys. Rev. Lett. 31, 51 (1973).
17. P.F. Meier, Solid State Commun. 27, 1163 (1978).
18. H.A. Mook and C.G. Shull, J. Appl. Phys. 37, 1034 (1966).
19. C.G. Shull and H.A. Mook, Phys. Rev. Lett. 16, 184 (1966).
20. K.G. Petzinger and R. Munjal, Phys. Rev. B 15, 1560 (1977).
21. B.D. Patterson and J. Keller, Hyperfine Interactions 6, 73 (1979).
22. B.N. Harmon and A.J. Freeman, Phys. Rev. B 10, 1979 (1974).
23. P. Jena, Hyperfine Interactions 6, 5 (1979) and Refs. quoted therein.
24. P. Raghavan, M. Senba, and R.S. Raghavan, Phys. Rev. Lett. 39, 1547 (1977).
25. M.B. Stearns, Phys. Lett. 47 A, 397 (1974).
26. D.E. Murnick, A.T. Fiory, and W.J. Kossler, Phys. Rev. Lett. 36, 100, (1976).

FREE RADICALS IN MUONIUM CHEMISTRY

Emil Roduner

Physikalisch-Chemisches Institut der Universität
Winterthurerstr. 190
CH-8057 Zürich

INTRODUCTION

The μSR technique using polarized positive muons in a magnetic field transverse to the muon spin direction has become an important method for studying kinetic and structural isotope effects as well as certain aspects of radiation chemistry. The standard experimental set-up is used to detect the three different classes of compounds which are distinguishable on the basis of their characteristic muon precession frequencies:

1. Muons substituted in diamagnetic molecules (e.g. MuOH) precess at their nuclear Larmor frequency, ν_μ=13.55kHz/G. Chemical shifts are usually not resolved.

2. Muonium ($Mu \equiv \mu^+ e^-$) is a two-spin-$\frac{1}{2}$ system. The variation of the four energy levels as a function of magnetic field strength is given by the Breit-Rabi diagram (fig.1). The dimensionless field parameter x is given by

$$x = \frac{\nu_e + \nu_\mu}{A} , \qquad (1)$$

where ν_e and ν_μ are the electron and muon Larmor frequencies and A is the isotropic hyperfine coupling constant (=4463 MHz for Mu). In general Mu is not prepared in one of the eigenstates. Therefore, the system will oscillate between such states. In μSR experiments we observe the evolution of muon spin polarization during these oscillations. For fields transverse to the muon polarization four transitions are allowed, giving rise to four charac-

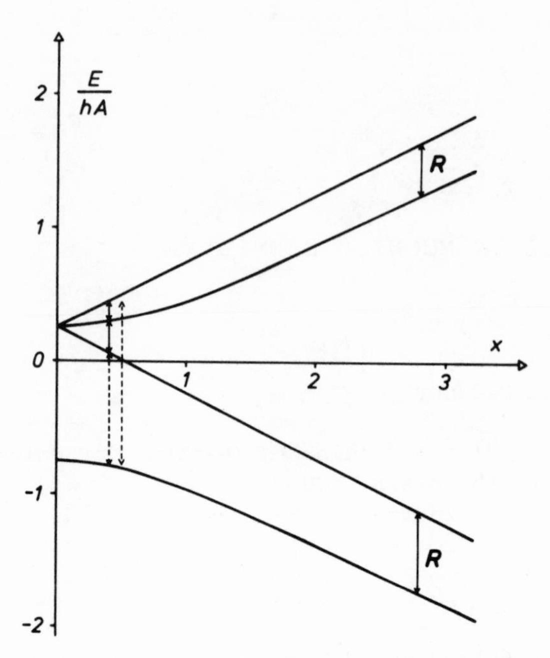

Fig. 1. Breit-Rabi diagram showing the energy levels of a two-spin-½ system as a function of magnetic field. Of the four transition frequencies allowed in low transverse fields, only the two denoted by full lines are resolvable for Mu. The transitions marked R are those detected for muonic radicals in high fields.

teristic μSR precession frequencies. Two of these, depicted by dotted lines in fig.1, are too high to be resolved. The remaining two, denoted by full lines, become degenerate for fields $\lesssim$ 20 G. This is the so-called "muonium frequency" (1.394 MHz/G) which is commonly used to characterize Mu.

3. Muonium substituted radicals are usually multi-spin systems. In high magnetic fields, however, the spins are decoupled. Here we can forget about the additional magnetic nuclei and think of the system as Mu with a reduced hyperfine interaction. The observable transitions are designated R in fig.1.

As a consequence of its production the muon enters the chemical target with high energy (MeV). Because of this radiolysis damage is to be expected along the muon track in liquids. μSR serves to investigate interaction of the muon/muonium with primary products from radiolysis. The relevant information is contained mainly in the initial amplitudes of the precession signals.

From a chemical point of view Mu is a light isotope of hydrogen since the Bohr radius and the ionization potential are within 0.5% the same as those of H. Mu is therefore expected to undergo reactions analogous to those well known from the chemistry of H. Any deviation arises primarily from the difference in mass which can influence rates of reactions. Tests of reaction rate theories rely to a great extent on the comparison of predicted and experimental kinetic isotope effects. The inclusion of Mu in the series of hydrogen isotopes (mass ratios 3:2:1:0.11) vastly increases the scope of such studies. Furthermore the lightness of Mu provides a means to investigate the contribution of tunnelling in chemical reactions.

Only recently has the nature of, and thence the means to observe muonic radical spectra become fully understood[1]. Since then radicals have been detected in liquid samples of a number of organic compounds. They are produced by addition of Mu to unsaturated molecules:

$$Mu + R_2C = CH_2 \rightarrow R_2\dot{C}\text{-}CH_2Mu. \qquad (2)$$

Comparison of their hyperfine coupling constants with those of their hydrogen analogs reveals large isotope effects.

At the conference account was given to all three major areas of muonium chemistry, i.e. kinetics, radiation chemistry and muonic radicals. Here we shall restrict ourselves to the last subject, since the first two have been covered extensively in recent reviews[2-4]. Theory for an understanding of muonic radicals is given here. First experimental results are reported and discussed.

THEORY

Hamiltonian and Eigenfunctions

Over 15 years ago it was postulated by A.M. Brodskii[5] that Mu may add to unsaturated molecules such as ethylene to form muonic radicals. Since then much effort in different laboratories has been devoted to low ($\lesssim$ 10 G) transverse field μSR experiments in the search for these species, with no success, however. The reason for this non-observation lies in the mistaken neglect of a term in the Hamiltonian describing the system. A muonic radical was viewed as a two-spin system analogous to Mu, but with a reduced hyperfine coupling constant[6]. Consequently the muonium precession frequency was sought in low fields, but with larger splitting than Mu at higher fields. However, additional magnetic nuclei, e.g. protons, are coupled to the unpaired electron in most organic radicals.

In accordance with convention we define the z-direction of an orthogonal coordinate system to be parallel to the external field B and the x-direction parallel to the muon polarization in the beam. The Hamiltonian for radicals in liquids is in frequency units:

$$\hat{H} = \nu_e \hat{S}_z - \nu_\mu \hat{I}_z^\mu - \sum_k^K \nu_k I_z^k + A_\mu \hat{S}\cdot\hat{I}^\mu + \sum_k^K A_k \hat{S}\cdot\hat{I}^k \quad (3)$$

where ν_e, ν_μ and ν_k are the Larmor precession frequencies for the electron, the muon and the k'th nucleus (ν_e = 2802.47 kHz/G for $g=g_e$, ν_μ = 13.55 kHz/G and ν_k = 4.258 kHz/G for protons). A_μ and A_k, the isotropic hyperfine coupling constants (hfc), are direct measures of the electron spin densities at the corresponding nuclei. Anisotropic contributions to the hyperfine interaction arising from direct dipole dipole interaction are averaged to zero through rapid tumbling of the molecules[7]. Similarly we employ the isotropic g-value for the unpaired electron which, however, may deviate slightly from g_e, the value for the free electron. Coupling between nuclei is small ($|J| \lesssim$ 20 Hz for most protons) and is neglected.

The orthonormal eigenkets to $\hat{H}$

$$|n\rangle = \sum_i c_{ni} |\chi_i\rangle \quad (4)$$

are conveniently expanded in the basis of product spin functions

$$|\chi_i\rangle = |\chi_i^\mu\rangle |\chi_i^e\rangle \prod_k |\chi_i^k\rangle \quad (5)$$

which are eigenkets to the Zeeman part of $\hat{H}$.

Selection Rules for μSR Transitions

The total number of eigenstates to $\hat{H}$ is $4\prod_k (2I^k+1)$. According to selection rules transitions are allowed between many but not all of them. If no polarization is lost during formation of the radical the amplitude of a transition $n \to m$ is[8]

$$P_{nm} = P_o \, |\langle m|\hat{\sigma}_q|n\rangle|^2, \quad (6)$$

where P_o is a normalization factor and $\hat{\sigma}_q$ is the suitable Pauli spin operator. Expressed in the basis of product states we have

$$\hat{\sigma}_q = \hat{\sigma}_z^{\mu} \otimes \underline{\underline{1}}^e \otimes \prod_k \underline{\underline{1}}^k \tag{7}$$

for q=o, i.e. for longitudinal fields and

$$\hat{\sigma}_q = \frac{1}{2} (\hat{\sigma}_+^{\mu} + \hat{\sigma}_-^{\mu}) \otimes \underline{\underline{1}}^e \otimes \prod_k \underline{\underline{1}}^k \tag{8}$$

for q = ±1, i.e. for transverse fields. Eq.6 illustrates the close analogy of μSR and magnetic resonance.

In zero field the Zeeman terms in the Hamiltonian vanish. The remainder of $\hat{H}$ commutes with $\hat{F}^2$ and with $\hat{F}_z$ $(\hat{F} = \hat{S} + \hat{I}^{\mu} + \sum \hat{I}^k)$. The eigenfunctions in low fields ($\nu_e << A_{\mu}, A_k$) are therefore characterized to a good approximation by the quantum numbers F_{τ}, M_F, where τ distinguishes between states of different energies but equal F, M_F. To deduce the selection rules we use the Wigner-Eckart theorem, which states [ref.9, p.489]:

$$\langle F'_{\tau}, M'_F | \hat{\sigma}_q | F_{\tau}, M_F \rangle = \frac{1}{\sqrt{2F'+1}} \langle F'_{\tau} | \hat{\sigma}_q | F_{\tau} \rangle \cdot \langle F, 1, M_F, q | F', M'_F \rangle. \tag{9}$$

For a non-vanishing transition neither the reduced matrix element $\langle F'_{\tau} | \hat{\sigma}_q | F_{\tau} \rangle$ (which is a continuous function) nor the Clebsch-Gordan coefficient $\langle F, 1, M_F, q | F', M'_F \rangle$ (which is a quantized factor) may vanish. The latter condition requires [ref. 9, p.478] that $M'_F = M_F + q$ and simultaneously $|F-1| \leq F' \leq F+1$. Thus the selection rules sought are

$$\Delta M_F = q, \qquad \Delta F = 0, \pm 1 \tag{10}$$

for μSR transitions in low magnetic fields.

In high fields ($\nu_e >> A_{\mu}, A_k$) the eigenfunctions to $\hat{H}$ become pure product functions (5) because of decoupling of the spins. Therefore they are characterized by the z-components of the individual angular momenta, m^s, m^{μ}, m^k. The transition operator $\hat{\sigma}_q$ acts upon the muon wavefunction only, and the simple high field selection rules

$$\Delta m^{\mu} = q, \quad \Delta m^s = \Delta m^k = 0 \tag{11}$$

are readily obtained from (6).

In intermediate fields neither F nor the individual m^i are good quantum numbers. $\hat{H}$ mixes product functions of equal $M = \Sigma m^i$. Thus M characterizes a wavefunction partly. Because of

$$|\hat{\sigma}_q|M>| = |M+q> \tag{12}$$

matrix elements of the form $<M'|\hat{\sigma}_q|M>$ are in general non-zero for

$$\Delta M = q \ . \tag{13}$$

It is only in high fields that the allowed transitions necessarily have considerable intensity, since then they occur between pure product functions. In intermediate and zero fields the intensities depend upon several terms which are products of the expansion coefficients corresponding to the relevant product functions. Accidental cancellation of terms may occur.

Numerical Example

A computer program was written to calculate eigenvalues and eigenvectors to $\hat{H}$ and to evaluate intensities based on equation (6). It is instructive to discuss the results on the transverse field example of a simple muon-electron-proton system with typical coupling constants. In the Breit-Rabi type diagram (fig.2, compare fig.1 for Mu) above $\approx$100 G all energies are essentially linear in B, i.e. the states are practically pure product states. It is clear that the muonium frequency, $\nu_- = \frac{1}{2}(\nu_e - \nu_\mu) = 1.394$ MHz/G, cannot occur. However, analogous transitions exist with $\Delta M_F = \pm 1$ within $F = \frac{3}{2}$ (see also fig.3). They are degenerate at $\nu = \frac{1}{3}(\nu_e - \nu_\mu - \nu_p) = 0.93$ MHz/G below $\approx$5 G. The allowed transitions are displayed in fig.3 and their calculated Fourier amplitudes are shown as stick plots in fig. 4. It is seen that the whole frequency spectrum collapses to two lines in high fields. (This is independent of the number of nuclei coupling with the unpaired electron.) At $\approx$20 G the spectrum is most complex. The total muon polarization is shared among 15 transitions. These degenerate to only three non-zero frequency transitions in zero field. Note that in the absence of an external field there is no directive force which determines the direction of the muon precession. For this reason the observable muon polarization oscillates back and forth in the x-direction, whereas no polarization should be observed in a telescope perpendicular to the initial muon polarization, i.e. in the y - z - plane.

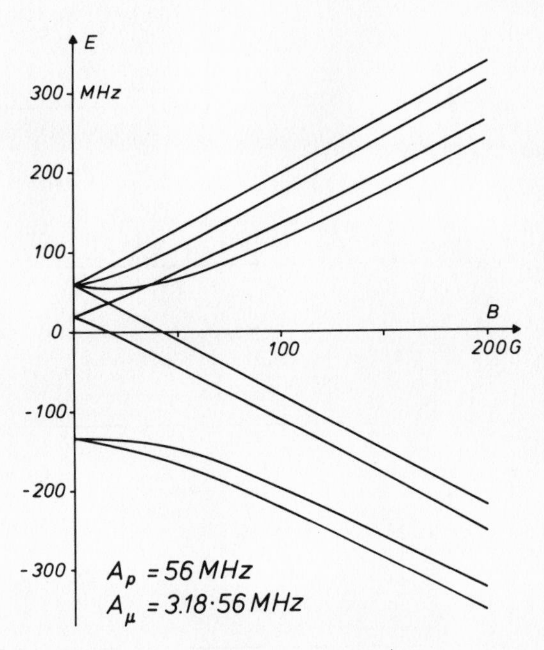

Fig. 2. Energy levels as a function of magnetic field for a muon-electron-proton system.

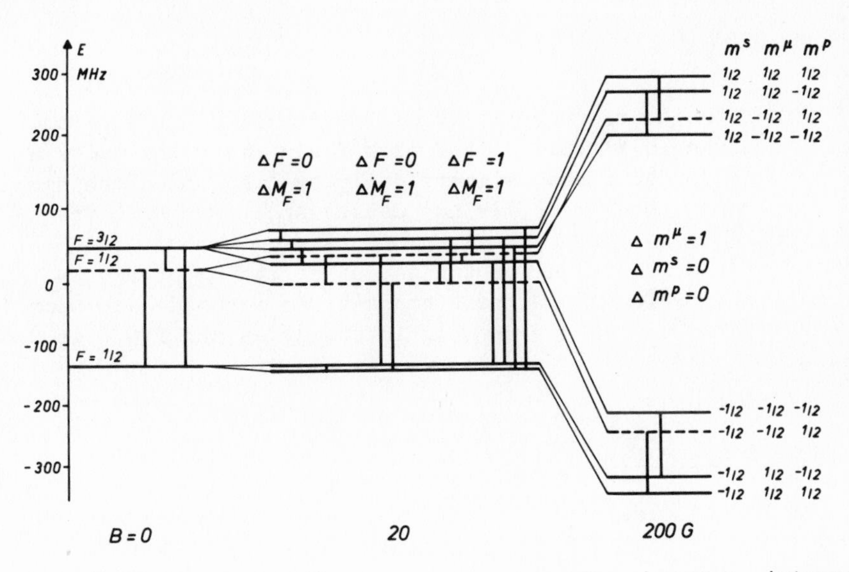

Fig. 3. μSR transitions for a μ - e - p system in zero, intermediate and high transverse magnetic fields.

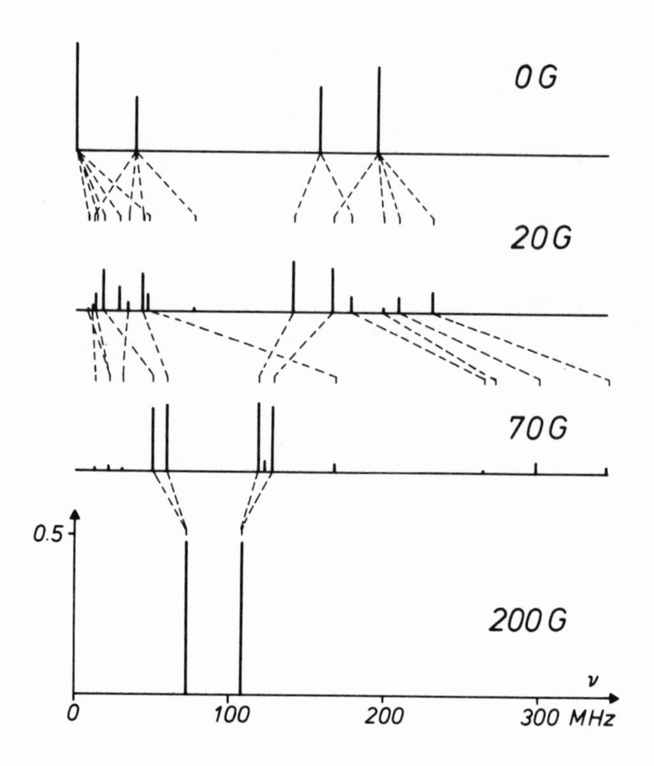

Fig. 4. Calculated μSR spectra for a μ - e - p system in different transverse magnetic fields (A_p= 56 MHz, A_μ = 3.18·56 MHz).

A continuous representation of the frequencies as a function of field is given in fig.5. In the high field limit the two radical frequencies are displaced symmetrically by $\frac{1}{2}$ A_μ about the signal of the bare muon. At B = 36.72·$|A_\mu|$ Gauss (A_μ in MHz) one frequency goes through zero due to crossing of states. Around 150 G the influence of the additional proton causes splitting of the two radical frequencies. Below 100 G those transitions forbidden in high field appear (broken lines) and grow in intensity as the field is decreased further.

Perturbation Treatment

In high fields the eigenvalues to $\hat{H}$ approach the values

$$E_n = \nu_e m^s - \nu_\mu m^\mu - \sum_k \nu_k m^k + A_\mu m^s m^\mu + \sum_k A_k m^s m^k. \tag{14}$$

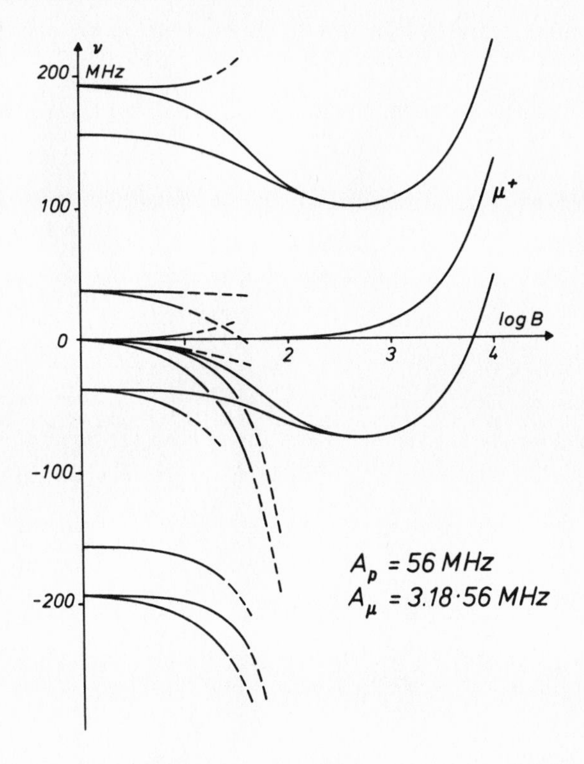

Fig. 5. Dependence of the μSR frequencies on a transverse magnetic field for the free μ^+ and a μ - e - p system. Broken lines indicate low intensities.

Using the selection rules (11) we immediately obtain the two transition frequencies

$$\nu_{nm} = \nu_\mu \pm \frac{1}{2} A_\mu \ , \tag{15}$$

which are the same as those encountered in ENDOR experiments of single spin $\frac{1}{2}$ nuclei[7]. Considerable deviations from (15) are to be seen in both experimental and exact numerical results for those fields, where most experiments have been performed. A more rigorous approach must be sought.

Eq.(15) is independent of nuclear terms in $\hat{H}$. To apply perturbation theory to the calculation of muon precession frequencies we may therefore start with the muon-electron Hamiltonian

$$\hat{H}^o = \nu_e \hat{S}_z - \nu_\mu \hat{I}^\mu_z + A_\mu \hat{S} \cdot \hat{I}^\mu \tag{16}$$

Table 1: Zero order transition frequencies and polarizations[a)]

$\nu_{12} = \nu_- - \Omega$	$P_{12} \sim c^2$
$\nu_{23} = \nu_- + \Omega$	$P_{23} \sim s^2$
$\nu_{14} = A_\mu + \nu_- + \Omega$	$P_{14} \sim s^2$
$\nu_{43} = -A_\mu + \nu_- - \Omega$	$P_{43} \sim c^2$

a) $\Omega = \frac{1}{2}\left\{ [(\nu_e + \nu_\mu)^2 + A_\mu{}^2]^{-\frac{1}{2}} - A_\mu \right\}$

$c^2 = \frac{1}{2}\left\{ 1 + (\nu_e + \nu_\mu)^2/[(\nu_e + \nu_\mu)^2 + A_\mu{}^2]^{\frac{1}{2}} \right\}, \quad s^2 = 1-c^2$

and introduce a set of N equivalent nuclei as the perturbation

$$\hat{H}' = -\sum_k^N \nu_k \hat{I}_z^k + \sum_k^N A_k \hat{S}\cdot\hat{I}^k = -\nu_k \hat{K}_z + A_k \hat{S}\cdot\hat{K}. \qquad (17)$$

This is a good approximation at least for $A_\mu > A_k$. In (17) we use K as a quantum number for the total nuclear angular momentum $\hat{K} = \sum_k \hat{I}^k$, and M_K for its z-component $\hat{K}_z$. K can take the values K_{Max}, $K_{Max} - 1$, $K_{Max} - 2$ .... 0 (or $\frac{1}{2}$) with different statistical probabilities. The eigenkets to $\hat{H}^o$ are products of muonium analogous eigenfunctions (see e.g. ref.6) and a nuclear wavefunction $|K, M_K\rangle$. Zero order transition frequencies ν_{nm}^o and their relative polarization P_{nm} are given in table 1. The coefficient s decreases rapidly with increasing field, so that in moderate and high fields only ν_{12} and ν_{43} are of significant intensity.

The first order corrections Δ' to ν_{12} and ν_{43} are

$$\Delta' = A_k M_k s^2, \qquad (18)$$

i.e. the two frequencies are split into symmetric multiplet patterns characteristic of magnetic resonance. The splitting increases with decreasing field because of its proportionality to s^2, and for a given field it grows with increasing A_k. These first order corrections give a good estimate of line broadening and splitting due to perturbing nuclei. For quantitative work second order contributions, given by more complicated expressions, cannot be neglected. The important result is that the magnitude of A_μ is obtained from the two radical frequencies (or centre of multiplets), accurate to the first order:

$$|A_\mu| = |\nu_{12}| \pm |\nu_{43}|. \tag{19}$$

Whether the "+" or the "-" applies is decided from the values of the individual frequencies (table 1; remember that one of them changes sign as seen in fig.5).

EXPERIMENTAL RESULTS

All experiments were carried out in the μE2 area of the Swiss Institute for Nuclear Research (SIN). The standard μSR experimental set-up with transverse magnetic fields $0.6 \leq B \leq 5$ kG was used. Substances of the highest commercially available grade (Fluka or Merck) were degassed via freeze - pump - thaw cycles, and then sealed in spherical glass bulbs of 25 or 40 mm diameter. For temperatures below ambient the samples were placed in a styrofoam block and cooled by passage of temperature regulated nitrogen gas. 4 - 20 million events were accumulated in two histograms. These were analyzed by Fourier transformation after correction for a constant background and the exponential decay.

Radicals are listed and compared with their hydrogen analogs in table 2. The structures proposed are based on the principle that the thermodynamically more stable radicals are formed, in analogy to the

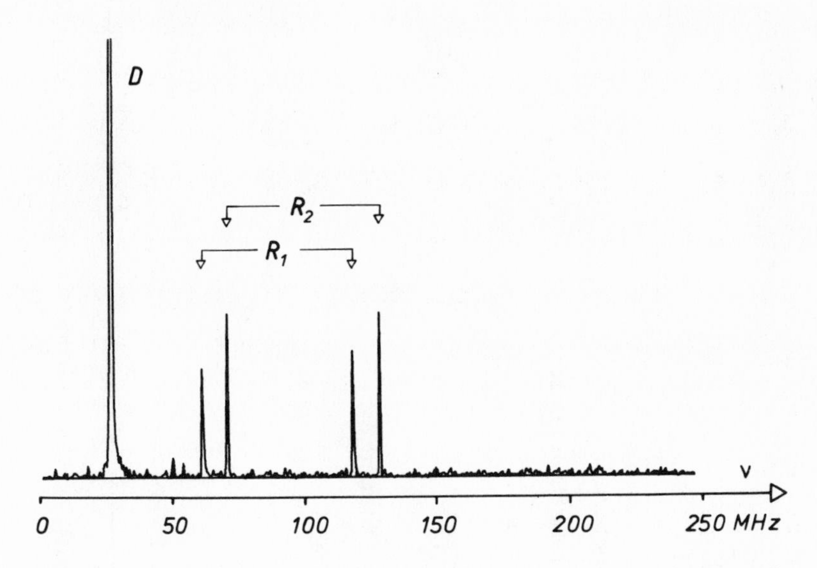

Fig. 6. Muon precession frequencies in isoprene at 2 kG and ambient temperature, D: muons in diamagnetic environments, R_1 and R_2 : muonic radicals.

Table 2: Hyperfine coupling constants of muonic radicals at ambient temperature and comparison with hydrogen analogs

substrate	radical	$A_\mu \cdot \mu_p / \mu_\mu$ [MHz]	A_p [MHz] [a]	$A_\mu \cdot \mu_p / A_p \cdot \mu_\mu$
isobutene	$(CH_3)_2\dot{C}CH_2Mu$	91.7	63.7 (203)	1.44
tetramethylethylene	$(CH_3)_2\dot{C}C(CH_3)_2Mu$	50.67	30.18 (298)	1.68
acetone	$(CH_3)_2\dot{C}$ O Mu	8.2	0.9 (300)	9.1
isoprene	$CH_2 = C(CH_3)\dot{C}HCH_2Mu$	56.8 or 62.7	37.8 (300)	1.50 - 1.66
	$CH_2Mu\dot{C}(CH_3)CH = CH_2$	62.7 or 56.8	42.91 (363)	1.32 - 1.46
1,3-pentadiene	$CH_2 = CH\dot{C}HCHMuCH_3$	53.1 or 57.4	39.74 (140)	1.34 - 1.44
	$CH_2Mu\dot{C}HCH = CHCH_3$	57.4 or 53.1		
benzene	cyclohexadienyl	161.6	133.7 (288)	1.21
benzene-d_6	cyclohexadienyl-d_6	163.4	135.9 (114)	1.20
toluene	o-methyl cyclohexadienyl	154.1 or 160.4	123.9 (300)	1.24 - 1.29
	p-methyl cyclohexadienyl	160.4 or 154.1		
thiophene	2-thiophenyl	106.5	95.3 (130)	1.12

a) values taken from Fischer and Hellwege[14]; temperature [K] in parantheses

chemistry of H. It is known that addition of thermal hydrogen isotopes to terminal olefins occurs almost exclusively (>95%) at the CH_2 group[10]. Addition does not occur at the central carbon atoms of 1,3 - butadiene[10]. Thus it seems reasonable that two muonic allyl radicals are formed in both isoprene (fig.6) and 1,3 -pentadiene. Furthermore muonic t-butyl radical is formed from isobutene.

H adds to the oxygen of acetone[11], and Mu is expected to do the same. Only one radical is formed in tetramethylethylene. The field dependence of the frequencies (fig.7) is as expected (table 1).

H is known to add to benzene to form the cyclohexadienyl radical[12]. μSR spectra of its muonic analogue are displayed in fig.8. Each line is split by 1.5 MHz at 1kG.

Di Gregorio et.al.[13] and Bennett and Mile[10] find that H adds to toluene in the ortho position. We observe two muonic radicals in toluene and assign them tentatively to the ortho and para addition products since these two positions usually show similar chemical behaviour.

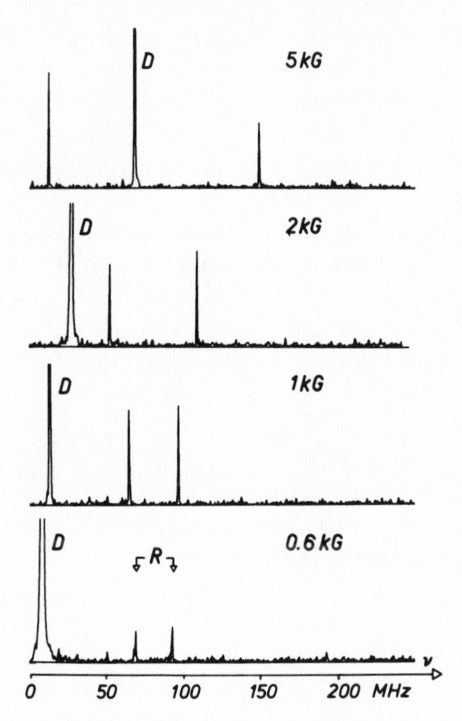

Fig. 7. Muon precession frequencies in tetramethylethylene at various magnetic fields and ambient temperature. D: muons in diamagnetic environments. R: muonic radical.

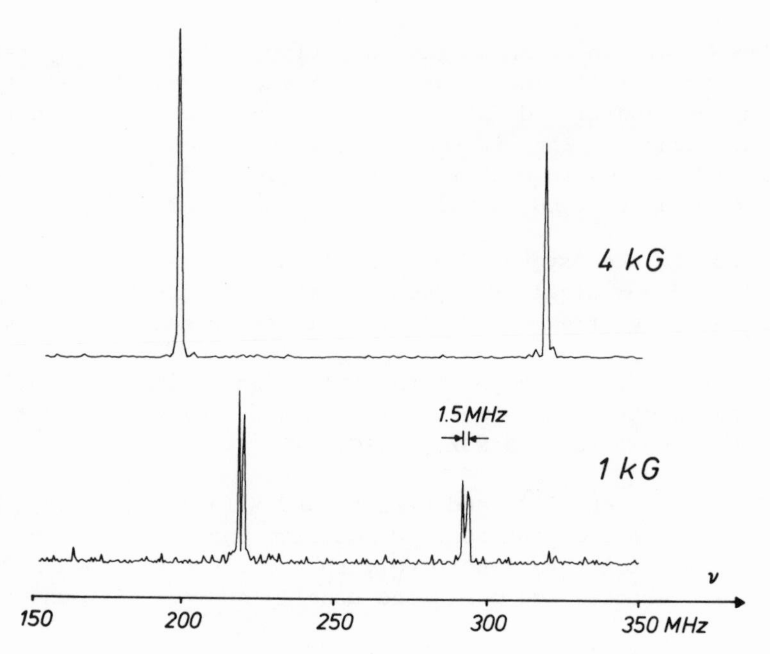

Fig. 8. Muon precession frequencies in benzene at two fields and ambient temperature.

In the absence of isotope effects the muon and proton hfc should be in the ratio of their magnetic moments, $\mu_p/\mu_\mu = 0.3141$. The discrepancies between the last two columns of table 2 clearly demonstrates the existence of isotope effects. In all cases the muon coupling constant is larger than predicted from the relative magnetic moments.

Due to the nature of radical addition (2), the hfc are of the "β" type, i.e. the muonium atom is bound to an atom adjacent to the radical centre (or conjugated system). Such constants are usually quite temperature dependent. This behaviour has been measured for the Mu adducts of isoprene and tetramethylethylene and is displayed in fig.9.

DISCUSSION

Comparison of the number of radicals observed in each substrate with the number of inequivalent possible sites of addition reveals some selectivity for this reaction. This is to be expected if the reaction occurs with a well behaved thermal Mu atom. This supports the view that the excess energy of the stopping muon is lost by the time of the Mu addition.

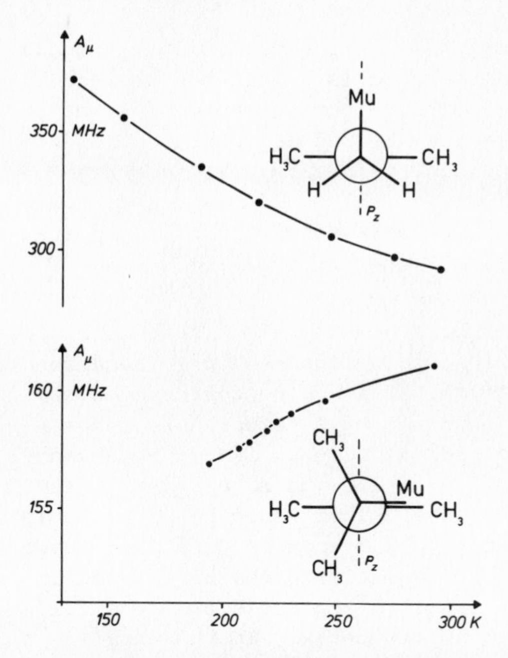

Fig. 9. Temperature dependencies of the hyperfine coupling constants in isobutene (upper) and in tetramethylethylene (lower). Note the different scales of the ordinates.

As mentioned above the muon is bound at the β-position in muonic radicals. β-coupling constants are described by

$$\langle A \rangle = B + A_o \langle \cos^2\theta \rangle \qquad (20)$$

where the first term, B, represents a contribution arising from spin polarization and the second term describes the angular dependence of the spin density at the β-nucleus. The definition of θ is given in a Newman projection where the projection is in the direction of the axis of internal rotation. The observed hfc is a weighted average over conformations with different θ. The weighting depends on the potential barrier to internal rotation, on the reduced moment of inertia, and on the temperature. In the high temperature limit there is free rotation with equal probabilities for all θ, and A becomes equal to $B + \frac{1}{2}A_o$. A will increase with decreasing temperature for an equilibrium conformation $-45^o<\theta_o<45^o$ and decrease for $45^o<\theta_o<135^o$.

The hfc for the nine equivalent protons in the t-butyl radical is practically temperature independent[15] because of the C_3 symmetry in the methyl groups. (This value corresponds to that for free rotation.) The strong negative temperature dependence for A_μ in the analogous muonic radical (fig.9) indicates an equilibrium conformation with $\theta_o(Mu) = 0$; i.e. Mu rather than H is eclipsed by the half-filled p_z orbital. This would be so if Mu is bulkier than H, so that larger steric hindrance occurs if the lighter isotope eclipses a methyl group. We estimate the difference in the effective radii (≙ van der Waals radii) of Mu and H from the difference in the zero point vibrational amplitudes x. In the harmonic approximation $<(x - x_o)^2>^{1/2}_{Mu} - <(x - x_o)^2>^{1/2}_{H}$ is 0.1 Å for a pure bending motion of Mu (H) with respect to the bulk of the radical, using a force constant $k_\delta \cdot \ell^2 = 0.5 \times 10^5$ dyn/cm. Thus the difference in the effective radii is considerable ($\approx$ 9% of the bond length) and makes an effect on the potential barrier plausible. The isotope effect $A_\mu \cdot \mu_p / A_p \cdot \mu_\mu$ for this radical at 215 K is 1.58, whereas the corresponding ratio for the deuterium analogue at 217 K, $A_d \cdot \mu_p / A_p \cdot \mu_d$, is 0.94 from data by Lloyd and Wood[16], in agreement with a preferred conformation $\theta_o(D) = 90^o$. Related isotope effects for partly deuterated alkyl radicals have been explained in the same way[17].

The radical formed by Mu addition to tetramethylethylene exhibits a positive temperature dependence as does its H analogue[18]. Thus, in terms of anisotropic rotational averaging the equilibrium conformation is $\theta_o(H,Mu) = 90^o$. Both the small temperature coefficient and the magnitude of A_μ are compatible with the view that Mu is almost as large as a methyl group, and that we have essentially free rotation over a very small barrier. This seems strange if one compares rotational barriers for related molecules (c.f. $\approx$20kJ/Mol for hexamethylethylene and $\approx$16kJ/Mol for 2,3-dimethylbutane[19]).

For the two radicals discussed above substitution of Mu by H does not much change the reduced moment of inertia for internal rotation. This is not so for the radical derived from acetone, $(CH_3)_2\dot{C}OMu$. Here, rotational motion of the hydroxyl group corresponds essentially to motion of Mu around the C-O axis. Replacement of Mu by H strongly changes the moment of inertia. This is most likely the reason for the observed isotope effect. $|A_p^{OH}|$ for $(CH_3)_2\dot{C}OH$ and for the related $H_2\dot{C}OH$ are both very small[14], so it is not unreasonable to assume similar barriers and equilibrium conformations. For $H_2\dot{C}OH$ the barrier is 17kJ/Mol and θ_o(H) is 90°[18]. This means that there is no free internal rotation but rather librational oscillation of OH(OMu) in these radicals.

For the methyl substituted allyl radicals derived from isoprene A_p (the one that corresponds to A_μ) takes the value for free rotation. A_μ is once more considerably greater than A_p, which indicates an equilibrium conformation with $|\theta_o|$ <45° for Mu.

Yim and Wood[20] conclude from INDO calculations that the cyclohexadienyl radical is probably non-planar and inverts rapidly. For the muonium substituted analogue (fig.10) the potential for this inversion is asymmetric with respect to the ring plane, again because Mu appears bulkier than H and tends to avoid the ring plane with its interacting hydrogen atoms. Thus Mu resides preferentially in the endo position with its higher coupling constant. The same argument holds for H in the cyclohexadienyl-d_6 radical, and for Mu in the cyclohexadienyl-d_6, the methyl-cyclohexadienyl and the thiophenyl radicals. The asymmetry of such a potential has been determined to be ≈600 J/Mol from NMR-studies of cycloheptatriene with one deuterium substituted in the methylene group[21]. Fig.10 gives the hfc's for the protons in cyclohexadienyl[20] and for the muon in its muonic analogue. Perturbation theory predicts that a coupling of 134 MHz (for the methylene proton) should lead to 1.4MHz splitting in each of the two μSR lines at 1 kG. The experimental verification of this (fig.9) is a proof for the proposed structure of the muonic radical.

SUMMARY

Theory for an understanding of μSR spectra of muonic radicals is given. μSR transitions obey selection rules analogous to those in magnetic resonance. It is shown that magnetic nuclei other than the muon do not influence the spectra in high transverse fields. Here

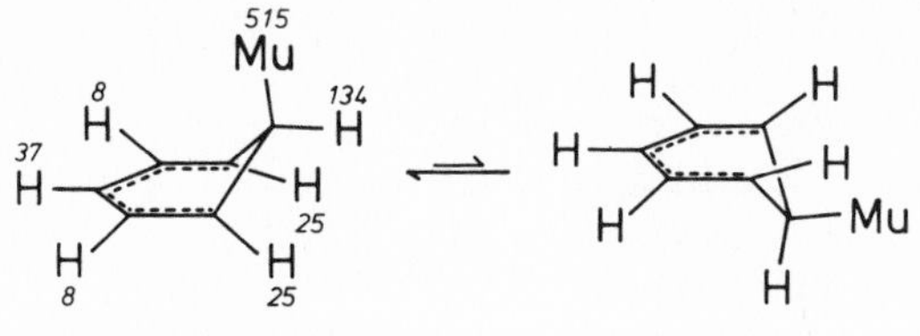

Fig. 10. Conformational dynamics of the muonic cyclohexadienyl radical. Numbers denote the hfc's (in MHz) for the muon, and for the protons in the hydrogen analogue.

muonic radicals are observed most easily, and the muon hyperfine coupling is obtained directly from the spectrum. In intermediate fields spectra become very complicated with an increasing number of magnetic nuclei. Muon polarization is scattered over many frequencies. This renders the detection of radicals difficult (there is recent evidence for radicals in low fields[22] where banding of frequencies occurs). Computer calculations can handle all fields, but a perturbative approach is developed to handle deviations from the high field limit analytically.

Comparison of hfc's between muonic radicals and their hydrogen counterparts reveals isotope effects exceeding the ratio of magnetic moments. They are interpreted in terms of intramolecular dynamics. Isotopic substitution can influence the barrier to internal rotation. Study of the temperature dependence of hfc's leads to information about radical conformations and internal dynamics.

Acknowledgements: This work is part of the author's doctoral thesis. Thanks are due to Prof. H. Fischer for his guidance and Drs. P.W. Percival and J. Hochmann for helpful discussions. Support by the Swiss National Foundation for Scientific Research and by SIN are gratefully acknowledged.

REFERENCES

1. E. Roduner, P. W. Percival, D. G. Fleming, J. Hochmann and H. Fischer, Muonium-Substituted Transient Radicals Observed by Muon Spin Rotation, Chem.Phys.Letts. 57:37 (1978).
2. D. G. Fleming, D. M. Garner, L. C. Vaz, D. C. Walker, J. H. Brewer and K. M. Crowe, Muonium Chemistry - a Review, in: "Positronium and Muonium Chemistry", H. J. Ache, ed., American Chemical Society, in press.
3. P. W. Percival, E. Roduner and H. Fischer, Radiation Chemistry and Reaction Kinetics of Muonium in Liquids, in: "Positronium and Muonium Chemistry", H. J. Ache, ed., American Chemical Society, in press.
4. P. W. Percival, Muonium Chemistry, Radiochimica Acta, Vol.55, in press.
5. A. M. Brodskii, Conditions of Formation of μ^+-Mesic Molecules, Zh.Exp.Teor.Fiz. 44:1612 (1963) [English transl. Soviet.Phys. JETP 17:1085 (1963)].
6. P. W. Percival and H. Fischer, Theory and Analysis of μ^+ Spin Polarization in Chemical Systems, Chem.Phys. 16:89 (1976).
7. J. E. Wertz and J. R. Bolton, "Electron Spin Resonance", McGraw-Hill, New York (1972).

8. E. Roduner and H. Fischer, The Evolution of Muon Spin Polarization in Muonic Radicals and Related Species, Chem.Phys.Letts., submitted.
9. A. Messiah, "Mécanique Quantique", Vol.II, Dunod, Paris (1964).
10. J. E. Bennett and B. Mile, Studies of Radical-Molecule Reactions Using a Rotating Cryostat, Faraday Trans.I, 69:1398 (1973).
11. R. A. Witter and P. Neta, On the Mode of Reaction of Hydrogen Atoms with Organic Compounds in Aqueons Solutions, J.Org.Chem. 38:484 (1973).
12. H. Fischer, Bildung von Cyclohexadienylradikalen durch Anlagerung von Wasserstoffatomen an festes Benzol, Z.Naturforsch. 17a:693 (1962).
13. S. DiGregorio, M. B. Yim and D. E. Wood, Aromatic Hydrogen Addition Radicals, J.Amer.Chem.Soc. 95:8455 (1973).
14. H. Fischer and K. H. Hellwege, eds. "Magnetic Properties of Free Radicals. Part b. Organic C-Centred Radicals", Landolt-Börnstein, New Series, Group II, Vol.9, Springer-Verlag, Berlin (1977).
15. H. Paul and H. Fischer, ESR freier Radikale bei photochemischen Reaktionen von Ketonen in Lösung, Helv.Chim.Acta 56:1575 (1973).
16. R. V. Lloyd and D. E. Wood, EPR Study of Conformations in the β-Halo-tert-butyl Radicals, J.Amer.Chem.Soc. 97:5986 (1975).
17. R. W. Fessenden, ESR Studies of Internal Rotation in Radicals, J.Chim.Phys. 45:1946 (1966).
18. P. J. Krusic, P. Meakin and P. J. Jesson, ESR Studies of Conformations and Hindered Internal Rotation in Transient Free Radicals, J.Phys.Chem. 75:3438 (1971).
19. H. Braun and W. Lüttke, Untersuchung der innern Rotation von Bicyclopropyl, Vinylcyclopropan, Butadien und einiger verwandter Verbindungen mit der Kraftfeld-Methode, J.Mol.Structure 31:97 (1976).
20. M. B. Yim and D. E. Wood, EPR and INDO Study of Fluorinated Cyclohexadienyl Radicals, J.Amer.Chem.Soc. 97:1004 (1975)
21. F. R. Jensen and L. A. Smith, The Structure and Interconversion of Cycloheptatriene, J.Amer.Chem.Soc. 86:956 (1964).
22. C. Bucci, G. Guidi, G. M. de'Munari, M. Manfredi, P. Podini, R. Tedeschi, P. R. Crippa, A. Vecli, Direct Evidence for Muonium Radicals in Water Solutions, Chem.Phys.Letts. 57:41 (1978).

INDEX